CAMBRIDGE LIBRARY COLLECTION

Books of enduring scholarly value

Travel and Exploration

The history of travel writing dates back to the Bible, Caesar, the Vikings and the Crusaders, and its many themes include war, trade, science and recreation. Explorers from Columbus to Cook charted lands not previously visited by Western travellers, and were followed by merchants, missionaries, and colonists, who wrote accounts of their experiences. The development of steam power in the nineteenth century provided opportunities for increasing numbers of 'ordinary' people to travel further, more economically, and more safely, and resulted in great enthusiasm for travel writing among the reading public. Works included in this series range from first-hand descriptions of previously unrecorded places, to literary accounts of the strange habits of foreigners, to examples of the burgeoning numbers of guidebooks produced to satisfy the needs of a new kind of traveller - the tourist.

The Voyages of Captain Luke Foxe and Captain Thomas James in Search of a North-West Passage, in 1631-32

The publications of the Hakluyt Society (founded in 1846) made available edited (and sometimes translated) early accounts of exploration. The first series, which ran from 1847 to 1899, consists of 100 books containing published or previously unpublished works by authors from Christopher Columbus to Sir Francis Drake, and covering voyages to the New World, to China and Japan, to Russia and to Africa and India. Volumes 88–9 (1884) contain accounts of two captains' searches for a North-West Passage to Asia in 1631. Their explorations were beset by bad weather. Foxe circumnavigated Hudson's Bay before retreating, while James became ice-bound for the winter, losing several members of his crew before retuning to England a year after Foxe. No new attempts were made for another century, as their accounts of the harrowing conditions they endured discouraged further voyages of exploration for the desired trade route.

Cambridge University Press has long been a pioneer in the reissuing of out-of-print titles from its own backlist, producing digital reprints of books that are still sought after by scholars and students but could not be reprinted economically using traditional technology. The Cambridge Library Collection extends this activity to a wider range of books which are still of importance to researchers and professionals, either for the source material they contain, or as landmarks in the history of their academic discipline.

Drawing from the world-renowned collections in the Cambridge University Library, and guided by the advice of experts in each subject area, Cambridge University Press is using state-of-the-art scanning machines in its own Printing House to capture the content of each book selected for inclusion. The files are processed to give a consistently clear, crisp image, and the books finished to the high quality standard for which the Press is recognised around the world. The latest print-on-demand technology ensures that the books will remain available indefinitely, and that orders for single or multiple copies can quickly be supplied.

The Cambridge Library Collection will bring back to life books of enduring scholarly value (including out-of-copyright works originally issued by other publishers) across a wide range of disciplines in the humanities and social sciences and in science and technology.

The Voyages of Captain Luke Foxe and Captain Thomas James in Search of a North-West Passage, in 1631-32

With Narratives of the Earlier North-West Voyages of Frobisher, Davis and Others

VOLUME 1

EDITED BY MILLER CHRISTY

CAMBRIDGE
UNIVERSITY PRESS

CAMBRIDGE UNIVERSITY PRESS

Cambridge, New York, Melbourne, Madrid, Cape Town, Singapore,
São Paolo, Delhi, Dubai, Tokyo

Published in the United States of America by Cambridge University Press, New York

www.cambridge.org
Information on this title: www.cambridge.org/9781108013550

This edition first published 1894
This digitally printed version 2010

ISBN 978-1-108-01355-0 Paperback

WORKS ISSUED BY

The Hakluyt Society.

———o———

THE VOYAGES

OF

CAPTAIN LUKE FOXE

AND

CAPTAIN THOMAS JAMES.

VOL. I.

No. LXXXVIII.

ADMIRAL SIR THOMAS BUTTON, KNIGHT.

DIED 1634.

From an original oil painting in the possession of Mr. G. M. Traherne, of Coedriglan Park, Glamorganshire.

THE VOYAGES

OF

CAPTAIN LUKE FOXE

OF HULL,

AND

CAPTAIN THOMAS JAMES

OF BRISTOL,

IN SEARCH OF A NORTH-WEST PASSAGE,

IN 1631-32;

WITH NARRATIVES OF THE EARLIER NORTH-WEST VOYAGES OF FROBISHER, DAVIS, WEYMOUTH, HALL, KNIGHT, HUDSON, BUTTON, GIBBONS, BYLOT, BAFFIN, HAWKRIDGE, AND OTHERS.

Edited, with Notes and an Introduction,

BY

MILLER CHRISTY, F.L.S.

IN TWO VOLUMES.—VOL. I.

LONDON:
PRINTED FOR THE HAKLUYT SOCIETY,
4, LINCOLN'S INN FIELDS, W.C.

M.DCCC.XCIV.

COUNCIL

OF

THE HAKLUYT SOCIETY.

CONTENTS OF VOLUME I.

*** *For conclusion of "North-West Fox", see Volume II.*

LIST OF MAPS AND ILLUSTRATIONS.

EDITOR'S PREFACE.

THE present volumes consist of a reprint of the narratives of two English navigators who sailed in search of a North-West Passage in the year 1631—namely, the *North-West Fox, or Fox from the North-West Passage*, by Captain Luke Foxe, of Hull, published in 1635, and *The Strange and Dangerous Voyage of Captain Thomas James* (of Bristol), published in 1633. Inasmuch, however, as Foxe prefixed to his own narrative abstracted accounts of all previous voyages northwestwards with which he was acquainted, his work became much more than a narrative of his own voyage. Most of these accounts of earlier voyages given by Foxe are merely abstracted from the pages of Hakluyt or Purchas ; but several of them are original and of considerable value.

Both works (which are of much interest in their way, though differing widely from each other in many respects) are fully described and discussed in the Introduction.

The advisability of making the two volumes of the present work approximately equal in thickness has caused the somewhat-inconvenient division in the middle of the reprint of the *North-West Fox*.

The first volume will be found to contain my Introduction (which treats, not only of the voyages of Foxe and James themselves, but also of the earlier voyages which led up to them), together with the first portion of the *North-West Fox*, containing Foxe's account of the north-west voyages preceding his own.

The second volume contains the remainder of the *North-West Fox* (namely, that portion relating to Foxe's own voyage), together with the whole of Captain James's work, and an Appendix containing certain interesting documents referred to in the Introduction.

It remains for me to thank the many kind friends who, in the course of my work as Editor, have so liberally responded to my inquiries and requests for information.

In the first place, I have especially to thank Mr. E. Delmar Morgan, late Honorary Secretary to the Hakluyt Society, who has from the first rendered me very valuable assistance in every possible way.

I am also very much indebted to Mr. G. H. Pope, Treasurer of the Merchant Venturers' Society of Bristol, who has both copied from the originals, and corrected the proof of, the interesting letters relating to James's voyage preserved at Bristol; and to my cousin, Miss C. Fell Smith, of Great Saling, Essex, who has made researches for me at the British Museum and elsewhere, and has rendered me much literary assistance.

My thanks are also due to Mr. Coles, of the Royal Geographical Society, who has given me advice on obscure geographical questions; to Prof. Skeat, who has assisted me with the derivation of various obsolete words; to Mr. J. G. Baker, F.R.S., of Kew, who has advised me with reference to several botanical points; to the Rev. Canon Austen of Whitby, who has searched the Parish Registers for the entry of Captain Foxe's death; to Captain W. J. L. Wharton, R.N., who has helped me with reference to several nautical matters; to the late Mr. John Taylor, of the Free Library, Bristol, who has sent me information relating to Captain James; to Mr. R. Hill Dawe, Town Clerk, Mr. W. Andrews, and Mr. T. Tindall Wildridge, all of Hull, for biographical information relating to Captain Foxe; to Mr. J. L. Wheatley, Town Clerk, Mr. John Storrie, and Mr. J. Ballinger, all of Cardiff, for much local and valuable information relating to Sir Thomas Button; to Captain Sir J. Sydney Webb, the late Mr. Inglis, and Mr. E. G. Weller, for facilities in searching the ancient records of the

Trinity House; to Mr. F. C. Danvers and Mr. W. Foster, of the Archives Department of the India Office, for facilities for examining the Records of the Old East India Company; to Dr. George H. Moore and Mr. Washington Eames, of the Lenox Library, New York, for information concerning, and for revising the proof of, the *Motives;* to Mr. Bernard Quaritch, for advice as to the present rarity and value of copies of the works of Foxe and James; to the Treasurer of the Inner Temple for permission to examine the Register of Admissions to that body; to Mr. G. W. Waddington, of Grosmont, Whitby, for information as to Foxe's marriage; to Mr. Ivor James, of the University College of South Wales and Monmouthshire, Cardiff, for information relating to Captain Thomas James; to Mr. C. H. Coote, of the British Museum, for help and advice upon several points; to Mr. Clements R. Markham, President of the Hakluyt Society, for timely assistance on various occasions; and to many other gentlemen who have assisted me in other ways.

In conclusion, I need only add that, as editor, I have not knowingly altered a single word in the narratives of either Foxe or James (bad as is the diction of the former) without clearly indicating any additions by inserting them within brackets [—], except in the case of certain obvious misprints, whether pointed out in Foxe's two pages of errata or not. With the punctuation, however (especially with that of Foxe, which is exceedingly erratic), I have taken great liberties, and have amended it throughout, in order to render the meaning clearer to modern readers.

With regard to the numerous foot-notes appended to Foxe's narrative, it is necessary to explain that those with the initial "*F.*" appended to them are Foxe's own, being explanations and remarks which appear as side-notes in his pages; those signed with a "*C.*" I am of course responsible for; while those distinguished by the words "*Foxe MS.*" or "*Master's MS.*" are extracted from the MS. Journals mentioned in the Introduction (p. cviii).

The notes to James's narrative are all mine.

None of the original documents introduced herein have, to the best of my belief, ever before been printed in full, except those contained in Appendix B and Sir Thomas Button's letter of Feb. 16th, 1629-30 (p. lxv), which appeared in Mr. G. T. Clark's *Account of Admirals Sir Robert Mansell and Sir Thomas Button*.

During the progress of my work as editor, I have used the latest Admiralty Charts of "Hudson's Bay and Strait" (No. 863), published June 28th, 1884, with corrections to August 1890, and of the "Harbours and Anchorages in Hudson's Bay and Strait" (No. 1221), published in 1886. All names I have used are taken from these charts. From the former has been prepared the large chart of Hudson's Bay, showing the routes of the early navigators, which is placed at the end of this volume.

All references (both in the Introduction and notes) to pages on which occur certain passages in the works of Foxe and James are to the pages of this reprint, and not to the original editions unless otherwise stated.

MILLER CHRISTY.

CHELMSFORD,
November 1893.

ERRATA, ADDENDA ET CORRIGENDA.

P. vi, ll. 15 and 16. *Delete* the words "as seems to be the case" (see pp. cviii and cix, and *note*).

P. xxviii. Button's Instructions: see also Postscript, p. ccxxiv.

P. xxxix, last line but two of *note*. *Delete* "John Evelyn, the Diarist".

P. xl, l. 2. *For* "twenty-seven" *read* "twenty-four".

P. xli, l. 13. *For* "all goods brought" *read* "all goods which might be brought".

P. xli, l. 20. *For* "on goods passing" *read* "on goods which might pass".

P. xlvii, *note*. Lord Carew's letter: This has been printed in *Letters from George Lord Carew* (Camden Society, 1860), p. 27.

P. l, *note*. Certificate concerning Hawkridge: see Postscript, p. ccxxvii.

Pp. xlviii and xlix. Date of Hawkridge's Voyage: see Postscript, p. ccxxvi.

P. liii, last line but one of *note*. *For* "narratives" *read* "narrative".

P. lvii. *For* "Trinity House" *read* "the Trinity House at Hull".

P. lvii, l. 15. Foxe's wife's name: see Postscript, p. ccxxviii.

P. lviii, last line but two of *note*. *For* "original ed., p. 228", *read* "see p. 370".

P. lxxxiv, ll. 7 to 11. Pieces of cloth for Japan: see also Postscript, p. ccxxiv.

P. cix. Foxe's MS. Journals: see also Postscript, p. ccxxix.

P. cxxxi, last line. *For* "Alderman Thomas" *read* "Alderman Thomas James".

P. 11, l. 27. "litle Mr. Jeffery [Hudson]": see Postscript, p. ccxxviii.

P. 11, l. 28. "Mr. Evans, his Maiesties great Porter": see Postscript, p. ccxxviii.

P. 19. Voyages of the brothers Zeno: see also Postscript, p. ccxxii.

P. 79, l. 20. Davis's "owne words": see Hakluyt's *Voyages*, vol. iii, p. 118.

P. 80. Weymouth's voyage: see also Postscript, p. ccxxii.

P. 115, *note*. Colbert, Cobreth, or Coolbrand: see also Postscript, p. ccxxii.

P. 122, last line but one of *note*. *For* "south" *read* "south-west".

P. 167, note 1. Button's deserted ship: see also Introduction, p. ci; Postscript, p. ccxxx; and pp. 344 to 348.

P. 175, note 1. *Add* "— C."

P. 177, note 2. *For* "p. 258, original edition" *read* "see p. 426".

P. 178, *note*. *For* "(cf. North-West Foxe, original edition, p. 161)" *read* "see p. 241".

P. 203, *note*. *Delete* from "Knowing, as we do" to "clearly unable to do." (See also Introduction, p. x, *note*).

P. 201, *note*. Promoters of Gibbons's Voyage: see also Postscript, p. ccxxv.

P. 248, last line. *For* "[1619]" *read* "[? 1617]": see Postscript, p. ccxxvi.

P. 399, sixth line of note 2. *Delete the word* "who".

P. 402, note 5. *For* "p. 283" *read* "p. 282".

P. 452. Note on Nash: see also p. ccvii, *note*.

P. 492, fourth line of note 3. *For* "he" *read* "him".

P. 493, *note*. Cape Monmouth: see Postscript, p. ccxxx.

P. 512, note 1. *For* "see *post*" *read* "see p. 614".

P. 529. Winter's Forest: see Postscript, p. ccxxxi.

P. 537, note 1. James was probably Welsh (see Introduction, p. ccvii).

P. 564, note 2. *Delete* the words "One authority states that", and see Introduction, p. clxxxix.

P. 591, last line but two of *notes*. *For* "Hudson's Bay" *read* "Hudson's Bay Company".

P. 632, *note*. *For* "foregoing" *read* "following".

INTRODUCTION.

IN the whole history of Geographical Discovery, there is certainly no more striking or more interesting feature than that remarkable fever for Arctic exploration, which first showed itself in the voyages of the early Italian, Spanish, and Portuguese navigators; which, during the Middle Ages, affected more or less all the maritime nations of Western Europe, developing at last into a perfect craze for the discovery of a navigable sea-passage round the northern coasts of North America or Asia to the rich spice-bearing countries of the Far East; and which continued with increased, rather than abated, strength even to our own times.

During this long period, scores of different expeditions were sent out; dozens of fine vessels were lost; hundreds of bold explorers perished; and many thousands of pounds were fruitlessly expended. Yet, so intense was the enthusiasm among the nations of Western Europe, and so strong was their belief in the existence of the passage so ardently

b

sought, and in its high commercial value when found, that there was seldom a time when there were not forthcoming still more Adventurers who were ready to risk their lives, their vessels, or their wealth in order to undertake further explorations. And when, at last, in the year 1851, success did crown this long and persistent search for a North-West Passage, and the long-sought passage was proved to have an actual existence, it is somewhat humiliating to have to record that it was at the same time proved to have no practical value whatever from a commercial point of view.

Yet, in spite of a result which seems at first sight so disappointing, it is impossible to regard the vast amount of time and treasure expended as by any means altogether lost. Although it is perfectly true that the main end in view, when attained, was found to be valueless for the purposes for which it was sought, it is nevertheless equally true that many most worthy and valuable ends were attained by the long-continued search. In short, a scientific knowledge of the region affected, and of its products and peculiarities, was obtained, such as could hardly have been acquired by any other means than those employed; and these results, it may be unhesitatingly stated, are of far greater value than those which have been the net outcome of many purely military and naval expeditions which have involved the sacrifice of vastly greater amounts of wealth and human life.

Among the more notable of the earlier English voyages undertaken in search of a North-West

Passage are certainly those of Captain Luke Foxe and Captain Thomas James in the year 1631. Their respective narratives are remarkable as being almost, if not quite, the earliest separately-published English works describing voyages in search of a North-West Passage. None of the earlier voyagers seem to have published detailed accounts of their explorations in separate form, though we have narratives of most of them in the pages of Hakluyt and Purchas. The two volumes in question are, moreover, remarkable in other ways: they are in all respects companion volumes, as they describe voyages undertaken in the same year, to the same place, and with the same object. Further, the two captains met, by accident, at the scene of their explorations; their respective works were published within two years of one another, and in the same form and size; and both alike have now become scarce and valuable, a perfect copy of either being very rare, and worth a considerable sum. The reasons for reprinting them together in the series of the Hakluyt Society are, therefore, obvious.

I will take Captain Foxe's narrative first, as it is the more important of the two. The volume containing it is fully described hereafter from a bibliographer's point of view. It is, therefore, sufficient to state here that it is a small quarto, consisting of about 275 pages; that it bears the whimsical title of *The North-West Fox;* and that it was published in London in 1635. Considerably less than half the work, and that the latter half, is occupied by Foxe's account of his own voyage. The greater part of the book is filled with abstracts of the accounts given of their explorations by the principal navigators, some fifteen in number, who had preceded him in the search for a North-West Passage from the very earliest times. These narratives, which are, of course, mainly taken from the pages of Hakluyt and Purchas, render *The North-West Fox*, so far as the history of Arctic research is concerned, a miniature and abridged edition of those well-known works. Three, however, of these narratives are of special interest, inasmuch as they give us almost the whole of the information extant respecting the voyages they describe. With reference to these abstracts of earlier voyages, Foxe himself, in a passage (Preface to the Reader; see p. 9) which, were it a little less uncouthly worded, would be a really fine piece of rhetoric, boldly admits that they were taken chiefly out of Hakluyt and Purchas, and says: "I do confesse my selfe to be infinitly bound unto them and others for their paines." He had, he tells the reader, by these abstracts, brought the large and costly works of the above-named writers cheaply within the reach of all, besides adding "much that never came in print as yet, being very difficult to be had. . . . Nor [says he, with much force] doe I hold that man fit to take charge of voyages remote, especially north-east or westward, and be ignorant of those Abstracts and Journals following. . . . In them I have done my best, and whosoever will amend them, I shall take him for my friend."

James's book is also a small quarto, but it only contains about 150 pages. Its smaller size enabled it to be issued from the press two years earlier than Foxe's book, namely, in 1633. It consists entirely of a clear and interesting account of James's voyage, and is remarkable for the exaggerated account he gives of the dangers he escaped. From a geographical point of view, it has not much value. A bibliographical description of it appears hereafter.

In order clearly to understand the position occupied by Foxe and James with reference to the Arctic explorers who preceded them, it is necessary to turn back and to notice briefly the achievements of some of these. Inasmuch as neither Foxe nor James explored anything beyond Hudson's Bay and Hudson's Strait, it will only be necessary here to notice fully the voyages of those who sought a Passage *through the Bay or Strait.* On this account, I have, in what follows, said little or nothing of the achievements of the Cabots, of Sir Martin Frobisher, of Captain John Davis, of Knight and Hall, and of others whose names are as household words in the History of Arctic Discovery, but have merely reprinted what Foxe said of them, adding here and there a note to point out some unusually glaring error or some specially interesting feature. Most of their narratives have been already treated in the series of the Hakluyt Society by editors far more competent than myself.

The men whose voyages it is essential to notice

more or less fully here are Hudson, Button, Gibbons, Bylot, Baffin, Hawkridge, and Munk. Of the achievements of these explorers, therefore, extended notices will be found in what follows. In the case of three of them, we are indebted to the industry of Foxe for almost, if not quite, all we know concerning their voyages, and his accounts of them may be said to be the most interesting and valuable portions of his book, not even excepting his account of his own voyage; for, whereas concerning them we have no other source of information, of Foxe's own voyage we should have been able to gather full particulars from a copy of his own manuscript journal which remains to us, even if the original has been destroyed, as seems to be the case. Two of these voyages—those of Gibbons and Hawkridge—it is true, were of little consequence, and accomplished almost nothing in the way of discovery; but the third—that of Button—is of high importance, and Foxe deserves credit for having rescued almost everything we know about it.

The man to whom the credit of having been the discoverer of the huge inland sea now known as "Hudson's Bay" is, of course, "that worthy irrecoverable discoverer" (as Purchas calls him), Captain Henry Hudson, whose tragic fate is told in many a children's story-book. And it is certain that the credit of being the first to make known to the world the existence of this great bay belongs as rightly to Henry Hudson as the credit of having been the first to expound the law of natural selection belongs to the illustrious Darwin; but, just as there appear to have been evolutionists (of a kind) before Darwin, and discoverers of America before Columbus, so also there seem to have been explorers of Hudson's Bay before Hudson. Dr. G. M. Asher produces evidence to show that at least forty years before the date of Hudson's voyage the early Portuguese navigators had reached the entrance to Hudson's Bay, if they had not actually sailed into it.[1] It is certain also that Sir Martin Frobisher, on his third voyage north-west in 1578, entered Hudson's Strait by accident and sailed up it for no less than sixty leagues without interruption (see p. 58). Frobisher himself believed he could have sailed straight on into the Western Ocean, and he even declares he would have done so if he had not been concerned for the safety of his large fleet, most of the vessels of which had become separated from him. Moreover, his voyage had an immediate commercial, rather than a geographical, object. Davis, too, had sighted the entrance to Hudson's Strait on his third voyage to the north-west in 1587. His companion Janes says: "We passed by a very great gulf, the water whirling and roring, as it were the meeting of tides." This whirlpool Davis spoke of as "the Furious Overfall", by which name many later navigators have mentioned it. Again, it is quite clear

[1] *Henry Hudson, the Navigator: from the Original Documents, etc.* (Hakluyt Society, 8vo, 1860), pp. xcvi and clxxi.

that Weymouth, in 1602, sailed some considerable distance—he himself says 100 leagues—into the Strait (see p. 84), and it is certain that Hudson was fully aware of this when he started on his fourth and last voyage in 1610, if, indeed, that knowledge had not partly led him to undertake the voyage; for Hessel Gerritsz.[1] tells us that in 1609 Weymouth's log-books were in the possession of Peter Plancius, a well-known geographer of Amster-

[1] Hessel Gerritzoon (Gerritsz.), or Gerrard, an eminent Dutch geographer and cartographer, was born in the village of Assum late in the sixteenth century. In 1617, he was appointed cartographer to the Dutch East India Company, a post he held until his death early in 1634. He it was who, in 1612, engraved and published at Amsterdam the very interesting and now rare map of Hudson's discoveries on his fourth voyage, of which a facsimile is prefixed to Dr. Asher's most valuable work on Hudson. It was, without doubt, engraved from Hudson's own "card", which we know the mutineers brought home (see p. 149). On the back of the map was a brief but interesting account in Dutch of Hudson's voyage and of the steps taken to follow up his discoveries. Later in the same year (1612), the chart was reprinted as part of a tract in Latin containing a fuller account of Hudson's voyage. Another edition of this tract, translated into Dutch, also appeared in the same year, and a revised Latin edition appeared in the following year (1613). In all these editions the notice of Hudson varies somewhat. Other editions of less consequence followed, and the important work was at once translated into several other languages. The notice of Hudson's fourth voyage in the first edition of *Purchas his Pilgrimage* (1613, p. 624) is a slightly abridged translation of it into English. Dr. Asher has given a very complete notice of Gerritsz. and his tract (*Henry Hudson*, pp. xliii-xlix), and has printed translations of the four chief editions cited (*loc. cit.*, pp. 181-194). An excellent facsimile reprint, with an introduction by S. Muller Fz., was published in Amsterdam in 1878.

dam, who had submitted them to Hudson on his visit to that city in the year named.[1] Still, as stated above, the credit is in reality Hudson's, for he first explored the Bay and Strait, and thus made them known to the world at large.

Of Hudson's life and works we know nothing, except during a period of a little over four years—from April 19th, 1607, to June 21st, 1611, or thereabouts, when his life ended. During this short period, he performed the four voyages which have made his name famous. The first was made under the auspices of the Muscovy Company in 1607, in search of a north-east passage to China. The second, in the following year, had the same aims and promoters. The third, undertaken in 1609, at the expense of the Dutch East India Company, had the same object; but, meeting with ice in the neighbourhood of Nova Zembla, he turned westward in search of a north-west passage, in the course of which he explored the Hudson River. His fourth recorded voyage was that herein described (see pp. 114-162). All these voyages have been fully described and discussed by Hudson's chief biographer, Dr. Asher, in the work above mentioned. Dr. Asher has reprinted from Purchas[2] the narratives of

[1] See *Henry Hudson, the Navigator*, p. 186.

[2] The bibliography of Purchas's various works (which are very often quoted from in these volumes) is somewhat complicated, and is incorrectly given by most writers. Even Dr. Asher, who attempts an explanation (*Henry Hudson*, p. 277), is, I believe, not quite correct. The Rev. Samuel Purchas was born at

Hudson's four voyages, and has added all the information about these voyages to be gleaned from the works of Hessel Gerritsz. and others.

Thaxted, Essex, in November 1577, and was educated at Cambridge. In 1601, he was curate of Purleigh, Essex, where, on December 2nd, he married Jane, daughter of Vincent Lease, of Westhall, Suffolk. In 1604, he was instituted to the vicarage of Eastwood in Essex, which he resigned late in 1614, when he was translated to the rectory of St. Martin's, Ludgate. He died in London in or about September 1626 (not in 1628, as is invariably stated). The statement made by all his biographers to the effect that he died much impoverished owing to his devotion to geographical research, is entirely without foundation, as will be proved by a perusal of his will, which has been printed (*Trans. Essex Archæological Society*, vol. iv, p. 164). His works, and those of his friend Hakluyt, have formed the basis of all later histories of travel and discovery. Purchas's first work, published while he lived at Eastwood, appears to have been *Purchas his Pilgrimage, or Relations of the World and the Religions observed in all Ages and Places discovered from the Creation to this present* [*Time*], *etc.* (London, pott folio, 1613). It is chiefly a theological work, but becomes more geographical near the end. The information relating to Hudson, which is very brief, appears on p. 624, and is obviously taken from the third edition of Hessel Gerritsz.' work (Amsterdam, 1612). The second edition of the *Pilgrimage* (London, pott folio, 1614) had almost the same title-page as the preceding edition, but was "much enlarged, with additions through the whole work". Though containing much the same matter, it seems to be an entirely new edition, and to have been reset throughout. The information concerning Hudson, which is much fuller than in the first edition, occupies pp. 743-745. Here, for the first time, we meet with Purchas's well-known account of Hudson's voyage as we are now acquainted with it, though in much briefer form than that in which Purchas afterwards gave it to the world. Purchas had, as he says in a side-note, received the matter from Hakluyt, since the appearance of his first edition in the previous year, and had also had some

Hudson sailed on his fourth and last memorable voyage in April 1610 in the *Discovery*, a vessel of 55 tons burden. Sir Thomas Smith, Sir John

---

help from Sir Dudley Digges, to whom he gives five long lines of gross flattery. Three years later, appeared the third edition (London, pott folio, 1617), which had almost the same title, but was again much enlarged. Though the same work, there are many additions, and the whole appears to have been reset. The information concerning Hudson occupies pp. 924-926, and is identical with that in the previous edition, except in a few trifling particulars. Although a fourth edition of the *Pilgrimage* afterwards appeared, the next of Purchas's geographical works to appear was *his Pilgrimes*, which is an entirely distinct work from the former, and of much greater consequence. It appeared in four large volumes in 1625, and is entitled *Hakluytus Posthumus, or Pvrchas his Pilgrimes, contayning a History of the World in Sea Voyages and Land Trauells by Englishmen and others, etc.* (London, 4 vols. [called "parts", each containing 5 books], fcap. fo., 1625). In this Purchas gives (mainly in their own words) the narratives of many early explorers, most of whose narratives would probably have been lost to us but for his diligence. Here he gives in full the information relating to Hudson, the narrative of whose fourth voyage occupies pp. 596-610 of the third volume. In the following year appeared the already-mentioned fourth edition of the *Pilgrimage* (London, fcap. fo., 1626), which contains the narrative of Hudson's voyages practically identical with that in the two previous editions (1614 and 1617). This narrative, though so much briefer than that contained in the *Pilgrimes* (1625), contains several interesting items of information not contained in the latter, as the source whence Purchas obtained his information, the names of Hudson's chief adventurers, the date when the survivors reached the Irish coast, and other minor matters. This work, though commonly regarded as a fifth volume of the *Pilgrimes* (which appeared in the previous year), and though it matches that work in size (which the three earlier editions do not), has no necessary connection whatever with it, being an entirely separate and distinct work. The fact that the fourth

Wolstenholme (the Elder), Sir Dudley Digges (of all of whom we shall hear again shortly), and twenty others (whose names are given hereafter; see Appendix D), bore the expenses of the expedition. His three previous recorded voyages had all been undertaken with a view of discovering a passage to the north-*eastward.* The voyage we are now concerned with, however, was set forth with the express purpose of searching to the north-*west* for a passage. The intention was, no doubt, to follow up Weymouth's discovery, as already hinted (p. viii). Leaving London on April 17th, 1610, Hudson entered the Strait early in July. Sailing slowly through it to the westward, on August 3rd he reached its westernmost limit, and, passing between two capes, which he named Capes Wolstenholme and Digges, after his chief "adventurers", he

---

edition of the *Pilgrimage* appeared in the year following the publication of the *Pilgrimes*, and that the two works are uniform in size, is no doubt responsible for the error, as I cannot find anything in either work to show that the former was intended in any way to be regarded as a part of the latter. The confusion is further increased by the fact that, in 1619, Purchas had published a small, thick, pott 8vo. volume entitled *Purchas his Pilgrim: Microcosmus, or the Historie of Man, etc.*, of which a second edition (evidently the old sheets bound up with a fresh title and dedication) appeared in 1627. This work, however, has no relation to geographical history, and is purely religious. Moreover, Purchas, in his will (*loc. cit.*, p. 175), not only mentions the three works—that is, the *Pilgrimage*, the *Pilgrim*, and the *Pilgrimes*—separately, but he speaks distinctly of "one entire work of my Pilgrims in fower bookes", which shows that he did not regard the *Pilgrimage* as a fifth volume of the *Pilgrimes.*

entered the huge inland sea which now bears his name. At this point Hudson's own journal ends abruptly. The narrative is continued by his companion, Abacuck Prickett, who relates in a vague manner how he and his companions sailed southward along the eastern shore of Hudson's Bay, and cruised about among the islands in what is now known as James' Bay. At last, on November 1st, they hauled the ship aground for the winter at a spot which may with fair certainty be identified as the southernmost extremity of the small bay or inlet which is situated at the extreme south-east of James' Bay, and which contains the Hudson's Bay Company's post known as Rupert House.[1] Here they remained frozen up till the 18th of the following June (1611), when they were again able to proceed upon their voyage of discovery. In the meantime, however, serious quarrels had arisen between Hudson and some of his crew, which before long ended in open mutiny, and Hudson, with his son and six of his crew who had remained faithful to him, were put by force into the ship's shallop and cast adrift, never to be heard of more, as has so often been related. The mutineers then, with all possible speed, set sail on their return journey; but

[1] See p. 130 (note). The identification of the spot attempted above is rendered all the more certain by the statements made by Hessel Gerritsz., in at least three out of the four editions of his work published in the years 1612-13 (for translations, see Dr. Asher's *Henry Hudson*, pp. 181-194), to the effect that Hudson wintered in Lat. 52° N.

just retribution soon overtook them. Green, the ringleader, and three others, were slain in an encounter with some savages, and Juet died of starvation before England was reached in the month of September, their provisions having given out. The survivors, several of whom were afterwards engaged in Arctic voyages, seem for the most part to have succeeded in excusing themselves for the infamous proceedings in which they had been engaged or had not striven to prevent, and there is no very satisfactory record of punishment having been inflicted upon any of them, though Hessel Gerritsz. says that, on their return to England, they were thrown into prison, where, he says, "they will be kept until their captain shall be safely brought home". If this sentence was carried out, it is to be feared those who were imprisoned died in gaol; but I am not aware of any contemporary English record of their imprisonment.

Practically, the whole of our knowledge of Hudson and his voyages is derived from the various works of Purchas and Gerritsz. all of which were published within twenty years of Hudson's death, that is to say, before 1630.[1] During the two centuries and a half which have elapsed since then, nothing of importance in the shape of a contemporary record has been brought to light. It is with

[1] Foxe, in the present work, gives an account, abridged from Purchas, of Hudson's fourth voyage; but, though writing only twenty years after, he does not add any important additional information.

especial pleasure, therefore, that I am enabled to print herein (see Appendix A) some most interesting documents relating to Hudson's last voyage, which have recently been discovered among the Records preserved at the Trinity House. These documents form part of a large parchment-bound folio volume labelled "Trinity House, Transactions, 1609 to 1625", of which they occupy folios 11-13. They consist of copies or abstracts of depositions of all the eight survivors of Hudson's expedition (with the exception of a boy), taken at Trinity House on the 24th of October 1611—that is, within a month of their return to England—together with the opinion of the Corporation of Trinity House as to the reliability of the statements contained in them. Next we have what appear to be abstracts of portions of the log-book of the *Discovery*, both while she was outward bound and when she was on her return home after the mutiny. Then follow some "Grounds for a Coniecture [as to the Existence of a North-West Passage]", which were probably supplied by the survivors. Finally, two days later (that is, on October 26th, 1611), we have the conclusions of the Corporation "concerning that discovery which is made in the North-West", arrived at, apparently, after due consideration of the preceding documents and after careful examination of the men. The full text of these documents is printed hereafter (for the first time in complete form)[1]; while their chief points

[1] In an article in the *St. James's Gazette* for April 20th, 1887, Mr. W. J. Hardy called attention to and described these docu-

of interest and their bearing upon the information previously existing are pointed out in foot-notes appended to Foxe's narrative of Hudson's voyage (pp. 114-162). Taken as a whole, the documents cannot be said to throw any very important new light on the main events of Hudson's voyage. Their interest lies mainly in the fact that they corroborate or supplement (and in some cases contradict) the details given in Purchas's hitherto-uncorroborated narrative. As corroborative evidence they have considerable value.

That the narrative of Hudson's discovery excited the greatest interest, both among those who were specially eager for the discovery of a North-West Passage and among the general public, is certain. Although there is no record of what took place immediately after the return of the mutineers, there is evidence of the interest their narrative excited in the fact that another voyage, which proved to be of considerable importance, was at once projected, and,

---

ments as though they were previously unknown. Six years earlier, however, all the ancient records of Trinity House had been described in the *Eighth Report of the Royal Commission on Historical Manuscripts* (1881, pp. 235-237), wherein appears a very full abstract of the documents in question. This fact, however, I believe, has escaped the notice of all later writers on early Arctic voyages, largely, no doubt, because neither in the text nor index of the *Report* is there anything connecting the documents with Hudson's well-known and important voyage. They are simply spoken of as "the depositions of some sailors who had been on a voyage to discover the North-West Passage". This was certainly a grave oversight and omission on the part of the cataloguer.

with commendable promptitude, was dispatched at the earliest possible moment the following spring.[1] This was the expedition of Captain (afterwards Admiral Sir) Thomas Button, which sailed from London about the middle of April 1612.[2] The

[1] Purchas's account of the matter (*his Pilgrimage*, 2nd ed., 1614, p. 745) is that, after Hudson's voyage, the Adventurers, "with princely assistance, pursued the action in more royall fashion, with greater shipping, under the command of a worthy sea-man, seruant to Prince Henry, Captaine Thomas Button". Henry Prince of Wales died before Button's return home, namely on November 12th, 1612, aged only eighteen years and a half. Though so young, he seems to have taken a genuine interest in Arctic discovery, and his loss was keenly felt. An obituary reference to his memory which is made by Purchas in *his Pilgrimage* (4th ed., 1626, p. 819), contains so many idolatrous adjectives that one sentence occupies no less than nine long lines, and the whole is almost inconceivably absurd in the fulsomeness of its flattery.

[2] There is evidence that the preliminaries of Button's voyage had already been settled, and that a ship was being selected as early as January 1612, for in the MS. *Autobiography of Phineas Pett* occurs the following passage:

"About this time also [January 1612], I did accompany Capt. Thomas Button to make choice of a Shipp for ye nor west passage, in which Journey he was to be imployed by the appoyntment of ye Prince. . . . ."

Phineas Pett was a son of Peter Pett of Deptford Strand, one of Queen Elizabeth's shipwrights. He was born at Deptford on November 1st, 1570; was educated at Emanuel College, Cambridge; and was apprenticed to a shipwright at Deptford. He afterwards became one of the principal shipwrights to King James, and seems, from his remarks, to have been a trusted adviser of the Prince of Wales on nautical matters. His MS. *Autobiography*, now in the British Museum (*Harl. MSS.* No. 6279), is of interest in connection with the history of the Navy at this period.

objects of this voyage we may assume to have been twofold. First, there was the humane desire to learn, if possible, what had become of Hudson and his fellow-sufferers.[1] Secondly, there was the need to follow up the discoveries of Hudson, which gave a most alluring promise, though no certainty, of the early discovery of the much-sought passage. Hudson, it must be remembered, had sailed through a noble strait which led him for 600 miles in the desired direction, after which he had entered a vast open sea. The whole of the eastern side of this sea or bay we know that he explored; but the question he had to solve was, not what lay on the eastern side, but what there was to the *west;* and the western side of the bay we know he did not explore.[2] It was unquestionably Hudson's ardent desire to continue exploration to the westward, after his stock of provisions had run perilously low, that led to the tragic mutiny among his crew and cost

---

[1] It may be surmised that this was to some extent one of the objects of the promoters of Button's expedition, small as was the hope of its successful accomplishment. In three out of the four editions of Hessel Gerritsz.'s work, published while Button's expedition was actually out (*vide* Dr. Asher's *Henry Hudson*, pp. 181-194), there occurs the very positive statement that one of the objects of the voyage was to recover Hudson; but, on the other hand, we have the significant fact that no mention of Hudson occurs in the elaborate Letter of Instructions given by Prince Henry to Captain Button before his departure (see Appendix B).

[2] At least, it is tolerably certain that he did not. The point is fully discussed in a foot-note on p. 131.

him his life. Hudson probably saw enough to show him that the sea he had entered was not the main western ocean which he sought; but it must have seemed to him more probable than not that there were other seas or passages leading from it into that main western ocean. His anxiety to continue his discovery was, therefore, very natural. Further, the mutinous survivors of his expedition, on their homeward passage, ran upon a rock near Cape Digges and remained fast for about eight hours, when "a great floud (which they, by this accident, took first notice of) came from the westward and set them on flote" (see p. 150). This flood or tide was regarded at the time as strong evidence of an opening into the Western (or "South") Sea. The necessity of following up this tide to its source is pointedly spoken of in Button's Instructions.

We may learn how bright, to the promoters of Button's expedition, seemed the prospect of the early discovery of a navigable passage from the very interesting and valuable Letter of Instructions given to Button by his patron, Henry Prince of Wales, which is printed in full as Appendix B. In it, the Prince says: "We assure ourself by God's Grace you will not returne, without either the good newes of a passage, or sufficient assurance of an impossibility." He is further instructed to "spend as little time as maie be in [the search for a good haven or harbour for ships "on the back of America or some Island in the South Sea"] or in any other search, saving of the Passage, till you have dispatched the Pinnace wth advertisement of your Entrie into the South Sea, wch must be done as soone as you shalbe thereof assured."

Another very strong piece of evidence of the extreme confidence felt in the complete success of Button's expedition is afforded by the statement made by Foxe (original edition, p. 268) that "in the Voyage of Sir Thomas Button were about 160 adventurers". It would have been quite impossible to have got together so large a number of adventurers had not the very greatest confidence been felt in the success of Button's expedition. No list of the names of this very large number of adventurers has been preserved; but there can be no doubt whatever that all of them were among the 288 persons who were incorporated as "The Company of the Merchants of London Discoverers of the North-West Passage", under a Royal Charter granted in July 1612, three months after Captain Button's departure. This Charter, with its highly remarkable list of names, is printed in full hereafter (see Appendix D). From it, one may gather that the general opinion then was that the much-sought passage *had been already discovered*, and that it only remained for Button's expedition *to sail through it*; for those noblemen and gentlemen who bore the expenses of Hudson's expedition are repeatedly spoken of in the Charter as being "the first adventurers and discoverers of the North-West Passage".

It is clear also from what Purchas says that he fully believed that Hudson had discovered a passage.

In the first edition of *his Pilgrimage*, published in 1613, less than two years after the return of Hudson's expedition, he says (*l. c.*, p. 624) that Hudson had "gained more hope of this discovery of the South Sea by a northerly Passage than ever before". In the last edition of the same work, published in 1626, after the expeditions of Button, Bylot and Baffin, Munk, and Hawkridge had proved that Hudson had not really discovered a passage, Purchas says of Hudson (*l. c.*, p. 817) that, having passed between Capes Wolstenholme and Digges, he came "into a spacious sea, wherein he sailed above a hundred leagues south, confidently proud that he had won the passage". Even then, however, some side-notes which Purchas inserts seem to give grounds for the belief that he still thought it possible a passage might be found through Hudson's Bay or Strait.

Still stronger evidence of the confident expectations that prevailed of the complete success of Button's expedition may be gathered from the various editions of Hessel Gerritsz.'s work which were published while Button was actually away upon his voyage; and it is interesting to note how those expectations grew when it was found that he did not return the first year. In the first, which appeared soon after Button's departure, we are told that one vessel was ordered to sail through the passage, when found, while the other "shall be sent home with the news, which we are expecting". In the next edition occurs the passage: "We expect more certain news by the ships which have already been sent there; and even the much-desired report that they have passed through the Strait. These ships will thus obtain eternal fame and glory." In the next edition, the tone is still more hopeful and we read that, when the ships "have found the open [western] ocean, one of them is to return with the desired news. This ship is daily expected home." In the last edition, we may observe even greater confidence. We read that nothing had been heard of the return of Button's ships. "We may, therefore [says Gerritsz.], hope that they have passed beyond that Strait, and we do not think that we shall hear anything about them before they return to England from East India or China and Japan," etc. Unfortunately, however, all these bright hopes ended in disappointment.

Captain (afterwards Admiral Sir) Thomas Button was fourth son of Miles Button of Worlton, Glamorganshire. The date of his birth has not been recorded, but he entered the naval service as early as 1589. Before the year 1600, he had served in the West Indies, in Ireland, and elsewhere, and had won considerable distinction and a pension; but not much is known of him previous to his coming into prominence in connection with the present voyage. After his return, he was knighted and rendered valuable services to the Crown as Admiral of the King's Ships on the Irish Coast, and in other posts. Later, he became mixed up in quarrels with the Admiralty; and, although he easily cleared himself of the charges brought against him, he died much impoverished in April 1634.[1] The best accounts we have of him are Mr. G. T. Clark's *Account of Sir Robert Mansel, Kt., . . . and of Admiral Sir Thomas Button, Kt.* (Dowlais, 8vo, 1883), and Prof. J. K. Laughton's sketch in the *Dictionary of National Biography.* The only popular work which, so far as I know, gives a notice of Button is the *Penny Cyclopædia* (London, 1836). In the *Encyclopædia Britannica* (9th ed., 1881) he is, by some extraordinary oversight, spoken of as *Butler.*

The Buttons, though not originally Welsh, had long been seated in Wales, and had intermarried with many good Welsh families. Various members of the family, both before and after the time of Sir Thomas, held the office of sheriff of their county. Clearly, therefore, Sir Thomas Button was of good family and education.[2] I had recently

---

[1] In a letter written by a certain G. Garrard to Lord Strafford, dated May 1st, 1634, and printed in *The Earl of Strafforde's Letters and Dispatches* (London, fol., 1739, vol. i, p. 242), the writer says: "Sir Thomas Button died of a burning Fever quickly, much discontented that he lost his Imployment in the Irish Seas."

[2] It is interesting to note here that Button Gwinnett, one of the "signers" of the American Declaration of Independence, was a direct lineal descendant from Sir Thomas Button. Mrs. Button, widow of Robert Button (the last of the Buttons of Cotterell), was mother of Mrs. Gwinnett, who had Cotterell from her. Button Gwinnett was her son. He is said to have been born about 1732, probably at Cotterell, though Sanderson states (*Biography of the Signers of the Declaration of Independence*, p. xvi,

an opportunity of visiting the residence of his father, and therefore his probable birthplace. The name of Worlton is now totally forgotten in the district, and the place is known as Duffryn House. It lies in the parish of St. Nicholas, about six miles inland from Cardiff. The house is a moderate-sized and unpretentious, but picturesque, mansion, charmingly situated in the bottom of a richly-timbered sheltered valley. A small stream, which flows through the grounds and passes close to the house, formerly worked a mill, which has now disappeared. Portions of the present house may be as old as the date of Button's birth, but the greater part seems more modern. It is rumoured in the neighbourhood that the ghost of Sir Thomas, armed and in full uniform, sits astride of one of the casks in the wine-cellar, and that pilferers are, in consequence, deterred from any attempt at robbery!

There are in existence two reputed portraits of Admiral Sir Thomas Button. One of these now hangs in the Town Hall at Cardiff. It appears from a report made by Mr. J. Storrie, Curator of the Borough Museum, to the Mayor (*Borough Minutes*, No. 1659, p. 405), that this picture formerly hung, with another, in the old Town Hall, and after its demolition, was stored in the gallery of one of the courts, where damp destroyed its

---

Philadelphia, 4to, 1865) that "the place of his birth, or where he resided, seems lost in obscurity". He was killed in a duel in May 1777.

frame and damaged the painting. In the course of time, it was, through Mr. Storrie's intervention, cleaned, restored, reframed, and hung. The picture, however, now represents a judge, and its style assigns it to the end of last century. According to Mr. G. T. Clark, F.S.A., of Talygarn (*loc. cit.*, p. 109), it was formerly a portrait of Sir Thomas Button, but the Corporation, wishing to honour Mr. Justice Hardinge, who was a favourite in Cardiff and a noted Welsh judge at the end of last century, engaged an artist to paint his portrait on an old picture in its possession, and that the artist added the red cloak and wig, changed the bâton in the Admiral's hand to a roll of paper, and made other alterations. This may very likely be the case, for the portrait is certainly painted over an older one; but, in its present form at least, it bears no resemblance whatever to the other portrait of Button.

This other portrait has a much more satisfactory history. It is a large oil-painting, measuring 34 by 39 inches, in perfect preservation, and it adorns the walls of the library of Mr. G. M. Traherne at Coedriglan Park, which is about five miles from Cardiff and in the parish of St. Nicholas. There is no clue whatever as to the artist, but his work proves his competence. The picture was acquired by purchase from the descendants of the Admiral, some seventy years ago, by an ancestor of the present Mr. Traherne, and has ever since hung in its present position. It represents an erect, almost full-faced, three-quarter length figure, with chestnut-brown hair and a decidedly pleasing expression. He wears a white ruff round his neck, and white frills at his wrists. His shoulders and chest are protected by what looks like a gorget of metal of a dark colour. The body is clothed in a tight-fitting doublet of greyish-brown material, finely embroidered, buttoned down the front, and with tight sleeves of the same material. The legs, which are shown almost to the knees, are encased in very loose breeches of dark green material, apparently velvet, embroidered in places. Across the body is a greyish-white sash of some light material, probably silk, tied in a large bow-knot over the right shoulder. At his side is a sword. Both arms are slightly extended from the sides, the left-hand resting on a terrestrial globe, the right holding what looks like a short staff or truncheon, one end of which is against the thigh. At the top corners are two smaller pictures, that on the left depicting a ship at sea in full sail, that on the right, apparently a landscape, but it is rather indistinct. At the bottom is the motto, "*Non mihi sed Patriæ*". That the portrait represents a navigator is obvious; while I have the authority of Mr. George Scharf, C.B., Keeper of the National Portrait Gallery, for stating that it is admirably painted, and that it is unquestionably of the date of Sir Thomas Button. Therefore, although there is no absolute proof that it represents the Admiral, there seems to be no good reason to doubt that it does. The motto (which may be translated, "Not for myself, but for my country") does not seem to have been that of the Button family; but this in no way proves that the portrait does not represent Sir Thomas, who may have adopted it personally.[1] The portrait, which has been reproduced by photography by the kind permission of Mr. Traherne, forms the frontispiece to this volume.

The instructions under which Button sailed were printed in full by Rundall (*Narratives of Voyages towards the North-West*, pp. 81-85). Rundall's work being now out of print, and the Instructions of great interest, they are herein reprinted in full (see Appendix B), together with the "Letter of Credence", dated Westminster, April 12th, 1612, given to Button by King James, which Rundall omits.[2]

---

[1] It was borne by the Springs, baronets (cr. 1641; ext. 1769), of Pakenham, Suffolk, and by the Hippisleys, baronets (cr. 1796), of Warfield Grove, Berks; but I cannot trace any connection between those families and that of Button.

[2] Both the Letter of Credence and the Instructions were printed in the *Athenæum* for 1834, p. 205, "through the courtesy of a gentleman who has a rich collection of such treasures. . . . . Both documents [it is stated] are beautifully written on vellum; the former is richly emblazoned: the latter has the seal and signature of the Prince affixed." Rundall (who evidently did not know of its previous publication), in 1849, printed the Instructions (see above) "from a rare *facsimile* of the original MS.", prepared at the expense of Mr. P. A. Hanrott (*l. c.*, p. iv), and still preserved in the British Museum (*Grenville Library*, No. 7218). It is a lithographed facsimile about 20 inches by 12, and apparently done about 1830. Rundall makes no allusion to the Letter of Credence, of which he probably knew nothing. In 1853, both

The Instructions were drawn up by Henry Prince of Wales, and are dated April 5th, 1612. The most noticeable feature in these instructions is (as has already been mentioned) the confident belief they indicate that nothing was left but for Button to sail through the Passage. The pith of the directions he received was that he was to hasten to Digges Island, "remembering that the waie [there] is alreadie beaten", and thence to "stand over to the opposite Maine [land] in the Latitude of some 58 degrees, where, riding at some headland, observe well the flood; if it come in southwest, then you maie be sure the passage is that waie; yf [it come in] from the north or northwest, your course must be to stand vpp into it." Among other things Button was instructed to do were to pay due regard to religious observances;

---

documents were again printed by Petherham in his *Bibliographical Miscellany* (No. 1, Nov. 15, 1853, pp. 5-8). From this work we may gather that the gentleman by whose courtesy the *Athenæum* was enabled to print the documents was Mr. Hanrott, who seems to have been fortunate enough to secure a copy of Birch's *Life of Henry Prince of Wales* (London, 8vo., 1760), containing the original documents engrossed on vellum. Mr. Hanrott had reproduced in facsimile a few copies of the Instructions, from one of which Rundall printed. Petherham states (*loc. cit.*, No. 3, Jan. 20, 1854, p. 22) that, at the sale of Mr Hanrott's library in 1833, the copy of Birch's *Life* (which formed lot 783 of part 1 of the catalogue) was sold for £2 15*s.* to Mr. Thorpe, bookseller; but the present possessor of this interesting volume is unknown. The version of the Instructions printed in the Appendix hereto follows Mr. Hanrott's facsimile, which differs slightly from the version given in the *Athenæum.*

to note which of his men appeared most ardent in the cause; to keep accurate and full journals; to make frequent observations; to follow a certain course on his outward voyage; to spend no time on the search for anything, except the passage, until he had discovered it; and, finally, when he had discovered it, to find some convenient haven on "the back part of America" or on some island in the western ocean, and then to send his pinnace back homeward through the passage with news of his success.

Button sailed about the middle of April 1612,[1] in two ships, victualled for eighteen months. He himself, as "Admiral" or "General" of the expedition, commanded the *Resolution*, and he had with him a relative named Gibbons, and a friend named Hawkridge (both of whom afterwards commanded expeditions which had similar aims, but failed to attain them); while Captain John Ingram commanded Hudson's old vessel, the *Discovery*. Prickett, too, accompanied Button, and he also had with him Robert Bylot or Bileth, who had been with Hudson (see pp. 164 and 203). Button passed through Hudson's Strait and anchored, as instructed, at Digges Island, where he put together a pinnace and lost five men, who were killed by the savages. Thence he stood to the westward, discovering the land he named Cary's Swan's-nest, and afterwards, in accordance with his instructions

[1] See p. 164, *note*.

he sailed on westwards (across what is now known as Hudson's Bay, but which was for long more appropriately known as Button's Bay),[1] falling in with the western shore of the bay in about latitude 60° 40′ N. Thus he was the first to cross the bay from east to west and to discover its western shore, which he named New Wales (see p. 170). The particular spot at which he encountered it, he named Hopes Checked, because he was by it disappointed in his hopes of sailing straight on into the Western Ocean. He then coasted southward; entered Port Nelson on August 15th; wintered there amid great hardships, losing his own ship, the *Resolution*[2]; and, on getting free again in the following spring, searched to the north-westward for the Passage, exploring in the *Discovery* the whole western shore of the bay as high as latitude 65° N., in the inlet afterwards know as Sir Thomas Roe's Welcome, which point he reached on July 29th, 1613. Then, his search having proved fruitless, he turned south-eastward, and, coasting along the southern part of Southampton Island, he spent a short time at the western end of Hudson's Strait, where he desired to further investigate the tide which Hudson's survivors had

[1] See p. 168, *note*.

[2] This fact is not mentioned in Foxe's narrative of the voyage, but Purchas says that Button "was forced to quit the Great Ship" (see p. 169, *note*), and he is supported in his statement bv an inscription on a board which Button erected, and which Foxe discovered in August 1631 (see hereafter).

observed there (see p. 150). This he admits he should have done at the outset, but his excuse was that, in acting otherwise, he was only carrying out his very definite instructions.[1] The result of his investigations greatly strengthened his hopes of a passage being ultimately found in this direction, but the season was too advanced for further search, it being now the end of August. He then revisited Digges Island, and thence proceeded homewards, disappointed in his high hopes, but still, like others, not finally convinced of the uselessness of further search.[2]

The all-too-meagre information which Foxe gives us about Button's voyage shows us that it was conducted with energy and discretion, and that (without fulfilling its main object) it was remarkably successful. By it, the discovery of the coast-line of the whole of Hudson's Bay was practically completed, with the exception of that part between Cape Henrietta Maria and Port Nelson, where, however, no hopes of a passage lay.

With reference to Button's observations on the variation of the needle, it is of interest now to note that Purchas makes the following comment on those observations (*his Pilgrimage*, 4th ed., 1626, p. 819):—

"This seemed strange: that, in this voyage, as he searched many leagues East and West, he found the Variation of the Compasse to rise and fall in admirable

---

[1] See p. 199.   [2] See p. 197, *note*.

proportion, as if the true Magneticall Pole might be discovered."

The variation of the needle had been first observed long before, namely by Columbus on the 13th of September 1492; and, although a certain amount of regularity in the variation had been observed by Frobisher on his voyages to Meta Incognita, according to William Borough's *Discours of the Variation of the Cumpas or Magneticall Needle* (London, 4to, 1581), the observations were thought of so little moment that, in his work published in the same year (*The Newe Attractive*, London, 4to, 1581), Robert Norman declared that there was no regularity whatever in the variation. From Purchas's remarks, therefore, it may be inferred that it was not until Button's voyage that the fact was clearly recognised; but it was not until 1831 that the North Magnetic Pole, or place of vertical dip, was first reached by Sir James Ross in about 70° 5′ N., 96° 45′ W.

Of Button's voyage, Rundall pertinently writes[1]:—

"The proceedings of this voyage are involved in what appears to be needless mystery. Purchas complains he could not obtain any information on the subject; and Mr. Briggs was also kept, to a great degree, in the dark, although he was eminent for his scientific acquirements, deeply interested in the success of the enterprise, and intimately acquainted with the navigator. For what is

---

[1] *Narratives of Voyages towards the North-West in Search of a Passage to Cathay and India*, 1496 *to* 1631 (Hakluyt Society, 8vo, 1849), p. 81.

known respecting the proceedings, thanks are due to the inquisitiveness and industry of Luke Fox, who sought and obtained information from some of the companions of Button, if not from the navigator himself, and also from Sir Thomas Roe, an energetic promoter of the north-west project [of whom we shall hear again shortly]. The information thus acquired was first printed in the *North-West Foxe*, A.D. 1635."

Foxe's own account of how he came by the information is as follows (see p. 162) :—

"Concerning this voyage, there cannot be much expected from me, seeing that I have met with none of the Journalls thereof. It appeareth that they have been concealed, for what reasons I know not; but it is fitting that such things should be made extant as may in any way redound to the good of the Common-wealth; and therefore I can but communicate what I have received from Abacuck Pricket, who was in the same voyage, and [from] others by Relation, and from Sir Thomas Roe [as regards the latter portion of the voyage]."

It seems from what Foxe says that, in addition to the two persons named above, he also received information as to Button's voyage from Captain Hawkridge, who sailed with him, and of whom also we shall hear again presently. The accounts given to Foxe by these three form a fairly-clear and connected narrative of Button's expedition; but why the full and official narrative should have been withheld is not altogether obvious. That it was purposely withheld, and that it has never since been brought to light, is certain. Purchas says that, on Button's return, his doings were "by him kept secret by some intent of his" (*his Pilgrimes*, part iii,

p. 843). A little later, however, Purchas tells us (*l. c.*, p. 848) that he had "solicited him for his noates, and had received from him gentle entertainment and kind promises" (cf. p. 240). He adds that, at the time these promises were made, Sir Thomas was detained in London by urgent affairs, but had undertaken, on his return home, to send his journals to Purchas. After that, instead of proceeding home, he was called away from England, so that the indefatigable Purchas never received the journals he so much and so advisedly coveted. "I cannot [he sadly says] communicate that which I could not receive"; but he added that, if he received the journals later on, he would insert them out of place, rather than not at all.

Foxe says (see p. 200) that, on the return of the captains of the ships who sailed under Button, their "journals were taken from them, and therefore who doth desire any further satisfaction from this voyage must seeke it from Sir Thomas Button". It will be seen hereafter, from a letter written by Button on February 16th, 1629-30, that on that date his journals were still in his own possession; and one may gather from what he says that it had not been through his wish that they had been kept secret. Probably some explanation of the mystery is to be found in the fact, already mentioned, that, on July 26th, 1612, about three months after Button sailed, the King had granted to the Adventurers who had sent forth Hudson and afterwards Button, a Royal Charter incorporating them as "The Company of the Merchants Discoverers of the North West Passage", with Prince Henry as "Supreme Protector". We may fairly assume that it was in consequence of instructions given by the Company incorporated under this Charter that the journals of Button and his companions were kept secret; but, inasmuch as Button did not discover the Passage, the members of the Company derived no advantage from their selfish desire to keep the information to themselves. It does not seem at all improbable that Button's journals may even yet be discovered in the possession of his descendants or in some public institution.

At this point, it will be convenient to make a digression, in order to speak of the Company abovementioned, and to trace its development.

Late in December 1851, there was sold by Messrs. Sotheby and Wilkinson the second portion of the library of Mr. Edward Drummond Hay, which contained a copy of that very rare work by John Davis, entitled *The Worldes Hydrographicall Description*, supposed to be a presentation copy from the author to Henry Prince of Wales, whose reputed arms appeared upon the cover. In this volume was a folio sheet, supposed at first to be in the handwriting of John Davis, entitled *Motives Inducing a Project for the Discoverie of the North Pole Terrestrial; the Streights of Anian into the South Sea; and the Coasts thereof* (*Athenæum*, Dec. 27th, 1851, p. 1377).[1] It was, however, after-

[1] It was purchased by Mr. Henry Stevens for £21 for the

wards pointed out (*Athenæum*, Jan. 3rd, 1852, p. 19) that the work could not have been a presentation copy to Prince Henry, who was only sixteen months old on the day of its publication, May 27th, 1595; while the inserted MS. could not have been in Davis's handwriting, for Davis had died in Dec. 1605, when Prince Henry was only about ten years old. More probably, the MS. was of the year 1610, that date appearing on the back of it. If this document was *not* by Davis, the question arises, *Who did write it?* It seems possible that it was written by Sir Dudley Digges, who was largely interested in the voyage undertaken by Hudson in the year in which the MS. is dated, and who, in 1611, published a work (see p. 119) entitled *Of the Circumference of the Earth; or a Treatise of the North-East* [should be *West*] *Passage*. Digges, moreover, was afterwards prominently connected with the Company of Discoverers of the North-West Passage, but there is no actual proof whatever that he was author of the *Motives*.

This MS. (which has never before been printed in an accessible form) appears as Appendix C. It is obviously addressed to the Prince of Wales, and,

---

Lenox Library, New York, where it is still preserved I am indebted to Mr. George H. Moore, Librarian of the Library, for this information, and for pointing out that the *Motives* was printed in No. 1 of the *Bibliographical Miscellany*, Nov. 15th, 1853, which seems to have been an appendage to the catalogue of second-hand books for sale by the late Mr. John Petherham, and which can only be obtained with great difficulty.

from its contents, one may note that the complaint that "The country is going to the dogs" is by no means a modern cry. The document commences with a dissertation on the advantages of commerce, especially to island kingdoms. It then proceeds to state that the profits of English merchants, trading with neighbouring kingdoms, had sunk so low that it was necessary to open trade relations with countries more remote; and that one of the most hopeful means of doing this was the discovery of a northern passage to China, Japan, and other countries in the "South Sea". It then proceeds to suggest that the King should create an order of knighthood, or some other dignity, to be conferred on those Adventurers who would provide funds and form a Company to promote the discovery of such a Passage, as other sovereigns had done with other objects in view. It is next proposed that the Adventurers should be incorporated into a Company with a view of discovering and "planting" the North-West Passage. These Adventurers, it is urged, should be men either of noble birth or high position; and, for every £100 adventured, it is suggested that they shall receive one thousand acres of land wherever the "Plantation" shall be located, with other benefits at the discretion of the Prince. In conclusion, it is stated that the document contains an outline merely; that the details and a scheme of management were thereafter to be communicated to the Prince by word of mouth; and that the writer, in order to enable the Prince better to understand the project, had left with

Mr. Wright,[1] at the Royal Library at St. James's, a small terrestrial globe.

From this summary of the contents of the *Motives* it will be obvious that (whoever the author may have been) we have in this document the germs of the project afterwards developed in much greater detail in the Charter of the Company of Discoverers of the North-West Passage, already alluded to.

The Charter in question, which is preserved at the Public Record Office (*Signs Manual, James I*, vol. ii, no. 30), has never yet appeared in print though portions of it are incorrectly given by Dr. Asher (*Henry Hudson*, pp. 255-256). It is, however, of considerable interest in connection with the History of Early Arctic Research, no less than five voyages of greater or less importance having been promoted by those who were incorporated under it between the years 1610 and 1616, namely, those of Hudson (1610-11), Button (1612-13), Gibbons (1614), Bylot and Baffin (1615), and the same again in 1616.[2] I have thought, therefore, that it is well worth while to print it entire, in spite of its great length (see Appendix D).

The Charter commences by reciting that, whereas

---

[1] Edward Wright was born at Garveston, Norfolk, in 1560, and died in 1616. He was an eminent mathematician, a friend of Briggs (see *post*), and preceptor to Henry Prince of Wales. He is the last named of the 288 persons incorporated as the Company of Discoverers of the North-West Passage.

[2] Probably, also, Hawkridge's voyage, in 1619, was partly promoted by the same Company, but of this there is no very clear record (see p. 249, and also Rundall's *Voyages North-Westwards*, p. 150).

twenty-three Adventurers, in conjunction with the Muscovy and East India Companies, had, in April 1610, equipped the *Discovery*, under the command of Henry Hudson, for the search for a North-West Passage to the South Sea; and, whereas Hudson had discovered a strait whereby the Adventurers hoped to open a trade with China, Japan, and other countries on those seas, the King, for the better accomplishment and discovery thereof, and according as he had been petitioned by the Prince of Wales and others, did incorporate the first Adventurers and some 265 others,[1] with the Prince as "Supreme Protector" under himself, into a "body corporate and politique" by the name of "The Governor and Company of the Merchants of London, Discoverers of the North-West Passage", with a Common Seal, a Coat of Arms, and all the rights and privileges of a corporate body. The direction of the Company's

[1] I have already (p. xx) alluded to this long list of names as "highly remarkable". It contains the name of nearly every person who is known to have taken a special interest in Arctic research at the time, together with those of an extraordinarily large number of eminent people and of persons belonging to well-known families, as a perusal of the list will at once show. It begins with many of the most distinguished statesmen of the day, with the Prince of Wales at their head, and enumerates altogether 25 Peers or Peeresses, 37 Knights, Baronets, or Court Officials, 38 Esquires, and 188 Merchants, or 288 persons in all. Among them, we meet with the two Wolstenholmes and Henry Briggs (who were afterwards connected with Foxe's voyage); Captain Thomas Button himself, and his companions Ingram, Gibbons, Bylot, and Prickett (but not Hawkridge); Richard Hakluyt, the celebrated author of the *Voyages;* John Evelyn, the Diarist; Sir James Lancaster, Alderman Francis Jones, Sir Thomas Smith, and a great many other well-known persons.

affairs was to be vested in a Governor and a Committee, or Board, of Directors, twenty-seven in number, who were to be elected annually in October, with certain prescribed ceremonies. Inasmuch as Sir Thomas Smith, Sir Dudley Digges, and John Wolstenholme, Esquire, had been the prime movers in the matter, the first-named was appointed first Governor, while the other two were appointed first Directors, together with a number of others who are named. The Company is empowered to set forth ships, to impose fines, and to sell merchandise. The exclusive right to trade and traffic through the Passage is granted to the Company for ever, and the subjects of the King and all foreign potentates are expressly prohibited from entering it; any patent to the contrary granted by Queen Elizabeth notwithstanding.[1] Lest any doubt should arise as to the boundaries of the Passage, it is very broadly defined as having its entrance between Desolation[2] and Labrador, including Davis' and Hudson's Straits, and extending to any seas lying northward, westward, or north-westward, either in America or Asia, between the entrance and the South Sea. Various customs

[1] This is, doubtless, an allusion to the fact that Queen Elizabeth had granted to Adrian Gilbert and others, Letters Patent, dated Westminster, February 6th, 1583-4, incorporating them as "The Fellowship for the Discoverie of the North-West Passage", with various exclusive rights and privileges similar to those granted by the Charter of 1612. This document is printed and discussed in Admiral Markham's *Voyages and Works of John Davis* (London, Hakluyt Society, 1880).

[2] That is, the southern part of Greenland (see p. 122).

regulations follow, together with provisions for the admission of sons and apprentices of members and of new members, it being provided that for the first seven years all admission-fines paid by the former should go for the use of the original Adventurers who set forth Hudson, in consideration of the expenses to which they had been put. New members were to pay a fine of one hundred marks. The King, in order to show his approval of their effort to extend the commerce of his kingdom and to add to his royal dignity, granted to the original Adventurers all customs and duties, which would otherwise have accrued to him, on all goods brought from or through the Passage between December the 15th, 1618, and December 15th, 1619, that being the seventh year after the Company's formation, by which time it was thought the trade through the Passage would be considerable. Similarly, the King granted to Captain Button and his crews the customs and duties payable on goods passing through the Passage between December 15th, 1616, and December 15th, 1617, that being the fifth year after the formation of the Company; one half thereof to go to Captain Button, and the other half to be divided among his Company in the proportions decided upon by Sir Thomas Smith, Sir Dudley Digges, and John Wolstenholme, Esquire. And, in order that the captains and crews of the Company's ships should not conceal from the Company any necessary and desirable information as to discoveries made or trade carried on upon any voyage, the Company was empowered to examine on oath any person employed on any voyage, and, in the event of any such person declining to give satisfactory information, to commit him to any prison in the City of London, there to remain until he consented to give the desired information. And, inasmuch as previous voyages had been marred by reason of mutiny and disorder on the part of the crew, the Company's captains and commanders were (by special request of the Company itself) authorised to exercise martial law, and to arrest and punish any who became mutinous. The document (which is a parchment about 28 inches by 24) ends with a brief summary of its own provisions.

Perhaps the most noticeable feature contained in this Charter is the evidence it affords of the degree of confidence—amounting almost to absolute certainty—with which the discovery of a North-West Passage by the expedition under Captain Button was looked forward to at the time, in consequence of the hopes raised by Hudson's discoveries. Keen must have been the disappointment when, not only Button, but those that followed him, failed to realise these high hopes.

At this point, it is necessary to remind the reader that we have brought the narrative of the search for a Passage up to the time of the return to England of Button's expedition in the autumn of 1613.

Never was the cause of Arctic Discovery more ardently prosecuted than at this period. As the return of Hudson's expedition in 1611 was at once followed by the dispatch, in 1612, of that under

Button, so now the return of the latter in 1613 was followed without the loss of a moment by the dispatch, in 1614, of yet another expedition having the same object in view. The sanguine hopes of the discovery of a passage raised by Hudson's voyage had not been realised by the results of Button's expedition. Indeed, they had been, if anything, lessened by the latter's failure ; but still, we know that Button himself believed that, if his own discoveries were properly followed up, the desired passage would yet be found. Accordingly, in the year following his return, yet another expedition was set forth. It was commanded by Captain Gibbons, who was a near relative of Button, whom he had accompanied as a volunteer.

Gibbons sailed in the *Discovery*, which had been the consort in the previous year of Button's vessel the *Resolution*, and which was also Hudson's old vessel. Purchas (*his Pilgrimage*, 2nd ed., 1614, p. 746) makes the following reference to Gibbons's voyage, which shows that he, at least, regarded the results of Button's expedition as satisfactory, even though they had not by any means come up to the expectations of himself and others (see p. xx). Speaking of Button at Digges Island, Purchas says that he there "found the comming in of the great and strong tide from the north-west, which feedes both those huge bayes, and leaves great assurance of nothing now left but a little sayling to the north-west for the finding of that Passage ; or reason to looke no further for it. For which the first adventurers, to their great charge, are now setting out the third voyage (which God prosper) this instant moneth of March 1614, in the good and luckie ship called the *Discovery*, and some other [ship],[1] which are now employed in hope of perfecting [the discovery of a Passage, under the command of Capt. Gibbons]." This, it should be observed, was written whilst Gibbons was actually preparing for his voyage. In the third and fourth editions of the same work (1617, p. 926, and 1626, p. 819), both of which were published after Gibbons's return, the passage reads as follows:

"The comming in of the floud from the North-West, giving them [*i.e.*, the Adventurers] hopes of a Passage, in March 1614, Captaine Gibbins was employed on this Discouery, in the *Discouery* (so was the ship called), but without any great discouery that I have heard of."

With the exception of the above references, we know nothing of Gibbons's voyage, beyond the few meagre lines of information given us by Foxe (p. 201), who says that he was victualled for twelve months; that, in attempting to enter Hudson's Strait, he was caught among the ice and driven into a bay (nicknamed by his crew, "Gibbons' Hole") on the coast of Labrador; that he lay there ten weeks in great danger; and that, when he got out, the season was lost and he had to return home. Thus, notwithstanding a very high testimony to his

[1] This is probably an error. There is no other record of Gibbons having had two vessels.

ability as a seaman which Button had given him (see p. 195), he most signally failed to accomplish anything whatever, and the source of the tide Button had reported was no more shown than if Gibbons had stayed at home. From what Purchas says, it is clear that the expense of Gibbons's expedition was borne by the enterprising Company of Discoverers of the North-West Passage, the history of whose incorporation by Royal Charter has already been given.

On the return of Gibbons's expedition in the autumn of 1614, the Adventurers had advanced no further forward than they had at the time of Button's return a year previously. Gibbons's ignominious failure had caused the total loss of a year of valuable time; and the further investigation of the tide Button had reported was not advanced in the slightest degree. It was necessary, therefore, that another expedition should be undertaken with the same object, and this the enterprise of the Company of North-West Adventurers was fully equal to. Without any loss of time, another expedition was planned and dispatched. As Purchas says (*his Pilgrimage*, 3rd ed., 1617, p. 926, and 4th ed., 1626, p. 819), after the return of Gibbons, the Adventurers, "persisting in their purpose, the next yeere, Robert Byleth, one which had been in [the] three former voyages, was sent forth in that ship as Master." "That ship" was the *Discovery*, which was thus making her fifth recorded Arctic voyage. Associated with Bylot, as "pilot", was William Baffin, a navigator of exceptional ability and character, of whom a notice appears in the body of the work (see p. 202). He it was who wrote the account of the voyage which Purchas printed and which Foxe has herein abstracted (see pp. 202-222).

Bylot and Baffin sailed on the 15th day of March 1615, and sighted Resolution Island on May 27th. Coasting thence along the northern shore of Hudson's Strait, they named various places which had not been named in previous voyages, and arrived at the western end of the Strait early in July. The whole of July and August was spent in cruising about among the ice and the islands at the western end of the Strait, and the southern end of what afterwards came to be known as Foxe Channel. Although the amount of original discovery effected on this voyage was but small, the coasts of the regions visited were for the first time laid down with tolerable accuracy on a chart; and we may fairly say (making use of an expression used by Foxe with reference to Weymouth and Davis) that Bylot and Baffin "lighted" Foxe into his Channel.

The general result of the expedition must, however, have been extremely disappointing to the Adventurers. Although Bylot and Baffin took great pains to ascertain the source of Button's reported tide, they came to the conclusion that it was not "the true channel tide", as Button had thought it to be. Baffin, in the clearest language, after his return home, declared his belief (see p. 220) that, if there was a passage at all, it was to be found by way of Davis

Strait, and not through Hudson's Strait. This opinion, though not literally, was practically, correct; and thus at last were the bright hopes of success, raised by Hudson's discoveries four years before, dashed to the ground.

Although Baffin's opinion that no Passage was to be found through Hudson's Strait was not so generally accepted as to altogether put a stop to further search in that direction at a later date, it seems at the time to have been accepted by the Company of North-West Adventurers; for they, at least, undertook no further search through Hudson's Strait, so far as we know definitely. But their energy and enterprise were still unexhausted. By the following spring, therefore, they had prepared yet another expedition, which was to sail once more in the old *Discovery* (which thus made her sixth and last recorded Arctic voyage), with Bylot again in command and Baffin as "pilot". The object of the expedition was to follow up the discoveries made by Captain John Davis in 1585-87, and to explore the further extremity of Davis Strait, as Baffin had recommended.[1]

---

[1] We find a contemporary reference to the disappointment felt in the result of Bylot and Baffin's first voyage, and to the hopes entertained of success on their second, in a letter, dated London, January 24th, 1615-16, written by George Lord Carew of Clopton to Sir Thomas Roe, and now preserved in the Public Record Office (*State Papers, Domestic, James I*, vol. lxxxvi, No. 16). The writer says:—"There is nothing this last soṁer performed, either by the Nor-West or North-East for the discovery of the Passage to the East Indies. I pray God that this next year may have better success."

This second voyage in 1616 was of far greater interest and importance than their first voyage in the previous year. Although, after its termination, Baffin was led to declare positively that, in the direction in which he had searched, there was "no passage, nor hope of a passage" (an opinion in which, as later discoveries have shown, he was not correct), still the voyage was well conducted and remarkably successful, the whole of Baffin's Bay being discovered and described with commendable accuracy. At the same time, it is not necessary here to notice further the result of the voyage. Foxe's account of it, abridged from *Purchas his Pilgrimes*, occupies pp. 223-239. Those desiring fuller information must consult Mr. C. R. Markham's *Voyages of William Baffin, 1612-1622* (London, Hakluyt Society, 1881).

The years 1617 and 1618 seem to have been a resting-time for those who were most closely associated with the search for the North-West Passage, and no voyage of exploration is recorded in those years. Very early in the following year, however, through the efforts of Sir John Wolstenholme (who had been knighted three years earlier, no doubt in recognition of his splendid services in the cause of Arctic research), another expedition was equipped and set forth in search of the Passage, under the command of Captain Hawkridge. It long remained unknown in what year, and under whose auspices, this voyage was made; but the researches of Rundall (*Voyages North-Westwards*, p. 150) showed that it took place in the year 1619; that Sir John

Wolstenholme was the prime mover; and that he succeeded in interesting a number of his friends in the venture. Whether or not the equipment of this voyage was another and a final effort made by the Company of Discoverers of the North-West Passage (of which Sir John Wolstenholme was a leading member), we are nowhere told; but, at any rate, we are safe in concluding that, even if the Company did not move in the matter in its corporate capacity, at least the friends who co-operated individually with Sir John Wolstenholme were mainly those who had been incorporated in that Company. However this may have been, Rundall shows (*loc. cit.*) that, on January 20th, 1619, Sir John Wolstenholme applied to the East India Company for assistance in sending out another expedition in search of a North-West Passage, stating that he desired "to raise meanes to furnishe forthe two pinnaces, w[ch] will cost *li.* 2000". He based his hopes of success on the old grounds, namely, "the great tyde of floode" which Hudson's survivors, Button, and Baffin, had all reported to exist in "Button's Bay". It is recorded that, in response to this application, and notwithstanding the opinion that had been expressed by Baffin in 1615 against further search through Hudson's Strait, the East India Company, with its customary enterprise, granted £200 towards the expenses of the voyage, or one-tenth of the total sum required.

Of Captain William Hawkridge, who commanded the expedition, very little is known It will be remembered that he had accompanied Button in 1612-13—it is said as a volunteer (see p. 167).[1] In spite of his good reputation as a seaman, and the experience he may be supposed to have gained under Button, his voyage proved almost as great a failure as that of Button's other companion, Gibbons (see p. 201), and produced absolutely no results of value.

---

[1] It seems probable that Hawkridge was a man of some property; for, among the volumes of "Transactions from 1609-1625" (fo. 95), preserved at the Trinity House, I have discovered a record of the fact that in after years he was unfortunate enough to lose by capture a ship and cargo (both his own), valued at £2,000, and that he was held captive in Algiers for ransom. The document runs as follows:

"Being requested to Certefie o^{r} knowledge of Capt. William Hockerage, now in Algier, a Captive, where he vndergoeth the vnhappie condicõn of a Miserable Slave, being taken wth 12 saile of Turke Men [of] Warre: touchinge his qualetye, hee hath had the com̃and of severall Shipps, both in the Seruice of the East Indye Companie, as also in other voyages: his ransome [is] sett at £250, wch hee is never able to paye wthout some charitable course be taken for him. And, beside this his ransome of £250, he hath lost in his Shipp: the Shipp wholie his owne: £2,000 in the value of his shipp and goods, wch great losse of his, together wth his most miserable and lamentable condicõn, is motiue to moove all xpian hartes to comiserat his case and minister to his libertye; and this much of our knowledge.

Witnes o^{r} handes this 24th of September 1631

Robt. Salmon,
T. Best,
Samu: Doves, Mr.,
&c."

Certificates for public begging (as they may be called), similar to the above, were not uncommonly given at the date in question by the Corporation of Trinity House, and other bodies. The result of this appeal is not known.

Of his proceedings, we have no account whatever, except the by-no-means satisfactory narrative given by Foxe (see pp. 248-259), which, he says, "hath come to my hands by manuscript or relation". This narrative shows that Hawkridge was either wanting in ability or very unfortunate. Although he arrived at the entrance of Hudson's Strait on June 29th, it was not until about July 22nd that he was able to enter, having first been impounded in Lumley's Inlet, to the north, and then driven much too far to the south. Having entered, he proceeded to the western extremity of the Strait, in and near which he remained until August 16th, cruising aimlessly about, first in one direction and then in another, in a manner which the very vague narrative does not enable us to follow, and along coasts which cannot be identified. He does not seem to have covered any ground which Bylot and Baffin had not carefully described in 1615, and he certainly made no fresh discoveries. He passed Resolution Island, on his homeward voyage, on September 7th. We should have lost very little had Foxe failed to preserve for us the account of this abortive voyage.[1]

This ignominious ending to another well-equipped expedition seems, at last, to have damped the ardour of those who had for years lavishly expended both their money and their time in promoting expeditions in search of a North-West Passage. From the time

[1] There seems also to be some inexplicable confusion in the latter part of the narrative, as is fully explained on pp. 256-57.

of the departure of Captain John Knight in April 1606, to the return of Captain Hawkridge in September 1619, a period of thirteen years,[1] there had

[1] I am speaking here of English voyages alone. Another year may be added, if the voyage of Captain Jens Munk is included. He was sent out by Christian IV, King of Denmark, with two ships, having a combined crew of 64 persons, and he sailed from Copenhagen on May 9th, 1619. On June 30th, he sighted Cape Farewell, and, on July 15th (after having entered Frobisher's Bay in error), he entered Hudson's Strait, upon which he bestowed the now-forgotten name of Fretum Christian. In the Strait, he met with very great trouble on account of the ice, and he did not reach its western extremity, Digges Island, until August the 20th. From this point, he seems to have sailed across Hudson's Bay straight to Port Churchill, where he laid up for the winter about September 9th, naming the place (which had not previously been named) Munk's Winter Harbour. Barrow (*Arctic Voyages before 1818*, p. 231) and others have identified his wintering-place with Chesterfield Inlet, much further north, but there is overwhelming evidence that it was Port Churchill. The early part of the winter was spent in fair comfort, the crews indulging largely in shooting; but the latter part of the winter and the spring brought the utmost privation, due apparently in the main to a too liberal use of spirits, and to a want of sufficient physical exercise. From the beginning of February onwards, deaths became of almost daily occurrence. On May 28th, only seven survived, and they were unable to bury their dead companions. By the 18th of June, all had died except Munk and two others. These gradually recovered, and, on the 18th of July, the three survivors set sail homewards in the smaller of the two vessels. They passed Resolution Island on August 18th, and eventually reached the coast of Norway in safety on September 21st. Munk accomplished nothing in the way of geographical discovery, but he is one of the pioneers of the search for a North-West Passage, and his narrative is, perhaps, the most tragic ever written. He appears to have known nothing of the voyage made by Captain Hawkridge in the same year, and does not once allude to it. Nor does it seem that, up to the date of publishing his book, Luke Foxe had ever heard of Munk's

never been a time (if we except the two years 1617 and 1618) when there had not been a well-equipped English expedition out searching for the Passage or just returned from the search, or when some fresh expedition was not being fitted out to start at the earliest possible moment. And, although considerably more than two centuries had yet to elapse before the object of all this searching was to be attained and a Passage was to be at last discovered, there need be no hesitation in pronouncing this the Golden Age of Arctic Research; for, never before nor since (except during the search for Franklin), was a Passage so ardently and so persistently sought. One can hardly wonder that, as all this searching had failed to reveal a Passage, the ardour of the searchers should have been damped, and that there should have followed a period of twelve years during which not a single attempt was made to discover a Passage, notwithstanding the fact that all the previous voyages had failed to prove the *non-existence* of the Passage, as completely as they had failed to prove its *existence*.

This brings us to the year 1631, in which both Foxe and James set forth. As Foxe's work was the

---

voyage, as he does not allude to it in any way whatever. Indeed, up to the present time, no adequate account of Munk's expedition has ever appeared in English. A reprint of his own narrative appeared in Copenhagen in 1883, with an admirable introduction by Mr. P. Lauridsen. I am glad to be able to announce that steps are now being taken to bring out an English translation of Munk's narratives, under the editorship of Mr. E. Delmar Morgan and myself.

most important of the two, as affecting any future discoveries, it will be well to take its author first.

Of the origin and personal history of Captain Luke Foxe, very little indeed is known. It is almost correct to say of him, as Mr. C. R. Markham says of Baffin (*Voyages of Baffin*, p. i), that "his previous history is an absolute blank"; and the same remark applies to many other early navigators. They were probably men, born of no particular family, who had risen to positions of responsibility by sheer hard work and their own competence as seamen. Mr. R. W. Corlass states that his father, Richard Foxe, of Kingston-upon-Hull, "was originally a sailor, and afterwards an assistant at the Trinity House, Hull." He brought up his son, who was born in the parish of St. Mary on the 20th of October, 1586, to the same profession.[1]

Through the kindness of R. Hill Dawe, Esq., Town Clerk of Kingston-upon-Hull, the Burgess Roll of that Borough has been most obligingly searched, for the purposes of this work, by Mr. T. Tindall Wildridge, who has brought to light the following interesting references to members of the Foxe family. The extracts have been translated:

4 Edward VI (1550) *Richard Fox*, porter, admitted and sworn a burgess of this town by fine of 40*s*.

---

[1] *Sketches of Hull Authors*, by the late Reginald W. Corlass, edited by C. F. Corlass and Wm. Andrews, p. 30 (Hull, 8vo, 1879).

1 Philip and Mary (1554). *Lawrence Foxe*, mariner, was admitted and sworn a burgess of this town 24th day of September in the year aforesaid, by fine of 40*s.*; to be paid into the hands of the Chamberlain 10*s.*, and the remainder to be paid in the manner and form following, videl[t]: at the feast of St. Martin in winter next, 10*s.*, and 20*s.* at the feast of Pentecost now next ensuing. If resident, and otherwise not, etc. By his pledge, Brian Bowmer.

13 Elizabeth (1571). *Richard Fox*, mariner, 19th day of July in the year aforesaid, is admitted and sworn a burgess of this town by apprenticeship with Robert Ratcliffe, etc. If resident, and not otherwise, etc.

33 Elizabeth (1591). *John Foxe*, the same day [4th February], is admitted and sworn a burgess of this town by apprenticeship with Robert Hall, baker, defunct. If resident, etc.

5 James I (1607). *Luke Foxe*, son of Richard Foxe [ffoxe], sailor [naut'], 22nd October in the year aforesaid, admitted and sworn a burgess of this town by patrimony of his said father. If, etc.

8 James I (1610). *Thomas Foxe*, by apprenticeship with Thomas Ramsdale, sadler.

Here we have Thomas Foxe, admitted in 1610, who (as likely as not) was a brother of Captain Luke Foxe, who we find was himself admitted a burgess by patrimony on October 22nd, 1607. Then we meet with a certain John Foxe, admitted in 1591, of whom we know nothing; and before him stands Richard Foxe, father of our subject, and also described as a mariner, who was admitted in 1571; while before him stands another mariner, a certain Lawrence Foxe, admitted in 1554, who was very likely Luke

Foxe's grandfather. It is tolerably safe to surmise that the family had been long settled in Hull.

Mr. Tindall Wildridge has also been good enough to favour me privately with a few extracts from the registers of St. Mary's Church, Hull, from which it appears that Thomas Foxe, four years after his admission (namely on May 19th, 1614), married Alice Hudspith; while the year before his admission (namely on February 12th, 1590), John Foxe married Alice Reanard, and was buried on September 20th, 1613, when he is described as a "baker".

Further than this, nothing relating to Captain Luke Foxe's previous history seems obtainable, except what he himself tells us. "I had [he says, in his account of his "Preparations" (p. 262)] beene sea-bred from my boyes-time, and had beene in the Mediterranian, Spaine, France, Holland, Norway, Denmarke, and the Balticke Sea." It is clear that, in his earlier years, he had been employed in the coasting-trade, and had been on many voyages in small trading ships sailing from Hull to Whitby, Newcastle, Yarmouth, and London, or other east-coast ports; for, in his "Preface to the Reader", he replies to certain nameless detractors, who declared him to be unfit to undertake the command of the expedition, urging, as he says (p. 11), "that I am but a North-country coaster, and hath but been brought up in small Vessels." There can be no doubt that he was diligent in the duties of his calling; for, in 1606, when only 20 years of age, we find him applying (though unsuccessfully) to a well-known explorer for a post as mate, "presuming [as he says] upon some parts I had, as the use of the globes and other Mathematicke Instruments", and upon his general experience as a seaman. It, is clear, moreover, that he had risen to some sort of official position, probably in connection with Trinity House, before he set sail on his famous voyage in 1631, when in his forty-fifth year; for he says (p. 11) that, by certain other of his perhaps-envious detractors, "it was also cast into my dish that I was an officer of the Admiralty". Against this absurd objection, Foxe vigorously defends himself.

That Foxe was married, we know; for a passing reference to his wife occurs on fo. 62 of his MS. Journal (alluded to hereafter); but who his wife was, and what descendants (if any) he left, seems quite unknown.

The foregoing seems to be literally everything known of Foxe previous to the time of our meeting with his name in connection with Arctic exploration; and, even for information as to his early interests in this direction, we have to rely solely upon his own narrative.

That Foxe's interest in Arctic research was due, in the first instance, to the ardour and enthusiasm of youth, there can be no question; for, in the passage in his account of his "Preparations" (p. 261), in which he candidly states that he "was neither importuned nor intreated to this undertaking [*i.e.*, his voyage] by any[one], eyther noble or gentle", he continues: "the truth is that I had beene itching after it ever since 1606, when I should have gone Mate to John Knight." However, he confesses that he came, in after years, to recognise the fact that, in 1606, when he was but twenty, "my ambition soared a pitch higher than my abilitie, [and I] . . . thought myself to bee fit for the best imployment, and desired to be pluckt before I was ripe." This fact, we may infer, Knight recognised at the time; for, as Foxe says, "he durst not depend upon me in that place [*i.e.*, as mate] for the voyage, so as I did not proceed with him." After this disappointment, "I was [he tells us] still kept in Marine imployments along the Coast, and crossing the Sea, whereby I gained Experience; and also at the Returnes home of all Ships from thence, I enquired of the Masters, Mates and others that were that way imployed, whereby I gathered by Report and Discourse and Manuscripts, how farre they had proceeded, what they had done, and what [there still] was to doe."[1] In order to thoroughly familiarise himself with the more abstruse and scientific side of the navigator's occupation, he tells us he cultivated the acquaintance of a certain Mr. John Tappe, who, he informs us, had a wide acquaintance among navigators. Tappe introduced him to Mr. Thomas Sterne, globe-maker, of whom he says, "I have found [him] to have engrossed all

[1] It is clear, from certain passing remarks made by Foxe in the course of his narrative, that he had come into personal contact with Prickett (p. 164), Bylot and Baffin (original ed., p. 228), and Sir Thomas Button (*l. c.*, p. 249), among other of the early Arctic navigators.

those former voyages by relation, manuscripts, and maps; from whom I gathered much, and must needs say he is a very well deserving Practitioner".

This John Tapp (concerning whom no information, so far as I have been able to discover, is given in any biographical dictionary) seems to have been a bookseller. His name is variously spelled Tap, Tape, Tapp, and Tappe. He was admitted, along with eleven others, a member of the Stationers' Company on June 3rd, 1600, "by translation hither from the Companye of the Drapers", each paying a fee of 3*s.* 4*d.* The first work he brought out was entitled *The Maryners Book, containing many Godlye Prayers, Hymnes, Songs, etc.*, and it was published on October 2nd, 1600. In publishing, Tapp seems to have chiefly confined his attention to works on navigation and arithmetic. He was also an author and editor of such works. His first shop seems to have been "on Tower-hill, neere the Bulwarke Gate". Here, in 1602, he brought out the first edition of his *Seaman's Kalendar*, a small quarto volume which ran through various editions, the fifth appearing in 1615. It must have been a standard work and well thought of at the time; for an edition, with additions by Henry Phillips and others, appeared as late as 1696, or nearly one hundred years after its first appearance. This, however, was not Tapp's earliest work, for in 1596 he brought out a revised edition of *The Arte of Navigation*, by Martin Cortes, which had been originally published at Seville in 1556, but of which a translation by Richard Eden had appeared in London in 1561. Tapp published this through his friend Hugh Astley, who was one of those admitted with Tapp, and to whose shop at "St. Magnus Corner" he seems afterwards to have succeeded. It was situated "beneath St. Magnus Church", which, according to Strype's edition of Stow's *Survey of London and Westminster* (Book ii, p. 174), stood on the east side of Bridge Ward. Here he brought out the later editions of his *Seaman's Kalendar*, as well as at least two later editions of his *Arte of Navigation* (1609 and 1615). In 1613, he published his *Arithmetic, or Pathway to the Knowledge of the Ground of Art*, of which a second edition, brought up to date by one Peter Ray, appeared in 1658. I have been unable to ascertain the dates of his birth and death.

Of Foxe's other early friend, Thomas Sterne, I have been able to ascertain absolutely nothing, though no doubt he was a well-known man and a competent adviser in his day. It seems not unlikely that the abstracts of earlier Arctic voyages which Foxe published were those "engrossed" by Sterne, but he nowhere tells us that they were so (see p. 218).

But Foxe's chief friend and ally in his schemes of Arctic exploration seems to have been "that famous mathematician, Mr. Henry Briggs", whom he says he had known even before he became acquainted with Messrs. Tapp and Sterne, and to whom he was chiefly indebted for the final accomplishment of his desires. This Henry Briggs was also a Yorkshireman. He was born at Warley Wood, in the parish of Halifax, in February 1561, and became first Savilian Professor of Astronomy at Oxford in 1596, which office he held for twenty-three years. He promoted largely the use of logarithms, first explained by Lord Napier in 1614, and made a journey to Edinburgh in 1616 on purpose to confer with that famous discoverer.[1] Between 1602 and 1633, he published various important works on mathematics, arithmetic, navigation, and kindred subjects; but the work with which we are chiefly concerned in the present connection was his *Treatise on the North-West Passage to the South Sea, through the Continent of Virginia, and by Fretum Hudson*, which was signed by his initials only, and appeared as an appendix to the second edition of Edward Waterhouse's *Declaration of the State of the Colony and Affaires in Virginia*, issued in 1622 (London, 4to). This discourse by Briggs was reprinted by Purchas (*his Pilgrimes*, vol. iii, p. 852) in 1625, together with a map of North America (apparently supplied by Briggs); but, ten years later, Luke Fox declared Briggs' arguments "to be but the imagination of men" (see p. 239). It must, indeed, be con-

[1] Mr. C. R. Markham, in speaking of the introduction of Logarithms by Henry Briggs (*Journal of the Royal United Service Institution*, 1891), says, "No greater service has ever been done by one man to navigation, and Luke Fox, who was among the first explorers who reaped the benefit of it, might well name one of the [groups of] Islands he discovered in Hudson's Bay—Mr. Briggs his Mathematicks."

fessed that this tract contains some very erroneous notions; for its author argues that because "the sea wherein Master *Hudson* did winter (which was first discovered by him, and is therefore now called *Fretum Hudson*) doth stretch so farre towards the west that it lieth as farre westward as the Cape of Florida", and because it receives rivers from the mountains to the westward as also does Virginia, that therefore the two places "cannot be in probability so farre distant . . . as 200 Leagues"; and also because the place where Sir Thos. Button wintered was as far to the westward as the Cape of California, that therefore the main western coast of America, or some of its inlets, could not be far to the westward of Virginia, and could probably be reached by following up the rivers running from the westward, especially as the Indians of Virginia, as well as those of Canada and Florida, all agreed in their reports of a large sea to the westward, with large ships upon it; all of which speculations, as we now know, are very erroneous.

Nevertheless, it is certain that the fact that Foxe ultimately sailed upon his voyage of discovery is due, largely at least, to the interest Briggs took in the venture.[1] Foxe tells us that Briggs encouraged

---

[1] A contemporary estimate of Foxe's ability as a practical seaman, and reference to Briggs' connection with the enterprise, is to be found in a letter dated "Christ College [Cambridge], January 30th, 1630-31" (*Harleian MSS.*, No. 390, fo. 534; printed in the *Court and Times of Charles the First*, vol. ii, p. 94), in which the Rev. Joseph Mead, writing to Sir Martin Stuteville,

him in his ambition to search for the Passage, and also introduced him to Sir John Brooke, "whom both to doe their King service, and the Publick good, perswades with divers of their friends to come into the Adventure." As to the personal history of this gentleman, and his connection with Arctic research, I have been unable to obtain any definite information. He was probably that "John Brooke of Heckington, co. Lincoln, Knt.", who was created Baron Cobham on January 3rd, 1644-5, when the King restored to him all the privileges and dignities that had formerly belonged to his grandfather, George (Brooke) Lord Cobham (G. E. C.'s *Complete Peerage*, vol. ii, p. 324); but no further information concerning him seems obtainable. His name does not appear, so far as I can find, in Metcalfe's *Book of Knights*. It was after him that Foxe, on July 29th, 1631, named Brooke Cobham Island (now known as Marble Island), "thinking then [as he says] of the many furtherances this voyage received from that Honourable Knight,"

---

informs him that "The voyage intended for the North-West discouery is chiefly furthered by Mr. Brigges, and the man to be therein imployed is one Fox, a master of a ship of Hull, and, though never yet in those parts (so my author[ity] thinks), yet is held to be a very skilful seaman." It is sad to have to add that, at the date of this letter, Briggs had been dead for four days. In the next letter Mead wrote to his correspondent, which bears date "Christ Church, February 6th [1630-31]" (*Harleian MSS.*, No. 390, fo. 535; *ibid.*, p. 97) he says: "Mr. H. Brigges, of Oxford, the great mathematician, is lately dead, at seventy-four yeares of age."

which "furtherances" he then proceeds to enumerate with considerable detail (see p. 324).[1]

These three, then—Foxe, Henry Briggs, and Sir John Brooke—as the first-named tells us, next drew up and "exhibited a Petition to his Majesty [asking] for the lend of a Ship for the Voyage, and [his Majesty's] countenance to the Action".[2] From what Foxe himself says, it is clear that this must have been in December 1629, or early in January 1629-30. There is independent evidence of this; for, in a letter from Sir Thomas Button to Lord Dorchester, one of the Lords of the Admiralty, which is dated Cardiff, February 16th, 1629-30, and which is preserved in the Public Record Office (*State Papers, Domestic, Chas. I*, vol. clxi, No. 10), we find Button acknowledging the receipt of a letter from Lord Dorchester, dated January 14th, containing a copy of Foxe's petition, and asking, on the King's behalf, for Button's opinion with respect thereto. This letter seems to have been a month in transit from London to Cardiff, for Button (who wrote in reply to it on February 16th) says it came to hand on the 14th inst. In his reply (which is remarkable for its quaint, yet pertinent, wording), Button says he has looked into his journals and papers, which he thought would never have been made use of, and proceeds, in answer

[1] In his MS. journal, Foxe speaks of it as "Brooke's Island" and also as "White Island"; but, in the printed narrative, it is always Brooke Cobham, although Sir John Brooke seems to have had no right to use the title of Lord Cobham in 1631.

[2] The text of this petition seems to have been lost.

to the questions, to give his opinion concerning the feasibility of finding a Passage, as follows:

"For his Mat^{s} especiall service.

"To the Right Honorable and very much honored good Lord, the Lord Viscount Dorchester, principall Secretarye of State to his Maties Court, or else where, hast these,

"Right Honorable and my much Honored Good Lord,

"Yo$_{r}$ Lo^{s} letter of the 14th of Januarye concerninge the North west passage, wth the Coppie of Luke ffox his petition and others in that business, coming by the way of Bristoll, came to my hand but the 14th of this p'sent, att 6 of the Clocke at night; whereby, findinge his Ma^{ts} pleasure and the contents of their peticõn, the next day, the better to inhable my selfe to give satisfaction in a pointe of so highe a nature, I oũlooked my Journall and those notes and papers that long have laine by me, wch I thought would never have bin made use on,[1] consideringe that these later tymes amonge o^{r} nation rather studies howe to forgett althinges that may conduce to the good of posteritye by adventuringe six pence if they fynd not a greate and presentt benifitt to insew thereof.

"But, in as much as yet att lenght it pleaseth God to open the eies of som to looke after soe importantt a busines for the Honõ of his Matie and not only the comon good of this o^{r} kingedome but of all o^{r} neighbore nations, I shall, in answer of yor Lo^{ps} letter, and in most Humble obedience to my most royall masters command, deliv^{r} not only my opinion, but under corection my knowledge, gotten by the sharpest experience, of that designe of any man of my sorte livinge, not only in o^{r} owne kingedome, but in any other in these neighboringe partes.

"What yor lo. writes off that his Matie requires to be informed of by me is:—

---

[1] See p. xxxiv.

"First, whether there be any likelyhood or p'babilitye to compass the designe, yea or noe?

"To that I answer that my opinion is nowe as it ev^{r} hath bin sithence my retorne thence, and as I then deliv^{r}ed it with the perticuler reasons of it to my most royall master of most famouse memorie that then was, Kinge James; that, beinge undertaken in a fittinge waye and a dewe season, I made, and doe make, as full accoumpt of the sensiblenes of it as I doe of any knowne chanell that is best knowne to us in these Norther partes, and to be p'formed w^{t} as little dainger (and was soe approved by his Matie to be), whoe inforst as manye and as importants questions for his owne satisfactions as if all the best experienst mariners of the Christian World had convented them selves together to have drawne the intergatories. The same reasons have I deliv^{d} to many most Honorable and knowinge persons and to o^{r} best Mathamaticians, as Mr. Briggs, Mr. Evells [?], and others, w'thall the best masters and mariners of o^{r} kingdome, as alsoe to others, both Hollanders and French; and, in my discourse with any one of them all, they never went unsatisfied from me of the probabilitie of it; and for further accompt herein att presentt I can give non; but, if my Journall or any other my notes or papers (w'th ought else in me) may give his Matie any further or fuller satisfaction, when I waight on his Highnes (w'ch I hope will be much sooner than is fitt for them to advance; for to sett out to tymlye, is to faule [fall] to soone into that danger that to late a repentance cannot healpe them out on), I will do my best, out of my ould experience, to affurther the good of it, and p'vent the evells and inconveniences that pretending men of little experience, or none at all, may suddenly bringe uppon it; for I will bouldlye saye that whoe shalbe fitt to have the manedginge of this unparaleld busines ought first to be so religiouse as to hould his end the happiest that dyes for the glory of God, the Hono^{r} of his Kinge, and the publique good of his countrye, all w'ch in this designe have their sev'all and p'ticuler interest; and therefore he must not looke backe for feare of the dainger of either unknowne coastes, hideouse storms, dark and long continewed mistes, to lye amonge and all ways to see more landes and Ilands of Ice than he can see of sea, and oft tymes rocks under him in sight, when he shall within theire his ships lenght finde twentye fathom water ; and, to incounter this under favo[r], must he be well armed that shall undergoe this busines : for thrice sithence my beinge there hath it bin attempt'd, and, for ought I here, little (or, rather, I may bouldly say, noe) advauncem[t] given to the busines[1] ; therefore cannot be to much curiositie used to put it into a good and choise hand, w'ch I will hartilye praye may be most happilye lighted on, for wee live not in the adge to fynde that they are the most p'fitt w'ch makes the [most] gloriose shewe.

"The second pointe required is, whether it may prove of such benefitt and advantadge as is pretended?

"To that I must most humbly answer, that that received opinion of form[r] adges (as well as of these moderne tymes) both in manye other countryes abroade, as in o[r] owne kingdom, and amongst o[r] own marchants att home, mainetaines and makes good that pointe; therefore to that I can saye noe more, but that I will as hartilye praye that God may give a blessinge to the discoverye ; w'ch, in the first place, must be the im̃ediate introduction, to bring on and p'fitt the rest; of w[ch] hono[r] to be that most happie man weare my yeares sutable to such an undertaking, or my purse answerable to what in hart I would be most willinge to adventur ; I would be loathe any man livinge should undertake it sooner than my selfe, or adventure more towards it then I would ; but, beinge no otherwise usefull in myne owne power or abilitie then in my welwishing, and what other affurthrance may lye in mee ; yet what I formerlye suffred by my winteringe doth sufficientlye satisfye all reasonable and experienst men that to

[1] Button refers, of course, to the expeditions of Gibbons (1614), Bylot and Baffin (1615), and Hawkridge (1619).

runn the hazard or chardge of such a purpose can be to noe other end then the inevitable hazard of all; and therefore, either the passage wilbe found, or not to be hoped for, the first yeare, soe by that assurance the first chardge wilbe much the less, and the source certaine w'ch will effect it the sooner; for nowe there will be noe faulinge into Hudson's Bay, nor Buttons Baye, to mispend tyme, as both he and I[1] did to noe purpose, and that only by Instructions out of England; but as soone as he[2] comes to the west parte or Cape of Notingham Iland where he is to anchor, and, according to the sett of that tyde which he shall finde there, to direct his course; w'ch must be and is the only way to fynde that passadge, w'ch I doe as confidently beleave to be a passadge as I doe there is on either between Calis and Dover or between Holy Head and Ireland. This being all att p'sent that I can doe in answer of y^{r} lo^{s} letter or for his Ma$^{t's}$ information in this busines, intendinge to bringe upp my Journall and such other notes as I have leaft, when I come upp myselfe for his Ma^{ts} or yor lo^{s} further satisfaction (but to noe other hands) I most humblye take my leave, assuring yor l^{p} on my faith yor letter came noe sooner then when I write; and, if there be any error happens by it, the fault is not myne, whoe am and all wayes shalbe,

"yor lo^{s} most respective, trew, thankfull,
"and humble servant,

"From my house att Cardiffe, this
16th of Februarye 1629[-30]."

---

1 That is, "Hudson and I."

2 That is, Foxe, or whoever was to command the projected expedition.

It seems probable that, in presenting their petition to the King, Foxe and his friends had also asked for a grant of certain concessions or advantages in the event of their discovering the Passage; for, in a second lengthy letter which Sir Thomas Button wrote to Lord Dorchester and enclosed with the foregoing, he reminds the latter that there was in existence (see p. xxxviii) a former patent or charter, granted by King James in 1612, to those who sent forth Hudson and himself (Button); and he suggests that His Majesty, before granting any further concessions, should confer with Sir John Wolstenholme and Sir Dudley Digges, who were the chief, under Prince Henry, of those to whom the former charter was granted. Button then proceeds to give a most doleful account of his position and prospects, and to beseech Lord Dorchester to use his influence with the King and the Lord Treasurer to pay over to him that to which he was entitled. He states his case with great force and evident truth. He had, he says, been thirty-seven years in the service of the King, the King's father, and his predecessors, during which time he had, as Lord Dorchester would be aware, "never balkt nor avoided any imploymentt, were it never so remote or dangerouse, so [long as] it pretended to his Maties service or the Comon Wealthe's," as was shown by his voyages to the North-West, the West Indies, and elsewhere; yet, he says, owing to the non-payment of moneys due to him, he was in a worse position than he had been ten years previously, and had been compelled to "morgadge and forfeit" lands worth nearly £500 per annum, in order to satisfy his creditors, who were pressing him to pay his debts. Under these circumstances, and in order to avert ruin from himself, his wife, and seven children in his last days (which he protests he cannot avoid if not soon relieved) he prefers, he says, to lay himself at his lordship's feet, beseeching the payment of a portion, at least, of what was due to him, rather than to trust to "the lardge p'mises of such as I fynd produces noe other effectte then smoak". The whole text of this long appeal is not given here, as it does not bear very directly upon the subject in hand. It is painful to think that a man of Sir Thomas Button's services was allowed to suffer from such injustice; but the circumstances were by no means unusual in the days of King Charles I; and we shall hereafter see that Captain Foxe himself afterwards had to complain of exactly the same scandalous treatment.

A contemporary reference to Foxe's intended voyage, and to the opinion of Button respecting the proper way for the expedition to search for the Passage, is to be found in a letter, dated January 13th, 1630-31, and printed in the *Court and Times of Charles the First*, vol. ii, p. 90 (London, 8vo, 1848), in which Mr. John Pory informs Sir Thomas Puckering:

"that certain merchants of London were about to set out a new voyage for the discovery of the North-West Passage, one of the adventurers wherein asking in his hearing Sir

Thomas Button (who had formerly been employed in the like discovery) his opinion concerning the same, his answer was that, if they meant to do good therein, they must make choice of an honest and sufficient man to be employed therein, whom they must not tie up with any directions, but, relying upon his art and judgment, leave him at liberty to practise the same according to occasion; 'for', said he, 'the directions given to me at my setting out to my voyage for the like discovery, destroyed quite the success thereof; for, having sailed four hundred leagues into the Strait, and the tides of the Eastern Ocean ceasing, and the tides of the Western Ocean coming in strong, which I ought to have followed, I was, by my directions, diverted into a bay, where, being frozen up, I was forced to winter, and so leave the further pursuit of that discovery.'"

We may learn from Lord Dorchester's endorsement that Button's letter was received by his lordship at Newmarket on Feb. 27th. Then followed a long (and, to Foxe, undoubtedly, a most aggravating) delay of over three months, during which nothing seems to have been done, and no reply made to his petition. At last, however (the supply of official red-tape having apparently come to an end), a decision was arrived at; and, on June 8th, after more than six months' delay, the prayer of the petitioners was granted, as we learn from a very badly-written and almost illegible minute by Nicholas[1] of an answer given by Secretary Coke to Captain Foxe's petition to the King, which is preserved in the Public Record Office (*State Papers*,

---

[1] See p. 302, note 5

Domestic, Charles I, vol. clxviii, no. 28), and runs as follows:

Luke ffoxe. W[hethe]r the petitioners will imploy this shippe att their own charge, which . . . [?] Y^e^ Reprisal: leave may be given unto it, or otherwise she is to be sould for his Maties benefit, as at first was appointed.

. . . . [?] *The Charles.* This ship is ordered to be sould, yf she be not employed by those to whom she is lent.

Foxe himself says (p. 263) that the requests of his colleagues and himself were "graciously accepted and granted".

We find the sequel to the annoying delay which had taken place in granting Foxe's request, in a second petition which Foxe submitted to his Majesty in Council at Whitehall on July 2nd, or less than a month after the prayer of his first petition had been granted. In this second petition, Foxe stated that, although his Majesty had lent the *Charles* pinnace to the Adventurers for the North-West Passage, for a voyage which it had been intended to set forth that spring, it had been found impossible to get her ready in time; and, as the Adventurers had expended upon her the sum of £35, it was prayed that, until the time came to start in the following year, she might be used as a man-of-war or as a merchant vessel with letters of marque.

That Foxe found himself unable to complete his preparations in the year 1630 was very natural; for he ought to have started (and afterwards did start) not later than the month of *April*, whereas we have seen that, owing to the delay caused by official red-tape, his request for a ship was not granted until the middle of *June*. Fox and his colleagues, however, were discreet enough not to point out to those he was petitioning that the muddle was of their making; and an Order in Council was thereupon made granting the petitioners all that they asked (*State Papers, Domestic, Chas. I*, vol. clxx, no. 10). The order runs as follows :—

An order about his Ma[ties] Pinnace called the Charles.

At White hall, the second of July 1630.

Present :

Lorde Keeper.	Lo: Visc. Dorchester.
Lo: Treasurer.	Mr. Treasurer.
Erle of Bridgwat[r].	Mr. Secret: Coke.

Whereas a petition was this day presented to the Boarde by Captaine Luke Ffox, wherein hee doth humblie remonstrate that his Ma[stie] was gratiously pleased to lende his Pinnas called the Charles, wth the ordinance, munition, and fourniture necessarie, unto the Adventurers for the North-West Passage, to be set foorth this spring for the Discoverie thereof, but by reason the tyme of the yeere was so farre spent before the provisions could be made readie, the Adventurers, out of their speciall care to take a seasonable and fit tyme, that they may give his Ma[tie] a good account of their diligence therein, are resolved to deferre the voyage untill the next yeare ; having before the deferring thereof been at the charge of thirty-five pounds for bringing the said Pinnace from Chatham, wth docking, graving, and preparing her for the voyage intended ; In regarde whereof hee doth humblie sue that, if their ll'pps shall not thinke fit to appoint the said Pinnace for his Ma[ts] service, they wilbe pleased to grant permission to him and the aforesaid Adventurers to employ the said Pinnace along the coast, or otherwise, untill the season shall fit for the aforesaid intended voyage; the rather for that it is regarded that she is slow of saile, and, being a cast ship and appointed to be solde, may want manie repaires, and other carpenter's worke not yet knowne unto them, to make her serviceable for a voyage of discoverie; which wilbe founde by her employment, the Ship much betterde, and the charges, past and to come, defrayed. Their ll'pps, having taken this his humble suite into consideration, doe pray, require, and autorise the Lords and other Commissioners for the Admiraltie to cause this and his Maj^{s} said Pinnace with her tackle, ordinance, fourniture to be lent to the petitioner and the aforesaid Adventurers, to be set forth and employed either as a mere man of warre, or in marchandizing with le^{rs} of Marqe,[1] untill the season shalbe fit for the undertaking of the voyage for wch his Matie was gratiously pleased to lend the same unto them.

Ex F. DICKENSON.

This document is endorsed by Nicholas: "Nothing donne upon this order." Six weeks later, however, it seems that something was done; for we find the following entry in a book containing orders for issuing Letters of Marque, which is also preserved in the Public Record Office (*State Papers, Domestic, Charles I*, vol. cxxx, p. 49):

Charles (I. m.). The like was graunted unto Captaine Luke ffox and others to sett forth the Charles of London, of about 150 tonnes, whereof sd. Luke is Capt: Date 20° Aug. 1630.

---

[1] "Letters of Marque" constituted a commission authorising a privateer to make war upon, or seize the property of, another nation. They were abolished among European nations at the Treaty of Paris in 1856.

For that year, therefore, the expedition was abandoned. Foxe's own account of the matter (p. 263) is that, "the time of the yeare was so farre spent before wee could make our provision ready, as wee were forced to desist untill the yeare following."

In the meantime, Foxe tells us (p. 263), "There came home that honorable knight, Sir Thomas Roe, from his ambassage to the King of Sweden." This illustrious man was born at Low Leyton, in Essex, in 1580; was educated at Magdalen College, Oxford; and, having been Esquire of the body to Queen Elizabeth, he was knighted by James I in 1604. He became a most successful diplomatist, and represented England at the Court of the Grand Vizier. He also concluded the peace between Gustavus Adolphus and Poland. On his return, he would have been chosen by the Protestant section of the country as Foreign Secretary, but the influence of Archbishop Laud, and others, was too strong against him, and in 1632 he retired to his estate at Bulwick, in Northamptonshire, where he chiefly resided until his death, in 1644, carrying on a correspondence with friends in various parts of Europe. Letters of great interest, addressed to Queen Elizabeth of Bohemia, to Lord Dorchester, and to many others, are preserved among the State Papers. Foxe named after him *Sir Thomas Roe's Welcome Island*, a small island in the north-west of Hudson's bay, but the name is now applied to a Strait.

Sir Thomas Roe, being made acquainted with the design of the Adventurers, gave the matter his best support. The King then sent for Sir John Wolstenholme, who had long been known as the staunch friend and supporter of all the Arctic voyages of his time.

This Sir John Wolstenholme, Knight, is stated in Burke's *Extinct and Dormant Baronetcies of England* (London, 8vo, 2nd ed., 1844, p. 578) to have been the second son of a certain Mr. John Wolstenholme, younger son of an ancient Derbyshire family, who was born about the year 1520; came to London in the time of Edward VI; and, obtaining an office in the Custom House, died at Stanmore, Middlesex, where he is buried. His son John, of whom we are speaking, was born in 1562. He was knighted by Charles I on March 12th, 1616-17[1]; and, acquiring great wealth as "farmer of the Customs", he purchased Nostell Priory. No man of his time was more energetic in promoting Arctic research, or more liberal in providing funds for the purpose of dispatching exploring expeditions. Foxe says that his own expedition cost Sir John at least £400, and that he had contributed more liberally than anyone else to the funds of no less than eight previous voyages, which Foxe estimates had cost him at least £700, making his contributions to the cause of Arctic discovery at least £1,100, a considerable sum in those days. We find his name

---

[1] Metcalfe's *Book of Knights*, p. 169.

very prominently connected with Hudson's fourth voyage in 1610. As a result, he was appointed one of the first Directors of the Company of Discoverers of the North-West Passage in 1612, and, consequently, he took a leading part in directing the four voyages in search of a North-West Passage sent out by that Company. In 1619, as has been shown (p. 248), he was the leading promoter of Hawkridge's abortive voyage. In the minutes of the Court of Committees of the East India Company, preserved at the India Office, it is recorded that, at a Court held on May 4th, 1625,

"Mr Governor moved the Court on the behalf of Sr John Wolstenholme, that they wold bee pleased (in regard that he and other the Adventurers for Discovery of the North-West Passage are now sending a small shipp and a pinnace for that Service[1]) to write their Le^{rs} to their ffacto^{rs} at Bantam that, in case the said shipp and pinnace shold discover that passage, and so come to the Indies, that they

---

[1] Another reference to what was, doubtless, the same intended expedition occurs in a letter from Sir John Coke to Sir Edward Conway, dated London, March 20th, 1624-25, and now preserved at the Public Record Office (*State Papers*, *Dom.*, *Jas. I*, vol. clxxxv, no. 82), in which it is mentioned that the Duke of Buckingham had adventured the pinnace *Lion's Whelp* (which had been given to him by the King) for the discovery of the North-West Passage. Apparently, nothing came of the project, as there is, I believe, no record of an expedition resulting. It is, however, interesting to note from the above that the Company of Discoverers of the North-West Passage still, to some extent, retained a corporate existence as late as 1625. The account given by Rundall (*Voyages North-West*, p. 186) of Wolstenholme's application is incorrect.

might be comãunded not only to relieve them and use them kindly, but also to suffer them to lade their shipps from thence for England, with pepp[r]. One of the Comĩt-tees disliked the moc̃on, being of opinion it were better the adventure were lost than that the said passage shold be discovered, for that it will greatly p'udice the Companies benefitt and trade to the East Indies. But, after some debate and discourse had of the said moc̃on and of S[r] John's request, the Court was content in favo[r] of him to write their Le[rs], but with these reservations: First, that the said shipps shall not be suffered to lade any pepp[r] from thence, in case the Company have any shipps there of their owne, and ready at that time to returne for England, but that they shall bee first employed. Secondly, if the Company have not any Shipps there at their arrival in the Indies, that then they are content the said 2 small shipps shallbee fraighted with pepp[r] for England, so as they be laden for the Companies accompt, for w[ch] they are willing to allowe them after the rate of 20*li.* the Tunne for the freight of the Same."

On March 13th, 1628-29, when a Mr. Christopher Gardiner, who had travelled in "Cataya", and claimed a special knowledge of the way to search for a North-West Passage thither, offered his services to the Company, he was thanked for his offer, but was referred to Sir John Wolstenholme and Sir William Russell, "who (having a great affec̃on to find out this discovery, and upon which they have bestowed some moneys) were very likely . . . to hearken to his moc̃on." Foxe named after him Cape Wolstenholme's Ultimum Vale (now known as Cape Henrietta Maria), "for that [as he says] I do beleeve S[r] John Wolstenholme will not lay out any more monies in search of this Bay."[1] He died in 1639, and was buried at Stanmore, beneath a curious marble monument, erected in a church which he had himself built. The following facsimile of his autograph is taken from a document, dated 1619, and relating to the victualling of certain ships of the Royal Navy. It is signed by him and three others, and is preserved in the British Museum (*Add. MSS.*, 15,750, fo. 13).

Sir John Wolstenholme, Knight, was succeeded by his son, Sir John, who had also been knighted by Charles I, namely, on May 8th, 1633.[2] He was generally known as Sir John Wolstenholme the Younger, to distinguish him from his father of the same name, both having for six years contemporaneously enjoyed the honour of knighthood. He it was who was appointed Treasurer to Foxe's Expedition. "During the Civil War [says Burke, *loc. cit.*], the Parliament fined this gentleman and his father's partners, in the farm of the Customs, one hundred and fifty thousand pounds, to answer which great

---

[1] Hudson, in 1610, named after him Cape Wolstenholme (see p. 119), and Baffin, in 1616, named after him Wolstenholme Sound (see p. 228).

[2] Metcalfe's *Book of Knights*, p. 192.

demand, his whole estate was sold under a Statute of Bankruptcy, unjustly brought against him for his loyalty to the Crown. Few, indeed, during the disastrous conflicts of the times in which he lived suffered so severely in prosperity. He lost considerably more than a hundred thousand pounds, and his son Henry, and his brother-in-law, Sir Thomas Dallison, both fell fighting under the Royal Banner. To compensate, in some degree, for this devotion to his family, Charles II, at the Restoration, reappointed Sir John farmer of the Customs, and restored to him a patent of collector oatward in the Port of London, from which he had been sequestered during the Usurpation. He [the King] likewise made him a Baronet [in 1664-5]." Sir John Wolstenholme the Younger died in 1670, and was buried near his father at Stanmore. The following facsimile of his autograph is taken from a letter of his in the British Museum (*Addl. MSS.*, 18,979, fo. 84), written from Nostell on December 22nd, 1644, and addressed to "The right hon^{ble} fferdinando Lord ffairfax, Lord Generall of the North" in which he indignantly complains of the incivilities inflicted upon himself and his household as Papists, and of the theft of his cattle and horses, by the riotous and brutal soldiery under the command of Lord Fairfax, for which indignities he asks redress.

There were at least six later Baronets of this family, the last being a Sir Francis Wolstenholme, at whose death, in 1780, the title became extinct.

To Sir John Wolstenholme (the elder) and Sir Thomas Roe, then, the King committed Foxe's enterprise, with the command to "expediate it forward". With them were associated the Master and Wardens of the Trinity House, and (we may presume) Sir John Brooke, while Mr. John Wolstenholme ("young Sir John Wolstenholme, that now is", as Foxe says) was Treasurer. One may gather from a remark of Foxe's (p. 297, *note*) that there were many other Adventurers, "noble and gentle", but none of their names have been preserved.

We may picture Foxe, then, as being busily employed during the autumn of 1630, and early spring of 1631, in command of his ship on various coasting trips; in putting her seaworthiness to the test; in fitting her out with all necessary additions; and in continued thought, study, and research into all matters concerning his long-pondered project, now about to be put in execution.

In the meantime, however, Foxe's best friend, Mr. Henry Briggs, had died at Merton College, on January 26th, 1630-1. Foxe, with quaint humour, subsequently named after him a group of islands in Hudson's Bay, *Briggs his Mathematickes* (see p. 329), a name which is no longer retained. This was a well-deserved acknowledgment of the many favours Foxe had received from his benefactor. By the death of Mr. Briggs, as Foxe sadly remarks, "one half of the adventure fell away."[1]

Probably the last step taken by Foxe's supporters before he sailed was an application to the East India Company. We find it recorded in the minutes that, on March 30th, 1631, before a Court of Committees:

"Sr John Wolstenholme made a moc̃on this day to the Court that, whereas his Matie intends to send out a Pynnace for the Discovery of the North-West Passage, Hee therefore desired that, if that design shalbe effected (of which there is good hope), and that the said shipp shall come to Bantam and be found able to returne for England, That the Company would be pleased, by their Le^{rs} to their ffacto^{rs}, to require them to lade home the said shipp wth pepper ; or, otherwise, if the shipp shalbe defective and not fitt to returne, then to give their men passage into England in the Companies shipps. It was answered that the Company cannot admitt of the lading of the said shipp wth pepp^{r} otherwise then for their own accompt ; for, if they

---

[1] Since the foregoing pages were printed off, I have discovered the following passage in a document containing a number of miscellaneous notes and memoranda relating to Virginia, the Northern Plantations, the Whale-fishery, etc., 1609-1656, preserved at the Public Record Office (*Colonial Papers*, vol. i, no. 23, fo. 17):—

"1629. *South Sea.*—Luke Foxe, Mariner, petitioned his Mat^{y} for a small supply of money towards the discovery of a Passage by the North-West to the South Sea, Mr. Hudson first, and after him Sir Thomas Button, having discovered a great way, and given great hope of opening ye rest."

Written, as this apparently was, nearly thirty years after the event, one may well suspect some error in the statement that Foxe petitioned the King for money, as it is not corroborated by any of the official papers hereinbefore printed.

should, it would be a kind of allowing of private Trade, wch they may not p'mitt. But they are content to give a reasonable fraight for such pepper as the said ship shall bring; or, otherwise, if Sr John shall give the Company 6*s.* p. dollar here, he shall have the pepper to dispose [of] at his pleasure. And, upon these termes, the Court p'mised to write to their ffacto^{rs}. Sr John referred himself either to accept of fraight or to give them 6*s.* p. D^{r} as the Court shall determyne. But, there being not a full Court, it was deferred until their next meeting, when it is again to be p'pounded, and resolved as they should think fitt."

On April 6th, it is recorded that:—

"Sr John Wolstenholme in Court [again] made request for the Companies Le^{rs} to their ffacto^{rs} that, in case the ship bound for the North-West Passage shall discover the same and come to Bantam, the ffacto^{rs} may lade upon her 200 bags of pepper, for wch he wilbe content they shall allow him sixpence p. pound fraight; or, otherwise, that he may treate with the ffactors to lade the said quantity of pepper out of their owne meanes, and charge him (the said Sr John) wth Bills of Exchange for the same to repay here in England. This request being taken into considerac̃on and divers p'positions made for accomodac̃on thereof, and being certified both from Sr John and Sr Thomas Roe, [they being] p'sent in Court, that the intenc̃on of this moc̃on was only to defray some extraordinary expense incident to the voyage, if the said passage shalbe discovered (whereof Sr Thomas gave great hopes, having lately taken paynes and discovered more p'babilities then were formerly knowne); thereupon they were content to write their Le^{rs}, as was desired, but to this effect: That, if the Comp^{nie} have an overplus of Stocke and want shipps to bring it home, then the ffactors are to lade the two hundred baggs requested, and Sr John to referre himselfe to the Companie to pay fraight p^{r} Tonne (allowing 1500^{lbs} w^{t} to a Tonne, according to the rate that the Companies shipps lately sent do stand them in for freight). But, if the Companie have not any overplus of Stocke at such tyme as the said Shipp shall arrive, then Sr John is left at libertie to capitulate with the ffacto^{rs} to lade 200 baggs of pepper in the said shipp and to charge Sr John wth Bills of Exchange for the same, at such rate upon the D^{r} as they amongst themselves shall agree on. Mr. Ellam, at the like request of Sr John, being p'mitted to advise wth and assist Sr John to p'vide tenn or twelve Cloathes of such sorte and coullo^{rs} as he shall conceave wilbe most vendible at Japan, if the said ship in her passage about shall arrive there; wth wch resolution of the Court he rested satisfied, and desires that the said Le^{rs} might be forthwth writt accordingly."

On the 15th of the same month, we read that:—

"A Le^{r} desired by Sr John Wolstenholme by the Companie to their ffacto^{rs} at Bantam, to be sent in his shipp now intended to be sent out for discovery of a passage into the East Indies by the North-West, was this day read in Court, and by a general consent approved and subscribed."

All preliminary difficulties were now overcome, and the *Charles*, having been already kept waiting for very nearly a year, was now, "by all our cares, sheathed, cordaged, builded, and repaired, and all things made exactly ready against an appointed time."[1] Foxe says (p. 264) she was of his

[1] Many references to the earlier movements of the *Charles* pinnace during the years 1625 to 1629 are to be found among the *State Papers* (*Domestic Series, Chas. I*) in the Public Record Office. On October 4th, 1627, we find mention of a rumour that she had taken a good prize (*l. c.*, vol. lxxx, no. 42). Her commander at this time was one Captain Hugh Ledyard, who on October 18th following (*l. c.*, vol. lxxxii, no. 23) was highly praised for his abilities as a seaman by Sir Henry Mervyn, and was recommended for the command of a better ship. On

"own chusing, and the best for condition and quality, especially for this voyage, that the world could afford". As regards this point, the facts may be as he states them; but, seeing that the vessel had been "a cast ship, and appointed to be sold" for the King's benefit, before she was granted to Foxe and his friends, and that (even after the grant) she was still ordered to be sold if they did not employ her at their own expense (all of which may be gathered from the Admiralty Minute, already published, p. lxxii), it is tolerably clear that she was granted to the Adventurers because she was no longer of any value, and that, in speaking as he does of her, Foxe's desire to show the King's liberality overcame his veracity. However this may be, the *Charles* was a pinnace of seventy or eighty tons

---

December 31st, in the same year, her captain was ordered by the Commissioners of the Navy to bring her to Chatham for repairs (*l. c.*, vol. lxxxvii, no. 65). A year later, in December 1628, when she was under orders to convoy soldiers for the defence of the Channel Islands, we find her captain petitioning for promotion (*l. c.*, vol. cxxiii, nos. 25 and 56). In April 1629, we find her under a fresh commander, one Captain Thomas Beale, who recommends her for service on the Irish coast (*l. c.*, vol. cxli, no. 73); but, in the following month, we hear of her as laid up, and her captain prays the Lords of the Admiralty to transfer himself and his company to another ship (*l. c.*, vol. cxliii, no. 42); though in the succeeding month again she is recommended for the guard of the Medway (*l. c.*, vol. cxlv, no. 9); while in July she is ordered to Chatham, and arrangements are made for manning and victualling her (*l. c.*, vol. cxlvi, no. 34). At the latter date she seems to have been again under the command of Captain Ledyard.

burden,[1] and her crew consisted of twenty men and two boys. In selecting these, Foxe says his greatest care was to have his men of "godley conversation and such as (their yeares of time not exceeding 35) had gained good experience". He was particularly anxious to have men who had had experience of the extreme cold and hardships of the Northern seas, and he was grieved to find it apparently impossible to get a single individual to sail with him, either as man or officer, who had previously sailed on any voyage for discovering a North-West Passage.

The list of Foxe's crew, so far as it can be ascertained from his MS. or printed narratives, is here given. None of the men, so far as is known, ever afterwards took part in any other Arctic voyage or came otherwise into prominence.

CREW OF THE PINNACE "CHARLES", 1631.

Captain and Pilot .	Luke Foxe.
Master . . .	——? Dune.
Master's Mate . .	——? Yourin (or Urin).
Cheirurgion . .	————?
Quartermaster . .	Andrew Hume.
Do. . . .	Peter Nesfield.
Boatswain . .	————?
Boatswain's Mate .	John Coatesworth.
Carpenter . .	————?
Gunner . . .	————?
Cook . . .	————?
"Exposer" . .	——? Russell.

[1] There is some uncertainty as to her exact tonnage (see note on p. 264).

Cooper	——— ?
Seaman ?	Samuel Blades.
Do. ?	Tom Brown.
Do. ?	John Honting.[1]
Do.	——— ?
Do.	——— ?
Do.	——— ?
Do.	——— ?
Boy	——— ?
"My Boy"	——— ?

Foxe says that his men were all chosen for him by the officials of the Trinity House. In his MS. Journal (fo. 39), he says that he asked for the appointment of one man only, but this man (though very competent) was denied him.

Although Foxe's relations with the subordinate members of his crew were very amicable, with his Master and the Master's Mate they were the exact reverse.[2] These two men evidently had little or no interest in the objects of the voyage. It seems from various passages in Foxe's MS. that they had been shipped by the Trinity House officials, apparently at the instigation of Sir John Wolstenholme, and in direct opposition to Foxe's own wish. From several passages in Foxe's MS. it appears that there was a certain amount of ill-feeling between Wolstenholme and Foxe, though there is not a word in the printed narrative to indicate this. In various other matters it appears that the Trinity House officials (ap-

---

1 See *State Papers, Dom., Chas. I*, vol. cciii, no. 68.

2 For an explanation of the respective duties and positions of the Captain and Master of a ship, see notes on pp. 49 and 300.

parently at Wolstenholme's instigation) had acted contrary to Foxe's wishes. Indeed, we may gather enough from Foxe's MS. to feel certain that, in several respects, he was by no means so well satisfied with his outfit as he declares himself to be in his "Preparations" (p. 265). Foxe's character was such (see *post*) that it seems more than likely that he was a man with whom it was difficult to work without friction; but it was certainly unfortunate that he had not the choosing at least of his first and second officers; for (as he very aptly remarks in his Journal) his honour and life depended upon fulfilling his instructions, but the others were not any way engaged but in point of labour.

No inconsiderable portion of Foxe's MS. Journal is taken up with complaints as to the laziness and obstinacy of the Master, who was aided and abetted by the Mate. These complaints, with scarcely a single exception, are omitted from the printed narrative; but, in reading the MS. Journal, it is impossible to overlook Foxe's intense ardour in his search, or to avoid the conclusion that his chief officers were a most serious hindrance and anxiety to him. Time after time, when he was anxious to push on, they desired him to anchor, and he seldom put into the ice, in order to make progress, without their wishing timorously to stand away from it. So earnest and so numerous, indeed, are Foxe's complaints, that it is impossible to doubt their honesty. Foxe makes it quite clear that the Master was a lazy, timid, and obstinate man, who strove to hinder and annoy the

Captain rather than to promote the success of the expedition.

The Master's name nowhere appears, either in Foxe's MS. Journal or in his printed narrative.[1] It seems, indeed, as if, in the latter, Foxe wished, as far as possible, to ignore the Master's existence. It may be gathered from Foxe's MS. that the Master had formerly been in the service of the East India Company. The name of the Master's Mate was Yourin, Hurin, or Urin. In the earlier portion of the voyage, he caused Foxe great annoyance by always taking the side of the Master and opposing the Captain; but, later on, when the Master was more or less prevented by sickness from interfering, he became more manageable, and Foxe several times speaks with approval of his services (see p. 378).

Innumerable extracts from Foxe's MS. Journal, showing his contempt for the Master, have been introduced into the body of the work in the shape of foot-notes, and some others are here introduced. Foxe had a quaint bluff humour of his own, which frequently crops up in these touches, and renders some of them very comic. Thus he says, with an exasperation which was probably well-founded:

> "I am not to be thus moved with a drone who cannot wake watches together, and so dogged is he to be set out of his will, as not one word will he utter. I never yet heard him bid good morrow, or good night when he comes or goes to bed."

A few days later, he writes (p. 353):

---

[1] It may have been Dune (see *post*).

"Not one word passed between him and me; nor have we drank one to the other since he told me he cared not for me. *It Spareth Drink well.*"

Then, on July 7th, with a very short-lived touch of contrition, he adds (fo. 37):

"I would be loth to write anything to do him wrong, not knowing whether I may live to come home to answer for, and satisfy, the same; but I do protest before God that I will not wrong him in one syllable, [and] that I do hold him the most arrogant Bullkase [? bull-calf] that ever went or came as Master this way, and the most faint-heartedest man. His very men, to the Cook, take notice thereof and laugh. Three or four things, he is good for: To make us believe he doth take a great deal of pains in Calculation. . . . Eating and sleeping is the other. But he doth not take half so much to command as at first . . . And again doth he discourse of Sir John Worstenholme towards me, . . . [and Mr. Salmon of the Trinity house, till I feel] sorrow that I (being the man allowed choice of Mr. Briggs, and allowed by his Majesty as Captain and Pilot for the voyage, by the approbation of the Trinity house—having been examined there by the means of Sr John Worstenholme), I should not have had the selecting of the Master and Mait, nay and all the whole company, to the intent my Command might have been the greater, and obedience the more. Nor needed there any Master, but only two Maits, the one for the one watch, the other for the other, which should have been mine; whereas now they hold themselves together, as was to be doubted before; for, say they: We are shipped by the Master of the [Trinity] House, by the direction of Sir John Worstenholme. . . . [After stating that he had delivered to Wolstenholme an account of what was necessary for the voyage, which was by him disregarded, so that he had to take the unsuitable outfit provided for him by the officials of the Trinity House Foxe continues as follows:—] For . . . the Master and others are but secondaries to me; nor are they any way engaged further than in point of Labour; but I am engaged in Honor (which is the most precious of all to man in things of this life), having the eyes of Court and Commonwealth fixed upon me in expectation. . . . Hoping these may come to the hands of my honourable Friend (who can remember, I hope, I told him of them to be needful for a Discovery Voyage—especially this), I will write them down as in a catalogue:

"*Imprimis.*

"One shallop, to send to and again to Land, or otherwise.

"One three-man Fishing Boat, with cable, which might have been hauled over Ice, and have sail, as occasion had served, at all times [see p. 295].

"Item. 20 Bonne poyes [?].
,, 10 Iron Bune frockes [?].
,, 12 Ship owers [? oars; see also p. 306].
,, 1 Small murderad [?] for the Shallop's head.
,, 1 Seen net to catch Fish.
,, Cloakes laid in for the ship's company's need, especially if we had wintered.
,, 2 Stoowefs [? stoves] for our wintering, besides others."

On June 14th, we meet with the following contemptuous passage:

"At night we argued about the course to be held more northerly, as it was still thought that the same should be N.W. by N., but the ambitious Master at his going forth directed N.W., which I bore withal, there being more danger to the Voyage in crossing a proud fellow, than could insue in that direction."

On July 12th (fo. 40), after a quarrel with the Master, Foxe says:—

"He told me I was desperate and cared not, but he had his life to lose. I told him so had I and must be careful thereof, for the voyage's good. Upon these termes, Urin broke forth, and said he would take upon him to be careful also of his life. I told him until now I did not know he had brought it with him and did wish he had left it at home, rather than it should hinder the voyage; for it was his bodily labour I did expect to have, and that it was not his part to be so saucy."

The following entry in Foxe's Journal on Oct. 7th (fo. 77) shows still further the kind of relations that existed between him and the Master:

"Yesterday the Master would needs take off the Main Bonnett, and bid me be hanged because I said she would eas[il]y bear it, if he thought she went not too fast homewards. The last evening he came to me to have the mainsail [taken] in, and go all night in forecourse. I told him the wind was much lessened, and God defend I should consent. . . . He could not prevail, and therefore I knew his meaning; he would have been in bed, and so he went to bed. This day, I have not seen him. They say he is sick, as it doth fall out when he cannot have his will. Howsoever, [as] he stirs not abroad, I may now put out all sail, without controul, which I caused [to be done]."

After this thrust, we are prepared for the following entry in the MS. Journal under Sept. 4th:

"The Master came not to prayers nor dinner this day. The one is a greater miracle than the other."

Again, we read in the Journal (fo. 60), under the date August 23rd:

"This morning the Master awoke after 7 hours sleep. He is always in hand with me to anchor at night, having dangerous presages of what may happen; but it is for nothing but [*i.e.*, except] to sleep; for he sleeps at £7 per month; but God help me; [for] if I sleep, I am like to get nothing; and therefore I am glad to ease myself [by sleeping] in my watch, to be sure to look to his better. The other day, I saw the cook giving beer about in small cans, which was contrary to order, and I did examine my boy to know whether he did or not. The Master, hearing me, went betwixt decks to tell the company how I had examined the boy strickly, and what I had said. This he did to bring me into contempt and dislike with my company. Thereupon the Boatswain [who, from other evidence, clearly sided with the Master] called my Boy a long-tongued rascal. . . . God enable me to perform my desire to his glory and mine, and [to my] Friends' and Country's Comfort. I pray him daily for my health; for all labours, I bless him, as yet are pleasures. . . . If the drone would consider that now is the time, he would shake off his sluggishness. . . . For my part, I am well pleased and much bound to my Maker that [He] hath brought me here into these remote parts, where I and my Church have served him in some places where he was never served before. All Glory be to his Holy Name."

When the Master laments the want of a pinnace, Foxe reminds him (p. 306) it is his own fault; for he, in his estimate, had provided for one. He had better, he adds, have spent less in victualling the cabin, and have bought a three-man cobble or fishing boat, which might have been carried over the ice at pleasure, or sent ashore at any time to try the tide, the boat they took being so heavy that it was almost unserviceable.

On the 13th of July, things appear to have reached a desperate pitch, and Foxe writes in his Journal:

"God, for thy mercy sake, send what thou seest we stand in need of; for, if it thus hold, it will break my heart. I have no comfort of one or other; nor doth any man bear a part of [the] care with me. We lie fast here; he [the

Master] eats and drinks, and is well pleased to sleep. That this noble voyage should be lost for want of fitting associates! If these be captain's best seamen because, as he said, "they be East Indians both", he and they are nought for this voyage, or any other; but the same too late I do remember Mr. Briggs, who in his letter gave me warning not to carry an East Indiaman, a man of war, or one out of the King's Ship; but [that] I could not do withal. I write truth, compelled by grief; for these fellows think to spin out time, and have wages due, come home and take it, never meaning to come here again, as they have vowed. . . . Now are they well pleased while I grieve; for they are never pleased without every night at least the ship be tied to Ice."

Again he protests that he will adventure everything for the sake of discovery; that he owes it to the King and the nobility and gentry who sent him forth; that he will not take their "money, and eat up their provisions like a slave to cousin them, and so go home; but while Life lasts [he says] I will have the voyage proceeded; nor do I care for their grumbling."

Many passages, similar to the above, appear in Foxe's MS. Journal. That scarcely a word of complaint against the Captain appears in the Master's MS. Journal (which is also alluded to hereafter) is easily explained; for he probably dared not enter complaints against his Captain. We are thus deprived of the Master's defence; but there is no room for doubt that it would have been a very lame one.

Apart from the Master and the Master's Mate, Foxe has practically no complaints to make against his men. More than once in his MS. Journal, he declares that he always found the subordinate members of his crew ready and willing to assist him in prosecuting his voyage (see pp. 286, 291, and 342). This makes it appear more probable that his complaints against his officers were justifiable. In one place, it is true, he speaks of Andrew Hume, one of the quarter-masters, as siding with the Master and being saucy to him, but this entry is exceptional. Several times, however, he speaks of the men's proneness to intemperance, but he does not seem to have had much cause to complain on this account. In one place in his MS. (fo. 39), he says that "of 20 men and 3 boyes, there is but one that takes no tobacco: nay, the pipe [is] never from their nose". Foxe himself was a man of strict frugality and most temperate habits. He evidently had no toleration for those who indulged in tobacco—indeed, in one passage (see p. 361) he speaks of it as "a thing good for nothing".

The ship was victualled for eighteen months, and, according to Foxe's own published account (p. 264), there was absolutely no cause for complaint in reference to his outfit. His ship-stores were perfect; his surgeon was bountifully supplied with drugs and instruments; and his carpenter, gunner, and boatswain had everything they could possibly require. There are, however, several passages in Foxe's MS. Journal, which are omitted from his printed narrative, and which give good grounds for doubt as to whether things were really so satisfactory as he says they were.

Finally, says he, "Things [being] in this [state of] readiness, I was brought to his Majestie, where I received his Gracious favour, with a Mappe of all my Predecessors Discoveries, his Majesties Instructions, with a letter to the Emperour of Japan. The copies of all which Captain James had."

Foxe's Instructions from the King have, unfortunately, not been preserved. He himself does not print them, which is somewhat strange; they have not been preserved among the State Papers; nor does Captain James (who had copies of them) put their tenour upon record. It is clear, however, from what Foxe himself says (pp. 321, 360, 364, 381, 409, 426), that his instructions were in the main pretty much as follows:—To proceed first to the western end of Hudson's Strait; thence to explore to the north-westward from Salisbury and Nottingham Islands, whence both the survivors of Hudson's expedition and Button had observed a tide which the latter regarded as "the true channel tyde".[1] Failing success in this direction, he was to proceed to Cary Swan's Nest, and thence to set his course westerly, so as to fall in with the western side of Hudson's Bay in about Lat. 63° or 64°. Then he was to explore the coast southward to Port Nelson, following upon Button's tracks; and, after that, he was to explore eastward along the then-unknown

[1] Owing to the quantity of ice Foxe met with at his entering, he was forced to defer carrying out this part of his instructions until after he had finished that part he should have undertaken later.

southern shore of the Bay until he had reached Hudson's most westerly point at or near Cape Henrietta Maria. The whole of these instructions (with the one exception noted) Foxe carried out to the letter, finally completing the exploration of Hudson's Bay, and practically discovering the channel which is now known by his name.

Having reached this point, about the end of April 1631, with Foxe just ready to start upon his voyage, it is necessary to leave him for awhile, and to turn back and notice certain concurrent events.

Very early in Foxe's preparations towards his voyage—probably about the time he first presented his petition to the King for a ship—he had come into contact with a certain Captain Thomas James, of Bristol, who, he says (p. 263), "had so wrought with the marchants of the sayd citie, for to set forth one ship for the same designe, as they were willing to adventure [one]." James, it seems, had written to Briggs before his death, asking that, whichever found the Passage, Foxe and he should have equal honour. Briggs, after consulting Sir John Brooke, consented to this; but James, not content with this alone, rode from Bristol to Oxford and proffered to Briggs the further request that, as both ships were to go forth in the same year, he might go in the higher place. This, however, was refused, as was probable, considering that Foxe sailed under the immediate patronage of the King. As he proudly says in his MS. Journal:—"I was not sent by private merchants; but the King, the nobility, and gentry of the Realm, were my adventurers." In the end, Captain James sailed from Bristol in the *Maria*, a vessel of seventy tons, on the third of May 1631, two days before Foxe sailed from Deptford. As his proceedings are minutely described hereafter, it is quite unnecessary to describe his movements here.

We must now return to Captain Foxe. It will not be needful here to trace minutely his movements whilst engaged upon his voyage, as his proceedings may be closely followed in the body of the work; but it is desirable to briefly sketch his progress.

Foxe set sail from Deptford on May 5th, 1631, and, proceeding by way of the Orkneys, he entered Hudson's Strait on June 22nd. He then sailed slowly through the Strait, keeping near the north shore. He found the ice very troublesome, but met with no serious difficulties. On July 10th, he reached Salisbury Island. Here he was enclosed by the ice, and on the 12th he broke his cutwater by contact with a piece, which he says was "the first harm, and all I received [on my whole voyage]". Being prevented by the ice from searching to the north-westward from this point, according to his instructions (see p. xcvi) he turned south-west for the Bay, intending on his return to search in that direction. Passing Mansell's Island on the 17th, and Cary's Swan's Nest on the 21st, he turned north-westward; and, sailing along the south-western coast of Southampton Island, without any hindrance from ice, he reached, on the 27th, an island off Cape Fullerton. This Island he named Sir Thomas Roe's Welcome, though that name has ever since been applied to the Strait in which it lies. Instead of searching to the northward up this Strait, he turned southwards (probably because so instructed) and sailed down the western shore of the Bay (then called Button's: now Hudson's), which coast Sir Thomas Button had discovered and sailed along (though in a northerly direction) in 1612-13. Foxe says he did this in order "to search [for] the Passage diligently all the Bay about until I come to Hudson's [*i.e.*, James's] Bay." On the 29th, he reached a "white island" which he named Brooke Cobham (see p. lxiii), but which is now known as Marble Island. On the 31st, sailing down the coast, he went in among a group of small islands which he quaintly named "Briggs his Mathematicks", in honour of his friend Henry Briggs, the Mathematician (see pp. lx and 329). On August 2nd, he passed Button's "Checks" (see p. 331). On the 5th, he was in Churchill Bay, which he speaks of as "Hubart's Hope", and where (for some reason which is not obvious; see pp. cvi and 331) both James and he had great hopes of finding a passage westward. All this time Foxe had been enjoying very fine warm weather. He had kept as near as possible to the mainland, in order not to pass by any opening which might prove a passage. Nevertheless, he failed to observe Chesterfield and Rankin Inlets. It is clear from remarks he now and then inserts that he recognised he was "out of the road for finding a passage" (as he says), and his very careful search in the face of this conviction was creditable to him. Rounding Cape Churchill, and sailing on still southward, Foxe, on the 8th, reached Port Nelson, which he entered on the following day, in order to overhaul his ship, to set up a pinnace he had brought from England in pieces, to search for a new main-yard, and for other reasons (see p. 340). Here he remained till the 20th, finding extensive relics of Button's wintering, twenty years before, including an inscribed cross (see p. 348). Just as he was leaving, he also found half of an inscribed board (see p. 349), which he brought away, being under sail. In his book, Foxe does not give the fragment of inscription on this board; but it is given in his MS. Journal. It is headed with the Royal Arms and Royal monogram "I.R.", together with the Prince of Wales's Feathers and his monogram "H.P.", and it runs as follows:—

"In . the . right . and . to [the honour of God]
"and . our . dread . souera[igne, by the Grace of]
"God . King . of . Great . Bre[tagne, France and Ireland,]
"defendor . of . the . tru[e faith of Christ,]
"this.coast . of . New . W[ales was discovered and possession]
"therof . taken . and . i[t was so named by Thomas]
"Button . Gentleman . of [Cardiff, when sailing to]
"the . northwest . under . I[. . . . . , raised]
"this . memoriall . with . h[. . . . . .]
"Britainis . to . gether . with [. . . . . .]
"and . our . most . hopefull [Prince Henry of]
"Wales . heir . aparent . to [the Throne of]

"Brittaine . the . Great . and [at the expense of the]
"honorable . companie . incor[porated for discovering a north]
"west . passage[1] . and . mygrations . a [. . .]
"I . thus . erected . on . the . S[eventh (?) day]
"of . July . and . in . the . year [of our Lord]
"1613."

I have endeavoured to supply [between brackets] that portion of the inscription which may be presumed to have occupied the missing half of the board. On the other side of the board was this inscription :—

"[[2] To this P]orte, I came the 27th of August, 1612[3]
"[by foul] weather, where I wintered the
"[before] written time, and then, by reason o[f]
"[wants] and sickness amongst my Company, [I]
"[was] forst to leave my owen Ship, [and with the]
"[Pinnas to] procecute my discovery [for the North]
"[West] the day and yeare afores[aid.]

"*Per me* THOMAS BUTTON,
"Gentleman."

From Port Nelson, Foxe sailed eastward, thus commencing the exploration of the then-totally-unknown southern shore of the Bay, between Button's easternmost point at Port Nelson and Hudson's western-

[1] This proves that, although the Company's Charter was not granted until three months after Button sailed, all preliminaries had been settled before his departure.

[2] "There is a fly Paper in the MS. where this hath been made compleat, and from which fly this is also made perfect." (Note by the copyist in the copy of Foxe's Journal in the British Museum.)

[3] Prickett (p. 166) says the 13th; Hawkridge (p. 169) says the 15th.

most at Cape Henrietta Maria. It is evident, however, from his Journal, that he merely did this in accordance with his instructions; for he clearly recognised the uselessness of searching further for a passage in this direction. On August 27th, he declared definitely that all hope of finding the passage he sought was at an end. On the 29th, he encountered Captain James, from whom he was eager to hear what he had accomplished. Having gone on board James's ship and heard his story, Foxe made a number of contemptuous and ill-mannered remarks about his host (see p. 360). They parted on the 31st, and Foxe continued his voyage eastward. On September 2nd, he reached Cape Henrietta Maria, which he named Wolstenholme's Ultimum Vale. Here Foxe's work in this part of the Bay ended, he having joined his own discoveries with those of Hudson; and he had no course open but to sail northward, in order to explore the then-unexplored channel which was known to lie to the north of Nottingham's Island, but which the ice had prevented him from exploring on his outward voyage. He had, as he says (p. 365), a good deal of time left, and no instructions to search elsewhere. The ship was accordingly turned northwards. On the 7th he sighted the southern part of the Southampton Islands. Thence he proceeded north-easterly (though some confusion in his narrative makes it difficult to follow his exact course), and reached Mill Island, on which he landed, on the 15th. He then stood over for the North Main, which he encountered near a cape, which he named (and which is still known as) the King's Cape. From this point, he continued along the coast, still northward, up the channel now known as Foxe's Channel and along the western side of what is now called Fox Land. Here, again, he accomplished an interesting piece of original exploration; for, although Bylot and Baffin had begun the exploration of this region in 1615, they had not proceeded nearly so far as Foxe did. At this point, we first hear of sickness among Foxe's crew. On the 7th, he says: "Now the Master and three men more are downe. God better it!" On the 18th, we hear of others being ill. Proceeding up this coast, Foxe reached his most northerly point, in lat. 66° 47′ N., on the 22nd, and named it "Foxe's Farthest". Here he turned southward, on account of the sickness of his men. In running back, he named various bays and headlands; but, owing to some confusion in his printed narrative (which, on many points, contradicts his MS. Journal), it is difficult now to identify them, especially as the coast he thus explored has never since been visited.

On getting back into Hudson's Strait, he notes the disappearance of the ice which had so delayed him on his entrance in the previous July. On the 28th (being now fairly on his homeward voyage) he became more liberal with his provisions, and "appointed 4 beefe dayes in the week". On the 1st of October, Foxe exclaims: "The Lord, for his mercy sake, look upon us, for we are all in weak case." On the 8th, he says: "Our sick men are as yet able to do nothing." His troubles were, however, nearly ended, for on the 15th he passed Cape Chidleigh and emerged on to the open ocean, and on the 31st he found himself safe in the Downs. The words with which he closes his narrative are these :—

"I came into the Downes with all my men recovered and sound, not having lost one man nor boy, nor any manner of tackling, having been forth neere six months. All glory be to God!"

The fact that he could say this bears powerful testimony to the skill with which he had conducted the expedition. It was more than most other Arctic explorers of his day were able to say.

For thus returning in the same year in which he sailed, Foxe had to put up with a good deal of ridicule and abuse from those who were either partisans of Captain James or losers by his failure to discover a passage. It was probably on this account that he added to his book a sort of appendix, entitled, "The benefit ensuing by my coming home this year," in order to defend himself.[1] We may find still clearer evidence of the same thing in the tone of a letter from Sir Thomas Roe to the Merchant Venturers of Bristol who had sent out Captain James. This letter, which is endorsed "Sir Tho. Roe his letter concerninge the returne of Mr. Fox from the N.W. discouery", is preserved in

---

[1] See p. 415. Button, it will be remembered (p. lxviii), had strongly urged the desirability of not wintering in the Bay.

the series at Bristol, mentioned hereafter, and runs as follows:

"To the Right W^{ll}, and my worthie friends, the Mayor and burgesses of the Citty of Bristoll.

"My worthie friends,

"I haue received from you an ho^{ble} testimonie of your loue, soe farre aboue any merritt in mee that it makes my obligation the greater; which I value and esteeme as highlie as those did, whoe for eminent seruices were made citizens of Roome in the greatest glorie. That I acknowledge it soe late, you will not impute to any neglect; because, uppon the returne of Fox from the North-West discouerie, I desired to giue you some true information of the Voiadge, and particulerlie of Capt. James. But Fox hath absented himselfe, whether for shame, or (as hee pretends) to make his cause p'fect, I cannot judge. I haue onelie twice spoken with him, and hee hath not giuen upp his accompte, wch he shall doe before his Matie; but we haue examined the Maister and his mate in the Trinitie House,[1] and find that they haue erred from their instructions, not haueinge made their first search to the North-west passage beyond Cape Comfort, but runne downe from Mansfield's Iland into the Bay, wch they haue exactlie discouered and found that land of Sr Thomas Button in Hudson's discouerie southward all shut upp.[2] This errour proceeded from to much trust to

[1] No record of this examination is preserved among the existing "Transactions" of the Trinity House, which I have been kindly permitted to examine.

[2] It was unjust to blame Foxe for this, when (as we have seen) he was unable to do otherwise. Probably Sir Thomas Roe recognised this fact when he received fuller information. Moreover, Captain James had done exactly the same thing, and was therefore equally blamable, if blamable at all.

the speculations of Mr. Brigges,[1] and hath donne us noe other good but to know there is noe passage, nor tide, nor rising of water to the south. As they report, Captaine James fell uppon the same southwardlie search, hauing beene much troubled and beaten with Ice. They meete and left him about fiftie-and-seaven degrees, about Port Nelson, resolued to winter there, and desirous that they would stay with him; but (as they say), that being in August, they hoped they had time to amend their errour, and to search for the N.W. passage, being the freest time from Ice; but they committed a worse in my iudgement; for, finding the high floude and current about Nottinghams Iland, they say they followed it North-east, and that it led them to that coast to 67 degrees; and then, winter approaching, and (as they pretend) the land and tide both tending easterlie, they resolued that those waters, though they confess fower faddome high, came from Fretum Davies; and soe they returned, and now would say there is noe passage[2]; and they can say noe lesse for themselues. But I never knew men seeke a North-west passage on a North-east shore[3]; nor doe I beleiue that there is any channell from Fretum Davies, nor that soe much water as doth dilate itselfe into soe great a bay can make the tide rise fower faddome in any place that is not a streight, comming in as Hudson's Straight, where in noe place it riseth more. Soe that all o[r] hopes nowe rest uppon Captaine James, whoe will attempt it next yeere, and resolued to loose his life or returne with more honour; and, though hee cannot finde it (if it be not there), yett hee shall and hath gained

---

[1] Probably it was Briggs who had hoped to find a passage in "Hubbard's Hope" (see p. 331).

[2] Apparently this was the opinion of the Master and mate. Foxe argued to the contrary (see p. 442).

[3] He means that Foxe should have examined the *western* side of Foxe Channel, not the eastern, in which he was correct; but Foxe would have done this (see pp. 388 and 411) had not the sickness among his crew compelled him to turn back.

infinite reputacõn to haue taken this resolution, and not to come home like a sluggard and say *a lion rores or was in his way*.[1] And in this honour I thinke myselfe nowe to haue interest, being by your fauour made a member of your Cittie; soe that as I will preserue for you His Ma^ts^ gratious graunt of priuiledge, wherein you haue now noe concurrent nor emulator.[2] Soe I wilbe ever readie to doe you all seruice, as hee that is

"Yo^r^ most affectioned brother and servant,

"THOMAS ROE.

"London, the 28th of November 1631."

In reviewing the whole of the circumstances connected with Foxe's Voyage, we are compelled to admit that (in spite of the fact that he did not attain the main object he had in view, namely, the discovery of a North-West Passage) he acted in almost every case with good judgment and discretion. Mr. Clements Markham rightly says of Foxe that "he conducted the voyage with judgment and energy, and achieved an excellent piece of geographical work"[3]. Foxe himself, in his preface, admits that

---

[1] Here, again (as we now know), Sir Thomas, by imputation, very unjustly blames Foxe for not having wintered.

[2] Sir Thomas doubtless meant that, having had the freedom of the City conferred upon him (as mentioned elsewhere), he felt an especial interest in the success of James's expedition, and he would use his influence with the King that (in the event of James finding the passage) the privilege thereupon ensuing by Royal Grant, should not be lost to the citizens, who (now that Foxe had returned) had no competition to fear. The whole tone of the letter is very unfair to Foxe, who is blamed for having followed what we now know to have been the wiser course.

[3] *The Voyages of William Baffin, 1612-1622*, p. lv.

he had not the good fortune to discover the Passage, but praises God that he had "brought home the newes thereof," by which he meant that he had opened the way for its discovery. In this he was to a certain extent right, though his "newes" was not sufficiently promising to induce anyone to send out any other expedition until nearly a century later, when the search was resumed by the Hudson's Bay Company.

In the course of his voyage, Captain Foxe named not a few islands, capes, and other geographical features in Hudson's Bay and Strait. A list of these appears on the following page, those few names which have survived to the present day being printed in *italics*. The fanciful nature of many of these names goes far to account for the disuse of the greater number.

The original journal or log kept by Foxe, as well as that kept by his Master, are both, I believe, preserved in the collection of MSS. belonging to the late Sir Thomas Phillipps, now at Thirlstane House, Cheltenham, but I have been unable to definitely ascertain this. Fortunately, however, there are MS. copies of these journals in the British Museum; and very interesting they both are, differing from the published accounts in many minor details, and supplying additional information in not a few places. These manuscripts afford very little information as to their own history. They are written on foolscap folio paper bearing the water-mark 1813, and they are both bound into a single volume (*Add. MSS.*, no.

A List of Islands, Capes, etc., named by Captain Luke Foxe.

Name.	*Latitude, N.*	*Longitude, W.*	*Date of Naming.*	*Page.*
Sir Thos. Roe's Welcome Island.	64° 15′ ?	87° 55′	July 27, 1631	321
Brooke Cobham Island	62° 40′	91° 00′	" 29 "	324
Dunne Foxe Island	62° 30′ ?	92° 35′	" 30 "	328
Brigges his Mathematicks	62° 00′ ?	94° 00′	" 31 "	329
Cape Ramsden's Hull	57° 5′ ?	92° 45′	Aug. 17 "	347
New Yorkshire	55° 57′	83° 92′	" 29 "	357
Wolstenholme's Ultimum Vale.	55° 10′	82° 45′	Sept. 2 "	367
Ile Sleepe	57° 50′	80° 30′	" 4 "	369
Cape Linsey	63° 20′ ?	81° 55′	" 9 "	373
Point Peregrine	63° 40′ ?	81° 20′	" 12 "	376
Hurin's Throughlet	64° 3′	77° 57′	" 15 "	381
King Charles' Promontory (King's Cape).	64° 25′	77° 50′	" 18 "	383
Cape Maria (Queen's Cape)	64° 45′	78° 20′	" 18 "	383
Trinity Islands	64° 30′	78° 10′	" 18 "	383
Isle Cooke	64° 30′	78° 12′	" 18 "	383
Lord Weston's Portland	66° 37′ ?	—	" 20 "	388
Cape Dorchester	65° 35′ ?	—	" 22 "	393
Fox his Farthest	66° 47′	76° 55′	" 22 "	393
Pointe Barte	— ?	—	" 22 "	394
Point Carleton	— ?	—	" 22 "	394
Prince's Cradle (Sound)	64° 20′	—	" 23 "	395
Prince's Nurse (Island)	64° 19′	—	" 23 "	395
Cape Dorset	64° 16′ ?	76° 25′	" 24 "	396
Cape Cooke	64° 14′ ?	—	" 24 "	396
Isle Nicholas	64° 12′	76° 30′	" 24 "	397
Sackville Island	63° 00′ ?	—	" 25 "	399
Crowe Island	63° 00′ ?	—	" 25 "	399

19,302), which was "Purch^{d} of Messrs. Boone, 8th Jan. 1853 (Arley Castle Sale, Lot 1071/$_{3}$)", as stated on the fly-leaf.[1] The two journals are

[1] Arley Castle, which is in Staffordshire, six miles from Kidderminster and four from Bewdley, was the seat of the Earls of Mountnorris (Viscounts Valentia), and was built by George (Annesley), second and last Earl, who died in 1844. The sale commenced on December 6th, 1852, and lasted over the twelve succeeding days. It comprised a quantity of antique furniture, the Valentia Library of 4,000 volumes (including a copy of the *North-West Foxe*, with map), a valuable botanical library,

copied in the same handwriting, but whose it is there is nothing to show. Evidently the copyist

---

Oriental MSS., and the collection of Egyptian and Etruscan antiquities formed by the Earl of Mountnorris during many years of travel with Belzoni and Salt. The lots, of which the journals in question (both originals and copies) formed part, are thus described in the catalogue:—

"Lot 1070. North-West Passage. A curious collection of Voyages: the Lord Conwaie's Relation concerning the Passages in the late Northerne Expedition; Description of the Braselians; Yourin's [Foxe's] Journal, 1631, signed Cuth. Pudsey; and several others, in 1 vol."

"Lot 1071. Instructions to Admiral Hosier, and Manuscript Papers time of George I and II, folio; Voyage to the Sooloo Islands, 1762-3, 4to.; Capt. Luke Foxe's Journal, 1631, folio."

I have taken considerable pains to ascertain the history of these two lots. The first (1070) was acquired by Mr. Henry Stevens, who, in his *Recollections of Mr. James Lenox of New York* (London, 1886, pp. 167-172), has given an interesting account of the hurried journey by night to Arley Castle, which led to his becoming the purchaser of this and various other lots. Among these was an important MS. by Hakluyt, which ultimately went to the collection of the late Sir Thomas Phillipps, and has since been published (*Collections of the Maine Historical Society*, 2nd Series, 1877). The two MSS. described as Yourin's Journal remained in Mr. Stevens's possession until May 1854, when they were sold by auction, together with his entire library of books and MSS., by Messrs. Puttick and Simpson, then of 191, Piccadilly, the sale lasting from the 24th to the 29th of the month. It included all the three MSS. which made up lot 1070 at the Arley Castle sale. The Foxe MS. (which had been incorrectly assigned to Yourin or Urin, and which really consisted of two distinct journals) formed Lot 408, and is thus described in Mr. Stevens's catalogue:—

"408. *Fox (Captain Luke)—Journal of his voyage toward the North-West*, calf extra, gilt edges, by Bedford, folio."

"Original and possibly autograph Manuscript from Lord Valentia's Library, 34 leaves, very closely and neatly written; most impor-

had little knowledge of sea-terms, as he makes many obvious errors of transcription. That of the

---

tant, and very interesting. Capt. Luke Foxe was an early English Navigator, who attempted to explore the North-West Passage, upon which expedition he started, according to this MS., on 28 April 1631, the entries concluding the 26th October in the same year. The result of his exploration appeared on his return in his volume entitled the *North-West Fox*, which is based upon the information supplied by this Journal, but differs in many particulars."

"*Yourin's Journal and Dune's Sea Journal* (relating to the same voyage as the preceding), calf extra, gilt edges, by Bedford, folio."

"Original Manuscript, consisting of a Log book and Journal, together about 50 pages, closely written. This Manuscript has been in the collection of the celebrated antiquary, William Herbert, whose writing is at both beginning and end."

"*** The two preceding articles will be sold in one lot."

These two interesting MSS. were purchased by the late Sir Thomas Phillipps, in whose collection (as already stated) I believe they still are. Doubtless they are what Mr. Stevens describes them to be—namely, the original autograph sea-journals of Captain Luke Foxe and his Master. I regret exceedingly that I have been unable to obtain access to them. I have been obliged to make use, for the purposes of this work, of the copies of them now in the British Museum. It has already been stated (p. lxxxix) that the name of Foxe's Master is unknown. The statement in the Arley Castle catalogue, that "Yourin's" Journal was signed "Cuthbert Pudsey" (though ridiculous on the face of it), might at first be taken to indicate the Master's name; but this is due to an error of the cataloguer, who should have written that the "Description of the Brazelians" (which also formed part of Lot 1070) was signed by Pudsey, as shown by Mr. Stevens's catalogue, wherein this MS. formed Lot 813, and is described as "Cuthbert Pudsey's *Journal of a Residence in Brazil*, folio, 1636." The other MS. which formed part of Lot 1070 at the Arley Castle sale (namely, Lord Conway's account of "The late Northerne Expedition") might, from the connection, be supposed to relate to some voyage in search of a North-West Passage; but Mr. Stevens's catalogue, wherein it formed Lot 724, shows that

Master, which is placed first, and occupies forty-two folios, is entitled, "The Journal | of | the

---

it merely relates to some military expedition to the North of England during the Civil War. Although both sale-catalogues assign the Foxe Journals to Yourin (Stevens's in part only), this appears to be an error. The MS. seems to consist of the Journals of Captain Foxe and (judging from the MS. copy in the British Museum) of his Master, whereas Yourin or Urin was the Master's mate. Probably it was assigned to Yourin, under the impression that he was the Master, for his name occurs many times in Foxe's narrative, while the Master's does not once appear. Indeed, the only clue we have to it is the reference in the Stevens catalogue to "Dune's Sea Journal", of which we have no description, and of which there does not appear to be a copy in the volume in the British Museum containing copies of the other journals. William Herbert, bibliographer (who is mentioned in the Stevens catalogue as having probably owned the Journals before Lord Mountnorris), was born at Hitchin in 1718. He was, by trade, a seller of books, maps, and charts. He acquired a valuable collection of old books and MSS., which was dispersed after his death in 1795.

The second lot in question (1071), sold at the Arley Castle sale in 1852, was purchased by Messrs. Boone, and passed into the possession of the British Museum (as stated above). I have made extensive use of it as editor of this work. Although described in the catalogue as though it was an autograph document, it appears to be merely a modern copy of the Journals of Foxe and the Master, spoken of above. Indeed, a manuscript addition to a copy of the catalogue, which I have seen, so describes them, and the paper on which they are written is watermarked "1813". Probably the history of these copies is contained in a passage in Sir W. E. Parry's *Journal of a Second Voyage for the Discovery of a North-West Passage* (London, 4to, 1824-25), wherein, speaking of "Yourin's Journal", he says (vol. i, p. 20):—"This journal, which is no less remarkable for its perspicuity and accuracy than for the neatness with which it is penned, is in the possession of Lord Mountnorris. By his lordship's permission, a copy of this journal was obtained by Captain Sabine, to whom I am indebted for it." We may assume that the copy thus obtained by Sabine,

Master | of the Ship Charles, | Captn Luke Fox | who was sent in quest of a North-West Passage in the Year | 1631," and it is headed, "A true discourse of our Voyage, bound for the North-West Passage and so by Japan; with the Latitude of divers Places and the distance [and bearings; together] with the blowing of the Wind, [the Tides, and] the Currents. Written by me [? —— Dune]." The words within brackets are "altogether destroyed or obliterated" in the original, as we are told in a note by the transcriber, and have been supplied in the copy in red ink. A good many other words have also been obliterated in the first few leaves of the

---

and shown by him to Parry, was that now in the British Museum, as the date watermarked in the paper is about that at which Sabine is likely to have wanted a copy, though why it should have been returned to Lord Mountnorris (as it must have been, seeing it was sold at his sale) is not clear. Parry speaks of the Journal as though it were solely Yourin's, and he also evidently regarded him as having been the Master; but Parry must have had before him Foxe's MS. Journal also, as he could not have obtained from "Yourin's" Journal alone all the information necessary to enable him to compile his map of Foxe's Discoveries alluded to on page 376.

To sum up: It appears that, about the beginning of this century, the Earl of Mountnorris acquired (probably from the collection of William Herbert) the original MS. Journals of Captain Luke Foxe and the Master; that, in or soon after 1873, the Earl had a copy made of them for the use of Captain Sabine; that, at the sale of the Earl's collections in 1852, both originals and copies were sold; that the originals were bought by Mr. Henry Stevens, and the copies for the British Museum (where they still are), that Mr. Stevens, in 1854, resold the originals to the late Sir Thomas Phillipps; and that they are probably still among his MSS. now at Cheltenham.

original. The Journal ends abruptly on Sept. 21st.

Foxe's own Journal, which follows the above, and occupies 114 folios, is entitled, "The Journal | of | Luke Fox, | Captain of the Ship *Charles*, | who was sent in quest of a North-West Passage | in the Year | 1631," and it is headed, "Towards the North-West, April 1631."

These two narratives have been very carefully compared with the published work as here reprinted, and, where they differ from or supplement the information given in the latter, the fact of their doing so has been pointed out in a foot-note. In these foot-notes, the Master's statements are indicated by the addition of the words "*Master's MS.*", and Foxe's by the words "*Foxe MS.*" An examination of the foot-notes so distinguished will show the interest belonging to the MSS., which may be said to give the secret history of the voyage, as written on the spot, and to contain many facts and expressions which were suppressed in the more-carefully-considered printed narrative. The remarks on the ill-feeling between Foxe and his first officer form the chief case in point (see pp. lxxxvii-xciv).[1]

[1] In Foxe's own MS., there is, I believe, not a single observation for longitude. In the Master's, however, observations, both of latitude and longitude, are entered daily, except when the ship was in harbour or enclosed among ice. The latter are at times rather confusing, as he carries his meridian distances from several different points. Thus, on May 22nd, he gives the position as "Lat. pr. judgement 59° 58′, and Longitude, to the west of the

Of Foxe's life and occupation during the period between his return to England, in October 1631, and his death, in July 1635, we know but little; but it is certain that he was allowed to pass these closing years of his life in neglect and discontent.

As regards the expense of Foxe's expedition, we find an interesting document among the State Papers (*Domestic*, vol. ccvi, no. 24) preserved at the Public Record Office. It is an account of sums (amounting altogether to £1,320 11*s*. 3*d*.) paid by Trinity House in 1631 in connection with the outfitting of the ship *Charles* for the discovery of the North-West Passage, and runs as follows:—

"Cash debitor upon the Accompt of the shipp Charles, nowe imployed, furnished, and sett forth for the discovery of the Nor.-west passage. Anno 1631.

	£	s.	d.
"Imprimis.—ffor monies paide by the M[rs] of the Trinitie house to divers men for severall materialls, as their particuler bills of receipt may appeare . .	135	14	08

Main Island of Orkney, 51′, according to plano, but, according unto parallel, 1° 4½′." Thenceforward he gives two daily observations for longitude (the one generally about double the other) until July 18th, when he drops the latter and greater. He was then, apparently, near Cary's Swan's Nest, in about longitude 83° west from Greenwich, which he sets down as "Lo. 38° 17′, Lo. 77° 14′ W." After that he gives only one daily reckoning (apparently from Orkney) until he reaches Port Nelson, after which he reckons his distance eastward from there. Clearly the Master (whatever his other faults) was very diligent in keeping a reckoning, and did not deserve Foxe's sneering reference to his "pains in calculation" (see p. xc).

"More for moneys paid the Marriners (vizt two moneth's wages each man) p. way of impresse	51 11 04
"More for bills perfected, being assigned and directed to Mr. John Wolstenholme for payement, the particulers amounting to	938 05 03
"More for a bill presented by Mr. Bland and the Pully-maker, but as yet not perfected nor signed, both amounting to 195 or thereabouts	195 00 00
"Summa total ...	1320 11 03

"P. J. BEST Clerke to ye Corporation."

The earliest extant official reference to Foxe's expedition after its return is contained in some notes by Mr. Edward Nicholas, Secretary to the Lords Commissioners of the Admiralty, of business to be considered by their lordships on November 12th and 26th, 1631. The entries in question (*State Papers, Domestic, Charles I*, vol. cciii, nos. 32 and 82) run as follows :—

"*Concerning Admiralty Affairs.*

"12th Nov. 1631.—The Charles pinnace, is returned from ye North-West Passage. Y^{r} L^{pps} may be pleased to resolve whether she shall be taken again into ye King's service.

"26th Nov. 1631.—[A similar entry exactly]."

Although the result of their deliberations does not seem to have been recorded, there can be no doubt (as the *Charles* was subsequently granted to

Sir John Wolstenholme) it was decided not again to use her in the King's service.[1]

We hear no more of the *Charles* until the 24th of March, 1632, when we learn from a royal grant still preserved in the Public Record Office (*Signs Manual, Chas. I*, vol. xiii, no. 88) that the ship, with all her ordnance, tackling, and anchors, had been given unreservedly to Sir John Wolstenholme, in return for his outlay upon the voyage. The document runs as follows:—

[*Endorsed 24th March* 1631-32.]

"Charles R.

"Charles, by the grace of God, etc., To o^r^ Com̃issioners and officers of and for o^r^ Admiraltie and Navie, and to all and singular the officers of o^r^ Ordonance and to other o^r^ officers, ministers, and loving subiects whatsoever to whome it doth or may apperteine, and everie one of them greeting. Whereas we are informed by the humble petic̃on of o^r^ trustie and well-beloued seruant Sr John Wolstenholme, Knight, that, in obedience to o^r^ com̃and, he undertooke to sett forth a voyage for the discouerie of the Northwest passage in May last, wherein we were graciously pleased to aduenture and lend o^r^ ship called the Charles, wth ordonance, wch is since returned w^th^out discouerie, and hath humblie besought us that we would be graciouslie pleased to grant unto him the said ship, wth the tackle and furniture thereunto belonging. And, whereas we are also informed by you the Com^rs^ of o^r^ Admiraltie, to whom we referred the consideration of the same, that the said ship was

[1] In October following, however, we learn (*l. c.*, vol. ccxxiv, no. 5) that another vessel, on which the King bestowed the same name, had been built at Woolwich by Peter Pett (see p. xvii); and many later entries relating to this ship are to be found among the State Papers.

certified long since to be old and decayed, and unfitt for o^{r} service. We, being graciously pleased to condescend to this his humble suite, as well in consideracĩon of his obedience to o^{r} Com̃and as aforesaid, as for div's other good causes and consideracions as hereunto mooving, have given and granted, and by this p'ents doe give and grant, unto the said Sr John Wolstenholme, o^{r} said ship called the Charles, together wth the said seven small pieces of iron Ordonance and the old tackle and anchors thereunto belonging. To have, receive, and take the said ship, wth the said seven small pieces of Iron Ordinance and the old tackle and Anchors thereunto belonging, to him the said Sr John Wolstenholme and his assignes, to his and their owne proper use and behoofe, and of o^{r} free and Princelie guift, without anie accompt or other matter whatsoever to be therefore rendered, paid, yealded, or done. Wherefore we doe hereby give full power and authoritie to all and everie of you o^{r} Com'ers, officers, ministers, and subiects aforesaid whatsoever to whome it shall or may apperteine to deliver or cause to be delivered to the said Sr John Wolstenholme, or to such person or p'sons as shalbe by him thereunto appointed and authorised, the said ship called the Charles, together wth the said seven small pieces of yron Ordonance, and the said old Tackle and Anchors thereunto belonging, as aforesaid. And these p'nts or th'inrollm't [the inrollment] thereof shalbe to all and everie of you o^{r} Com'ers, officers, ministers, and loving subiects whatsoever to whom it doth or may apperteine, sufficient warrant and discharge in this behalfe. Given, etc.

"This conteyneth yor Ma^{ts} grant unto Sr John Wolstenholme, knt., of yor Ma^{ts} Ship called the Charles, (wch yor Ma^{ty} was pleased to adventure and lend for the discouerie of the Northwest passage), together wth seven small yron Ordonance and the old tackle and Anchors thereunto belonging, And is in consideracion that the said Sr John Wolstenholme, in obedience to yor Mat^{s} Com̃ands, undertook to sett forth the saide voyage. Yo[r] Mat[s] pleasure signified by Mr. Secretary

"WINDEBANK."

We hear nothing of Foxe himself until April 28th, 1632, six months after his return, when he seems still to have been in charge of the *Charles*, and we find him petitioning the Lords of the Admiralty for a second time for satisfaction for his long attendance in keeping the pinnace *Charles*, their lordships having (he says), on his former petition, desired Sir John Wolstenholme to give him satisfaction, because his Majesty had bestowed the pinnace on him; but Sir John Wolstenholme had told him that he had expended a great deal of money on the North-West search, and that until his Majesty paid him he could not pay him (Foxe), who therefore again prayed the Lords of the Admiralty for relief.[1] This petition, which occupies three-quarters of a page, is still at the Public Record Office (*State Papers, Domestic, Chas. I*, vol. ccxv, no. 97), and runs as follows:—

"To the right honn[able] the Lords Commissioners for the Admiraltie.

"The humble petition of Cap. Luke Foxe.

"Humblie sheweth,

"Whereas yo[r] pet[r] late humblie pet'ioned unto this Board for sattisfaction for his long attendance in keeping y[e]

---

[1] There seems room for doubt as to whether this money was really payable by the Admiralty. Apparently Wolstenholme should have paid it; for, when the King gave the *Charles* to the Adventurers, it was expressly stipulated that they should employ her at their own expense (see p. lxxii).

Charles pinnace, from y$_{e}$ date of his Maties Privie Seale (according to his Maties allowance and yor peto^{rs} qualitie), untill such tyme yor Lo^{ps} were pleased to give hime a discharge, yor Lo^{ps} pleased thereuppon, as apeareth by yor Lo^{ps} referrence, to desire Sr John Wolstenholme to take the said peticõn into consideracon and to give the pe'co^{r} iust satisfaction for his attendance in keeping ye said pynnace wch his Matie of his bounty bestowed on hime, wch yor petico^{r} hath accordinglie demanded at the tendringe of the said peticõn unto the said Sr John Wolstenholme, whoe answered yor pe'co^{r} thatt he was a great deale more mony out for the Northwest Voyadge. And thatt untill his Matie was pleased to pay hime all his said monies he would nor could nott pay the pe'co^{r}, although his Matie hath freelie given hime the said pinnace.

"Humblie beeseecheth,

"Yo^{r} Lo^{ps} to take the p'mises into Yo^{r} hono^{able} consideracons and be hono^{ablie} pleased to give p'sent order for yor pet^{rs} spedie reliefe (the said Sr John Wolstenholme haveinge 2 yeares since p'mised yor pet^{r} reall sattisfacõn), yor pet^{r} haveinge noe other p'sent hoapes or meanes to subsist by. And, wthout the same, he shal be liable to the perrill of arest and undoeinge, being brought into debt by this longe attendance of himselfe and 2 seruantes.

"And, as in dutie bound, he shall dailie pray, etc."

This petition is endorsed by Mr. Edward Nicholas, Secretary to the Admiralty: "R[eceived] 28° April 1632: Pet^{n} of Mr. ffox. I am to speake with Sr John Wolstenholme about this."

In this petition, it seems that Foxe did not claim any specific amount, and it does not appear that he received anything; for, on May 2nd, we find him sending in a "particular accompt" of moneys due to him and two servants for keeping the *Charles* from the time of the King's grant of her to that date, together with wages for the boatswain and gunner, amounting altogether to £160 14*s*. 6*d*. This document, which occupies three-quarters of a page, and may also be found at the Public Record Office (*State Papers, Domestic, Charles I*, vol. ccxvi, no. 7), runs as follows:

"A p'ticular Accompt of such monies wch are due unto mee for my charges, for my selfe and twoe seruants for victualls and wages for keepinge the Charles Pinnace Regis from the date of his Maties Priuie Seale to this p'sent, (besides whatt more will acrue from this said p'sent unto the day of my discharge), Viz^{t}

	£	s.	d.
"Ffor victuall for my selfe and two seruants at 8*d*. a peece p' diem according to his Ma^{tes} allowance, from the date of his Maties said Priuie Seale to this day is 777 daies, makinge 27 mo. ii daies, is	77	14	00
"Ffor wages for 27 months ii daies accordinge to his Ma^{ts} allowance as fol^{s}: Viz^{t} the Bottswaine at 23*s*. 4*d*. p' month; the gunn^{r} att 23*s*. 4*d*. p' month; one seruant att 14*s*. p' month, amounteth unto . . .	83	00	6
"Suma totalis ...	160	14	06

"May ye 2^{d} A^{o} 1632."

This is endorsed by Nicholas: "Capt. ffox his demaunds."

Whether or not this claim was ever satisfied, does not appear from the official papers ; but, from what Foxe himself says, we may infer that it was not. In the postscript at the end of his work (which was probably written rather more than three years after his return and about seven months before his death), he says : "For my part, I have now washt the Black-moore these five yeares,[1] having yet received neither sallery, wages, or reward, except what some few Gentlemen hath (I know not whether in curtesie or charitie) bestowed upon me ; having [here]tofore had my meanes taken from me in the time of warres betwixt France, Spaine, and us." Foxe was not alone in being thus circumstanced ; for we have already seen that an Arctic explorer of even greater prominence than he had been left in the same, or even in a worse, plight.

During all this period, we may picture Foxe as busily engaged in the preparation of his book. Why its appearance should have been so long delayed, it is impossible to learn with any certainty, though in all probability the cause was its author's want of skill as a literary workman. Whether or not this was the case, there appear to have been in the meantime many inquiries for the work from those who were curious to know what he had done ; for, in the preface, Foxe, affectedly speaking of his

[1] This is a figurative expression, and may be taken to mean that he had been engaged in a profitless occupation, as it were, trying to wash a black-a-moor white.

book as his "child", says "there were some [who] desired to have had him come abroad like the Lapwing,[1] before the shell was from over his head; but the Nurse, more careful of the tenderly-borne Infant, procrastinated the time, supposing that after birth it would strengthen apace in this temporate Clime, and in time be the better able to Journey". This is but a lame apology for so much delay. Captain James's narrative had appeared some two years earlier, though it must be admitted that it is a much smaller volume than Foxe's.

At last, however, the work was complete, and on December 15th, 1634, according to Professor E. Arber's *Transcript of the Registers of the Company of Stationers of London* (London, 4to, 1875-77, vol. iv, p. 305), Thomas Fawcett and Bernard Alsop

> "Entered for their Copie[right], vnder the handes of Master BAKER and master Rothwell warden, a Booke called *Northwest FFOX or FFOX from the Northwest passage*, with *an Abstract of the* [*earlier*] *Viages*, with *the Authours owne viage, being the sixteenth*, by Captain LUKE FFOX . . . . vj[d]."

---

[1] This passage seems to be in allusion to the fact that the young of the peewit or lapwing (*Vanellus cristatus*), and many allied species, run about and are able to feed themselves, led by the parent bird, directly they are hatched, probably, in some cases, with the upper portion of the egg-shell still adhering. Shakespeare has not overlooked this peculiarity; for he says, in *Hamlet*, Act v, Sc. 2:

> "This lapwing runs away with the shell on his head,"

as pointed out by Mr. Harting in his *Ornithology of Shakespeare*, p. 222.

B. Alsop and Tho. Fawcett are the gentlemen whose names appear on the title-page as printers and publishers[1]; and it is interesting to note that the work emanated from that renowned literary locality, Grub Street; for its printers are described as "dwelling in *Grub Street*". It is probable, however, that the work was not actually issued until early in the year 1635, that being the date which appears on the title-page; for we also find in Prof. Arber's *Transcript* (*loc. cit.*, p. 306), under date February 4th, 1635, the following additional entry:

> "Assigned ouer vnto Michaell Sparkes, by Vertue of a Note vnder the hands and seales of Bernard Alsop and Thomas ffawcett, and subscribed by both the wardens, All their estate, right, Title, and interest in their Copy[right] called *Northwest FOX or FOX from the Northwest, etc.*; but they are to haue the printing of it for euer, doeing it as well and as [cheaply as any other would] . . vj^{d}."

The reason for this assignment to another of the copyright in the work is not stated.

The work eventually appeared as a small "pott quarto" volume, measuring about 7½ inches by 6 inches. The title is unusually voluminous, even for

---

[1] We learn from the Addendum to Arber's *Transcript of the Register of the Stationers' Company* (vol. iii, pp. 699-704) that Bernard Alsop succeeded his deceased partner, Master Creede, about the year 1617. He took in as partner Thomas Fawcett, who is described as "a pore man", though "ye abler man and better workeman and better Governer". They were at one time suppressed, because Alsop was "unruly", and had printed something not authorised.

a period when long titles were the rule—when, in short, the title was usually made to answer also the purposes of the more modern "Table of Contents". In this particular instance, the title-page actually bears no less than 218 words. It is herein reproduced in almost exact facsimile (see p. 3).

Of Foxe's narrative as a piece of literary work, it is impossible to speak highly. As a matter of fact, from this point of view, his book is about as bad as it well could be.

It may be that Captain Luke Foxe was "the earliest of the known Hull authors", as stated by Sheahan in his *History and Description of the Town and Port of Kingston-upon-Hull* (London, 8vo, 1864, p. 628); but it is quite certain that his literary work reflected little or no credit upon his native place. His style is so faulty that it is scarcely correct to speak of it as "style" at all; for his punctuation is extremely defective; his spelling is bad, even for the period[1]; whilst his diction is ill-chosen and often slangy. As a natural consequence of these defects, Foxe's sentences are often so confused as to be almost, and sometimes quite, incomprehensible. In addition, we meet in the narrative with a constant straining after far-fetched witticisms. The text is, moreover, encumbered with an immense amount of indigestible information as to the direction of the

[1] Thus, *to lay to hull* always appears *to lay to Hull*, which Foxe perhaps intended as a play on the name of his native place; while *off* stands for *of*; *then* for *than*, *too* for *to*; etc., etc.

wind, the latitude, the state of the weather, his course, the hour of the day at which he did this thing or that, and many other minor observations, which a more skilled narrator would have thrown into tabular form and inserted as an appendix. He might thus have avoided burdening his text with a mass of small facts which detract seriously from its clearness and interest as a narrative. All this, in a work "printed by his Majesties Command", is hardly what one expects to meet with.[1]

Perfect copies of Foxe's book have now become very scarce, and are highly valued by bibliophiles. Sabin, in his *Dictionary of Books relating to America* (vol. vi, p. 561), says: "[Of] this very rare collection of early voyages towards making a discovery of the North-West Passage . . . a perfect copy is worth over a hundred dollars." At the present time, however, the work has a much higher value. Mr. Quaritch has been good

[1] Sir John Barrow, in his *Chronological History of Voyages into the Arctic Regions* (London, 8vo, 1818, p. 236), curtly dismisses Foxe's book with the following harsh, but just, criticism: "The narrative of the voyage is written by Fox himself, who affectedly assumes the name of the *North-West Fox*. . . . The style of his journal is so uncouth, and the jargon so obscure and comical, as in many places to be scarcely intelligible." Rundall (*Voyages towards the North-West*, 1849, p. 156), says of Foxe's narrative that it "exhibits many singularities in point of style. It contains many *euphuisms* calculated to excite a smile. . . . These affectations cannot, however, be considered the result of a frivolous mind. . . . The proceedings of Foxe . . . give evidence of his having been a man of no ordinary mind. They are characterised by decision, sound judgment, and skill."

enough to inform me that the average value of a good and complete copy is about £30, while an unusually fresh copy would be worth even more.[1] The Globe, which forms a frontispiece, through its having been printed on the fly-sheet, is very often missing.[2] The map of the Arctic regions, which

---

[1] There are three copies in the library of the British Museum. One of these (303. d. 19) is imperfect, wanting the map. Another (C. 32. e. 8) is similarly imperfect, but a facsimile hand-drawn map has been supplied. The third copy, however—that in the Grenville Library (G. 7167)—is perfect, having the map and also the frontispiece of the Globe, both in prime condition. The collation of the work seems to be as follows: Fly-sheet, with the signature "A" on recto, and the Globe (woodcut) on the verso; Title-page, with back blank; Address to the King [2 pp.]; Preface [6 pp.]; the Folding Map (engraved); Accounts of Earlier Voyages, pp. 1-171; Errata, [1 p.]; "My Preparations to the Voyage," pp. 169-172 (pp. 169-172 being repeated); Foxe's own Voyage, pp. 173-244; "To whom this may Concern," pp. 244-249; "The Benefit ensuing by my Comming home this year," pp. 250-251; Discourse on the certainty of there being a passage, pp. 252-[270]; Address to the Masters of Trinity House, [pp. 270-272]; License, [p. 272]; Errata, [p. 272]. There are not a few errors in the pagination. Pages 169-172 are often missing, as are also some few pages at the end. Sabin (vol. vi, p. 561), says: "After page 168, are two leaves paged 170, 172, 170, and blank, which are said to be cancelled leaves. There are several other errors in the pagination. After page 79, the next is page 100, and the *hiatus* is not supplied; the other errors are corrected by duplication; signature Bb is incorrectly paged 225-232, instead of 205-212."

[2] It was evidently printed from the identical block used in Davis's *Seaman's Secret* (Lond., 4to, 1607). The block of the reproduction given herein was prepared for Capt. Markham's *Voyages and Works of John Davis* (Hakluyt Society, 1880, p. 288), which contains a reprint of the *Seaman's Secret.*

measures 19 inches by 14 inches, and has in the bottom left-hand corner a design of a fox running off with a goose, is still more rare. Mr. C. H. Coote speaks of it (*Dictionary of National Biography*, vol. xx, p. 132) as "a large folded map of the Arctic regions, now rarely found in the book, but which is one of the most interesting and important documents in the history of Arctic exploration". Foxe's own remarks upon his "Polar Map, or Card", as he calls it, will be found on p. 12. Although he modestly apologises for it as being "but rough-hewn, like ship-wright's timber", it is undoubtedly a very valuable map, and is herewith reproduced in facsimile by photo-lithography.[1]

It is from Foxe's work that we gain almost our only insight into its author's personal character. From what has been said of the book, it might be inferred that his early education had been very limited. This was the inevitable result of his having been, as he tells us, "sea-bred from his boye's-time." Foxe was, in all probability, at least as well-informed and well-educated as the average of his class in his own time. He erred in not having sought the aid of someone more skilled than himself in the art of

[1] Mr. C. R. Markham, in discussing the misconceptions as to the extent and position of Baffin's Bay which arose through the suppression by Purchas of Baffin's map of his discoveries in 1616 (see p. 223), says (*The Voyages of William Baffin*, 1612-1622), that on Foxe's chart "Baffin's Bay is shown correctly, and it seems probable that this part of Fox's map may have been copied from the lost map of Baffin." This is more than likely, as we know that Foxe knew Baffin personally (see p. 370).

literature, whilst engaged in compiling his book. On this point, he says (p. 444) that the shortcomings of his book must be excused; for, "being no Scholler and having had no helpe, which I did know was very needful, but [I] was not able to buy it, and I was told it would not be had for nought, especially by [? *from*] the Scholler that was acquainted with the language of the Sea. . . . . Whereupon, not knowing otherwise how to proceed, I was enforced, with such Tackling, Cordage, and Raftage as I had, to Rigge and Tackle this ship myselfe."

Yet in spite of this obvious lack of education, Foxe stands revealed to us in the pages of his own narrative as a man of very considerable personal ability, and endowed with tremendous energy and perseverance. These points in his character are obvious upon almost every page, while not less obvious is the fact that he was a most able and experienced seaman. It is evident, moreover, that Foxe was thoroughly imbued with the strong religious feeling, and the fervent belief in the Divine leading which characterises the Yorkshire man and the time. Yet he is not without a touch of self-complacent superiority for that same reason, where he says: "For my part, I am well pleased and much bound to my Maker that hath brought me here into these remote parts, where I and my Church have served him in some places where he was never served before. All glory to his holy name." That Foxe was an excellent all-round observer, his Journal also shows. He

k

mentions no fewer than twenty-three species of plants, shrubs, and trees; twenty-one of mammals; twenty of birds; and several of fish. He also took note of the burial-places and weapons of the natives, though of the natives themselves he saw nothing.

The less agreeable feature in Foxe's character which his book reveals to us is his excessive self-conceit, and his blindness to the merits of others. This personal peculiarity is unpleasantly apparent in his references to Captain James. It may, however, be noted in his favour that he never descends in his narrative to real vulgarity or coarse abuse; and it must also be allowed in his favour that his conceit was the self-reliant and largely-unconscious conceit which may be seen in many, perhaps most, self-made men of limited education, who have accomplished some great work mainly by their own efforts and in spite of their early disadvantages. Sir John Barrow sums up Foxe's character as well and briefly as he dismisses his book when he says (*loc. cit.*, p. 236): "He was a keen shrewd Yorkshireman, and evidently a man of considerable talent, but conceited beyond measure."

Captain Luke Foxe died at Whitby, where he had probably gone to reside, on or about July 15th, 1635, and was buried there on the 20th. According to Lionel Charlton (*The History of Whitby and Whitby Abbey*, 4to, 1779, p. 315), he "is the first sea-captain of whose death particular mention is made" in the Whitby Registers. The approximate date of his death has not before been ascertained, Charlton merely stating that it took place in the month of July. I am enabled to give the above dates through the kindness of the Rev. Canon Austen, M.A., Rector of Whitby, who has examined the registers for me.

The general circumstances which led up to the voyage of Captain James were identical with those which led up to Foxe's expedition. The discoveries of the earlier explorers of Hudson's Bay occupy exactly the same position with regard to the expedition of Captain James as they do to that of Captain Foxe. It is totally unnecessary, therefore, to repeat anything that has been already said upon this subject.

James's personal history is, like Foxe's, involved in much obscurity. It is clear from his book that he differed widely from Foxe in his characteristics, and we know that, unlike Foxe, he came of good family and was well educated. The reference to "my fellow countrymen" on p. 460 all but proves his Welsh origin; but the place of his birth is doubtful, and the date of it can only be inferred from the legend round the portrait on his map, which goes to show that he was born in 1593. It is practically certain that he came of one of the many branches of a family of James which held property at many places in and around Bristol and in South Wales. Probably he was a near relative (most likely a nephew) of his contemporary, Alderman Thomas of Bristol, who died in or about

1619, and lies buried beneath a handsome monument in the Church of St. Mark. Other members of the family attained high positions in the civic and commercial life of Bristol. Mr. Ivor James of Cardiff, who has searched records and made enquiries with extreme diligence in every likely quarter, has, I believe, utterly failed to discover any more definite information concerning him. It is probable that he never married. That he received a good education is clear from Boyle's statement printed on p. clxxxviii, and might have been inferred from his book. Seyer (*Memoirs of Bristol*, vol. ii, p. 286) says "Adams, his contemporary, in his Calendar [a MS. I have failed to trace] calls him a man of great learning and experience in navigation, and well seen in the Mathematical Science." On July 16th, 1628, he was in command of the *Dragon*, of Bristol, a vessel of 200 tons belonging to himself and others, to which letters of marque were issued (*Calendar of State Papers, Domestic*, vol. cxv, p. 113). Where else he obtained his "experience in navigation" is unknown, but he would scarcely have been selected for his command had he not possessed such. Nicholls (*Bristol Biographies*, No. 2, p. 57) speaks of him as "a rich barrister-at-law of the Inner Temple", a description of him which is to some extent supported by the words of Nash (p. 451) and Boyle (p. clxxxviii).

The circumstances which led immediately to the voyage of Captain James were apparently very simple. So far as one can now gather, James's voyage was due solely to a wholesome feeling of rivalry between the merchants of Bristol and those of London. Foxe had, as we know, long cherished a design to undertake a voyage in search of the North-West Passage. He had, by his persistence, interested many influential persons in his scheme, and had at last gained all he wanted in the way of support and recognition. The merchants of the ancient city of Bristol, always to the front in enterprises of the kind, hearing of what he was doing, formed the highly creditable resolve to fit out a ship at their own expense to undertake a similar voyage in search of the passage. To Foxe, therefore, belongs the credit of having been the prime mover, directly or indirectly, in both voyages. It is true that James, in his "Preparations", says that his friends had long pressed him to undertake such a voyage, and that he at last consented, knowing the King's desire for the discovery of a passage (see p. 455); but it is difficult to explain the coincidence of the two expeditions sailing in the same year, without supposing that the Bristol merchants dispatched theirs when they did out of active rivalry with the London "Adventurers"; indeed, Foxe, in his "Preparations", distinctly says that they did.[1]

However this may have been, the Bristol mer-

[1] It has been stated by several writers that Captain James had had previous experience in Arctic navigation. Thus, Mr. Ivor James, in a letter to the *Athenæum* (Nov. 30th, 1889), says: "The voyage of 1631-32 was not his first in search of the North-West Passage." Mr. J. F. Nicholls (*Bristol Biographies*, No. 2,

chants offered to furnish Captain James with a ship. In his narrative, James does not say clearly that his voyage was promoted by the Company of Merchant Venturers of Bristol. In the letter he left at Charlton Island, however, he distinctly says (see p. 594) that his voyage was promoted by the Company, or by a number of its members. Moreover, a full copy of the lengthy correspondence that took place during the preparations for the voyage is preserved among the archives of that body in the Merchants' Hall at Bristol. I am indebted for copies of these letters to Mr. George H. Pope, Treasurer of the Company. I have printed them in full, and have left them very largely to tell their own tale.[1]

---

p. 64) and Mr. Ivor James (*The Source of the Ancient Mariner*, pp. 31 and 64) both express their belief that he had accompanied Sir Thomas Button in 1611-12; while, in *Bristol Past and Present* (of which Mr. Nicholls was part author), there occurs the positive statement (vol. iii, p. 301) that he did so. Mr. Nicholls, at least, rested his belief partly on a paragraph in which James himself says: "A great ship, *as by former experience I had found*, was unfit to be forced through the ice" (see p. 456). These words certainly give some support to the supposition that James had previously sailed in icy seas; but I believe that by "experience" he meant not *personal* experience, but rather conclusions that he had come to as a result of studying the proceedings of earlier Arctic explorers. There is no other passage in James's narrative, or in the letters from the Bristol merchants quoted hereafter, which affords the slightest ground for the belief that he had been previously engaged in the search for the Passage, and it is morally certain he had not been.

[1] Some of them were printed by Seyer, in his *Memoirs of Bristol* (vol. ii, pp. 277-286), but inaccurately, and with considerable

The Company of Merchant Venturers of Bristol still exists, though it no longer discharges its original functions. No history of it seems ever to have been published; and, as it was one of the earliest and most important of the incorporated societies of Merchant Adventurers, the following sketch of its history has been compiled mainly from information kindly supplied by Mr. George H. Pope.

The Fellowship, Company, or Society of Merchant Venturers is the only surviving representative of the many mediæval guilds which Bristol once possessed.

No exact record of its foundation remains, though traces of such a guild are discoverable in 1314; but it was fully organised as early as 1467, and in 1500 an elaborate code of laws (which still exists) was framed for its regulation.

At this date, it seems to have been little more than a trade committee of the Town Council of Bristol; but, in 1552, its independence was secured by royal charter, granted by King Edward VI, and it was incorporated as "The Master, Wardens, and Commonalty of Merchant Venturers of the City of Bristol". The first Master of the new Corporation was Edward Pryn, and the first Wardens were Thomas Hickes and Robert Butler.

Fresh charters were granted by Elizabeth, by

---

omissions which are here supplied, Mr. Pope having also most carefully corrected the proofs with the original documents. I believe the letters are here printed in proper sequence, but the dates of some of them are certainly wrong.

Charles I, and by Charles II. The most important was that of Charles I, which finally settled the constitution of the Company, by giving to the Master and Wardens "ten of the gravest and discreetest men" as assistants, thus completing the Court of thirteen members, which still forms the executive body, elected (as the Master and Wardens were elected long before the Company's incorporation) on the 10th November in every year. Its Master holds, during his year of office, a position second only to that of the Mayor and Sheriff of the city.

In the reign of Edward IV, the Company occupied, for the purposes of business, "the Chappell, and the Draughte Chameber apperteyninge thereto, in the Hows callyd Spiceris Halle, upon the Back of Bristowe;" but, in 1561, they acquired, and used thenceforth as their Common Hall, the desecrated Chapel of St. Clement, in Avon Marsh, upon the site of which they built, in 1701, the present Merchants' Hall.

From the time of Elizabeth down to a very recent period, the history of the Company has been the history of commercial and patriotic enterprise in Bristol, and the roll of the members' names is a list of the leading burgesses. The charters confine the privileges of membership to freemen of Bristol, which freemen now form but a small portion of the whole body of citizens.

The annals of the Society record adventures for the "planting" of Virginia and New England, with detailed rules for the conduct of the plantations;

the expedition with which we are herein concerned for the discovery of a north-west passage to India; contributions for the suppression of pirates (whether the "Spaniards and Dunkirkes" in 1597, or the "Turkes of Algier" a hundred years later) and for redeeming the English captives; statistics of the produce imported into Bristol during the seventeenth and eighteenth centuries; lists of the ships lost or sold to foreigners; particulars of the wharfage and other duties payable at the port; and many interesting facts beside, well worthy of the notice of a competent historian.

It is to the Society of Merchant Venturers that Bristol owes its wharves and quays. These were built by the Society, under leases of the river from the city, principally with borrowed money, and at an outlay which seems to have been long unremunerative.

About two hundred years ago, the Merchant Venturers became by purchase Lords of the Manor of Clifton. Much of this manor was subsequently sold. The wastes, which are known as "The Downs", were given by the Society to the citizens.

In 1696 and 1708, Edward Colston, a prominent member of the guild, constituted the Society trustee of the various lands with which he endowed an almshouse and a free boarding-school or hospital. The almshouse remains under the Society's control; but, in 1875, a scheme was framed for the management of the school by the Charity Commissioners. A few years, however, proved that the Commissioners had miscalculated the value of the property, and then the Merchant Venturers came to the rescue of the scheme.

Recognising that technical education is the most pressing need of the day, the ancient Society of Merchant Venturers, after careful inquiries, both in this country and abroad, planned and erected, in 1885, upon the site of the old Bristol Grammar School, in Unity and Denmark Streets, the range of buildings which now contains the Merchant Venturers' School. The Society has borne, out of its own private funds, the whole expense of founding this school, and continues to defray, from the same source, the cost of its maintenance. The Colston Trust has been thus relieved of a burden which was beyond its strength; and its whole income is now applicable for the benefit of the school which Colston himself established.

The Merchant Venturers' School has exceeded the highest expectations that were formed of its success. During the first year of its existence, nearly 1,000 students—men, women, and boys—received instruction within its walls; and, during the present year, the number has risen to nearly 1,700.

The Bristol merchants, having agreed upon their expedition, desired for it the approval and support of the King. For this purpose, they sent Captain James to London with a letter to Sir Thomas Roe,[1]

[1] No copy of this letter, apparently, has been preserved.

of whom we have heard before, as he was one of those who had charge of Foxe's expedition. James himself describes Sir Thomas Roe as the most learned and eminent English traveller of that day (see p. 455). The result of this action on the part of the merchants may be seen in the following letter from Sir Thomas Roe to the Mayor of Bristol, which is endorsed, "Sir Thomas Roe his letter to Mr. Mayor, to incite the marchants of Bristoll to goe forward in the enterprize," and of which a copy is preserved among the Company's records at Bristol, in the interesting series already alluded to (see pp. civ and cxxxiv). It runs:

"To the right Woor^{ll} and my very good brother, Mr John Tomlinson, Mayor of the Citty of Bristoll.

"Sir,

"His Matie haueinge gratiously pleased to take to his care the discouerie of the North-West passage to China, and giuen one of his shipps to that use, and comaunded mee to attend the dispatch of that busines, and to giue those direcc̄ons and instrucc̄ons wch were requisite, I haue nowe newely begunn to putt in execution his Royall order, and to prepare the shipp fitt for such a voyadge; when very opportunely and to my greate incouragement there came to mee Captaine James of yor Citty, propoundinge from you that you were resolved to make an adventure that way, and had fitted a shipp to that designe, provided that you might haue graunted equall priviledges wth those whoe were undertakers under his Matie; but hee, beinge there arrived in a tyme of his Ma^{ts} absence, thoughe hee hath pressed his busines as well as he could, I haue not presumed to giue him any determinate answere, not daringe to conclude my Maisters grace wthout his leaue and Communicac̄on. But his Matie is soe just a Prince, and so gratious to encourage and reward all vertuous accõns, that I am p'swaded hee will make noe difference betweene his subjects, but [will] indifferently proceed to animate you as well as any other, and that he wilbee glad that you shall goe forward, and that the accõn bee strengthned by yor endeavo^{r}. For my owne part, I haue noe other thoughts nor ends but such as are directed to serve the publique; and, therefore, as I doubt not the motion wilbee acceptable to his Matie, whose desires are that the worke bee done, soe I doe promise to the undertakers of yor Citty that I will moue him as effectually for them as for o^{r} selues, not wisheinge that any such publique utillitie should bee monopolised by any one Citty or Societie of men. As yet I knowe not of any Pattent or priviledge graunted to any; but as I haue seene a draught in paper, to encourage those that shall nowe adventure, wch I doubt not shalbee extended equally to you, but in what manner I cannot prescribe. This I suppose you shall obtayne at the least: that, yf yor shipp haue the fortune to make the passage first, you shall haue that priviledge wch is due unto you; and, yf wee make it, yet that you shall haue such Libertie of yor trade as yor adventure and industrie doth merritt; for, whether his Matie will shutt upp the trade to Companyes, or what his resoluc̃on is, I am not able to explayne. When his Matie shalbee retorned (wch wee expect wthin twelue daies) I will make you a cleere answere, and doe promise you that, before any Pattent bee taken out, you shall haue that satisfacc̃on wch may content you. In the meane tyme, you may bee pleased to proceed cheerefully, and to make ready against the Season, in full Confidence of his Ma^{ts} Justice and equal distribuc̃on of his favours; and to be assured that I wilbee a Careful Sollicito^{r} for you that you may receyue that measure wch soe generous undertakinge doth merritt. I haue acquainted this day my Lord Treasouro^{r} wth yor proposic̃on; whoe, thoughe hee will not oblige himselfe untill his Matie bee informed, hath declared his opinion that it is just that you haue equall share wth any whoe beare equall adventure, and that hee will doe you any favour, as beinge behouldinge to you for yor loue to chouse him Steward of yor Citty ; soe that I am Confident you shall haue due content. It wilbee alsoe convenient, yf you like it, that wee may hould some Correspondence in o^r^ seũall instruccõns for the manner of the search of the passage, and to communicate o^r^ Cards, and the workes and errors of other men, that wee may helpe one another, and make the more expedicõn, and the more exquisite discouerie. This, S^r^, I pray you present unto those of yor Citty whose names I knowe not, and that, yf I may here a resolucõn from them, I wilbee glad to doe them seruice. For yor perticuler, I receyved from you a kind letter, wch I haue not opportunitie to answere by reason of my infirmitie ; but against the next season of Venison I will provide you of Venison in such measure as I am able. So, desiringe you to remember my loue to yor wife, my sister[1], I will ever rest

"Yo^r^ very loveinge frynd and brother to serve you,

"THO: ROE.

"St. Martin's Lane, 25th of January 1630[-31]."

The Mayor having communicated the above to the Company, a reply, endorsed "The Companies letter to Sir Thomas Roe for ye obteyninge of Condicõns to encouradge them in the enterprize with Capt. James for discouery of the N.W. passage", was sent to Sir Thomas Roe. It runs as follows :—

---

[1] From this and other passages in Sir Thomas Roe's letters to Mr. John Tomlinson, Mayor of Bristol in 1630-31., it appears that the two men were brothers-in-law ; but I have not been able to corroborate this.

"To the ho^{ble} Sr Thomas Roe, knight—these.
In London."

"Ho^{ble} S^{r},

"Mr Mayor of this Citty hath comunicated to us a letter, wch hee lately receyved from you by the Conveyance of Captaine James, whereby doth appeare yor wonted noble disposic̄on and good affecc̄on to such enterprises as tend to the service of his Matie and the weale publicque: In perticuler, you are pleased to approue the dessigne of some marchants here, wch intend to ymploy the said Captaine for discouerie of the North-west passage. And, for that purpose, they haue intreated him to make a Jorney unto London to understand what condic̄ons may bee obtayned for their incouragement therein. Nowe wee perceyve that his Matie hath byn gratiously pleased to recom̄end the mannaginge of that busines to yor Care, wherein wee are much comforted and soe satisfied wth the noble expressions of yor favour in yor said letter, as that wee depend and rely on yor wisedom and goodnesse to procure such faire condic̄ons for us as others enioy, and as o^{r} allacritie in undertakinge the voyadge and proporc̄on of o^{r} Charge merritt, whereby you shall much oblige all the adventurers here, and p'ticulerly us whoe are

"Yo^{r} humble Servants,[1]

[Bristol.]
[? Feb. 2, 1630-31.]

Captain James had sent Sir Thomas Roe's letter to the merchants enclosed in a letter from himself,[2] which elicited from them the following answer, endorsed, "The Companies letter to Captayne James

---

[1] Neither signatures nor date are given in the copy of this letter.

[2] Of this letter, no copy has been preserved. It was dated January 26th (see *post*).

concerninge their desire of his care in prosecutinge the designe" :—

"To o[r] worthie and most respected Frynd, Capt. Thomas James, at the three Cupps in Bredstreet[1] these, in London.

"Captaine James,

"Wee haue receyved yo[rs] of the 26th ult[mo] [January], wth the inclosed from S[r] Thomas Roe, the coppie [w]hereof wee nowe send you herewith,[2] as alsoe of yo[r] [? *our*] answere thereto,[3] whereby you may perceyve what doth passe 'twixt us. He doth presuppose that wee haue already fitted a shipp, but you knowe the contrary, and that wee haue done noe more then to procure adventurers for 800*li.* or thereabouts, and that the fraightinge of a shipp is deferred till wee knowe from you the Priviledges wee shall obteyne (yf the benefitt of the discouery bee appropriated). But wee rather desire that it might continue free from [? *for*] all his Ma[ts] subjects p'miscuously. It wilbee sufficient hono[r] and advancement unto us to bee Instrumen[ts] of soe greate good. But, yf his Ma[tie] and the State in their wisedome shall thinke fitt to appropriate it, then wee must rely on S[r] Thomas Roe his faire offers and promise, and yo[r] provident Circumspection in o[r] behalfe. It wilbee very difficult for us to procure a fitt ship suddenly; for the White Angell[4] will not bee had, and the Will'm and John is nott arrived, nor expected this moneth. Therefore, wee advise you to bee very Cautious howe you engage us and yo[r] selfe to bee ready this springe, unlesse his Ma[ts] pleasure in the

---

[1] The "Three Cups", in Bread Street, was a well-known inn in the seventeenth century. It is mentioned by Hatton.

[2] From this, it appears that James was ignorant of the contents of the letter he forwarded, and had asked for a copy.

[3] Printed above (p. cxli).

[4] The *White Angel*, a vessel of 140 tons, was owned by Giles Elbridge (one of James's adventurers) and others. Letters of marque had been issued for her on July 7th, 1626 (vide *State Papers, Domestic Series*, vol. cxv, p. 113).

busines be deliũed wthin fewe daies, and that you are assured to gett a fitt ship in London yf wee faile here. This wee write not out of any disaffec̃con to the busines (for wee all continue very constant and well p'swaded therein); but it is onely to preuent any disreputac̃on or disparragem^{t} wch may befall both yor selfe and us yf wee should promise and not p'forme, or doe it unseasonably. Thus, leauinge the Successe of the Interprize to the blessinge of God, and comendinge o^{r} louinge salutac̃ons to yo^{r}selfe, wisheinge you all true happinės, and desiringe to bee advised of yor proceedings, wee rest

"Your affectionate Frynds,

"JOHN BARKER. JOHN TAYLER.
RICHARD LONGE. GILES ELBRIDGE."

"Bristol, Feb. 3, 1630[-31]."

"Sir,—I thanke you for yor kind letter directed to my selfe. Lett mee request you to recommend my humble respects to Mr. Cary[1] and his noble Consort, who are pleased to accept and esteem the toies I sent them much more then they meritt. It is their humanitie and noblenesse soe to doe. "Yours,

"JOHN BARKER."

Thus far, matters seem to have proceeded to the satisfaction of the merchants, for we next find them addressing Lord Weston, Lord High Treasurer, in a letter which Captain James must have delivered personally. It is endorsed, "The Companys letter to the Lord Treasourer in the behalfe of Capt. James, and to crave his Lo^{pps} favour for equall priviledge with others that shall attempt the enterprize", and runs as follows:—

---

[1] Probably Mr. Thomas Cary of the Bedchamber (see p. 569).

"To the right Ho^{ble}, o^{r} very good Lord, Richard Lo: Weston,[1] Lo: Highe Tre^{r} of England.

"Right Ho^{ble},

"The marchants of this Citty, beinge ambitious to serve his Matie and their Countrie, as alsoe to inlarge trafficque for their future ymployment, are willinge to sett forth a shipp this springe of the yeere for the discouery of the North-West passage into the South Sea, under the Command and Conduct of this bearer, Captaine Thomas James, a well-deservinge gent', very expert in the Arte of Navigation, valiant, and a good commander. Hee is nowe sent by them unto London, to inquire what Patent is graunted to those Gent^{s} wch haue undertaken the same designe there, that wee may obteyne equall priviledges wth them from his Royall Matie, accordinge to the proporc̃on of o^{r} adventure; haueinge also direcc̃on to attend yor Lo. when hee hath discerned howe the busines doth depend, and humbly to craue yor ho^{ble} favour in o^{r} behalfe, that wee may enioy such priviledges and Immũties (yf the passage bee discouered) as are graunted to any others: Whereby yor Lo^{p} shall increase the obligations of this Citty to yor hono^{r}, especially the Company of Marchants, yor humble servants. Thus, being confident of your honorable inclination to this our suite, we most humbly take leave, and do rest your Honours in all dutiful observance.

"JOHN TOMLINSON, Mayor. RICHARD LONGE.
HUMFRIE HOOKE. JOHN TAILER.
JOHN BARKER. GILES ELBRIDGE."

No reply to this letter has been preserved.

---

[1] This was the man who was created Earl of Portland in 1633, and after whom Foxe named a cape and James an island (see pp. 388 and 493). He seems to have been connected with Bristol, and Mr. Pope informs me that a portrait of him hangs in the Council House there.

We have already seen that there was associated with Sir Thomas Roe, in making the arrangements for Foxe's voyage, that energetic promoter of Arctic research, Sir John Wolstenholme; and this latter gentleman seems also to have helped forward the arrangements for Captain James's voyage. He had, it seems, addressed a letter to the Earl of Danby, asking his lordship to arrange, if possible, for Captain James to have a personal interview with the King.[1] The Earl of Danby's reply to this request is endorsed, "The Earle of Danby his letter to Sir John Worstenham, knighte, concerning his Ma^{ts} pleasure to speake with the Capteyne, whereby hee shall receyve grace and encouradgment for the undertakinge of the voyadge."

"To the Right Wo^{rll} Sr John Worstenham, knight.

"S^{r},

"Accordinge to yor desire, I haue moved his Matie concerninge Captaine James and the Towne of Bristoll, whoe is gratiously pleased to speake wth the man, and will then expresse himselfe both thankefull to the Towne of Bristoll for this present enterprize of discoueringe the North-West passage, as likewise for their good indeavo^{r} p'formed in this last warr with Spayne, more actiuely and wth better Successe then any other Port in those parts: where alsoe Captain James shall receyve grace and incouragement from his Ma^{ts} owne mouth for his worthie undertakinge of the voiadge intended. And Sunday next,[2] about

---

[1] No copy has been preserved of Wolstenholme's letter to the Earl of Danby (see p. 506), who was, at the time, a member of the Privy Council.

[2] February the 4th, 1630-31, was a Friday, so the following Sunday was the 6th of the month.

nyne of the Clocke in the morninge, I thinke the fittest tyme; yf you and hee cann bee in the presence or Privie Chamber, I will not faile to meete either of you there to accumplishe the effect hereof, God willinge.

"Yo^r^ very lovinge Frynd,

"DANBYE.

"St. James, Feb. 4, 1630[-31]."

We may presume that Captain James had an audience of the King at the appointed time; for, among the Company's records at Bristol, we find a copy of the following petition:—

"To the Kinge's most Exellent Matie.

"The humble peticõn of Thomas James, in the behalfe of yor M^r^chants Adventurers, Citizens of the Citty of Bristoll.

"Whereas yor Ma^ts^ most faithfull subjects, the Citizens of yor Citty of Bristoll, understandinge that yor Matie hath pleased not onely to desire the discovery of the North-west passage into the South Sea, but to grace and protect all those that shall indeavo^r^ to seeke it; whereby incouraged, yor said subjects, the m'chants of Bristoll, haue determined to sett out one good shipp, well furnished and provided for such a service, wch shalbe ready in the beginninge of May next. And, in full assurance of yor Ma^ts^ gratious disposicõn, and equall distribucõn of yor favors to all yor subjects that shall equallie pursue the waies of hono^r^ and the benifitt of your Ma^ts^ kingdomes, they haue presumed most humbly to peticõn yor Matie that you wilbee pleased to graunt unto them, the adventurers of the Citty of Bristoll, such equall shares and priviledges, both of Trade and Libertie, as you shall vouchsafe to graunt to others, the adventurers of the Citty of London, proporcõnably to their Charge and adventures. And they (as in dutie bound) will ever pray for yor Ma^tis^ longe and happie Raigne."

The Company in Bristol having received an account of the presenting of this petition, wrote as follows to Captain James:—

"Captaine James,

"Wee haue receyved yor last letters of the first and second present,[1] wth the Coppie of the petic̃on wch you exhibited to his Matie on o^{r} behalfe inclosed. Wee approve well the method thereof, and doe rejoyce that it was so gratiously accepted and interteyned by the Kinge.[2] The speedie procuringe of the King's Signature, wee consent wth you to bee necessary (if it bee his Maties pleasure to Confine the trade when the passage is discouered); and, untill wee may haue a severall distinct Patent for o^{r} selves, wee conceave it were expedient some Articles of agreement were interchangably made and sealed 'twixt us, providinge that, when the passage is found by either, the priviledge of trade, wth all other dependencies thereon, may be equall, accordinge to the proporc̃on of o^{r} adventure. We pray you forgett not to conferr and consult effectually wth Sr Thomas Button, and any other juditious men wch you cann learne haue byn formerly ymployed in this discovery, comparinge their mapps and cardes wth yo^{rs} and collectinge all their observac̃ons for yor better informac̃on and satisfacc̃on. Alsoe, we thinke it necessary (yf you approve thereof) that you enterteyne[3] twoe or three men, before yor retorne from London, of those wch were last ymployed in this Interprize, yf any may bee procured on reasonable termes, wch you deeme industrious and

---

[1] No copies of these letters are to be found among the others at Bristol.

[2] Presumably the copy of the Petition and the account of its reception by the King were sent in a later letter than either of the two here acknowledged; for James, writing on February 1st or 2nd, could hardly tell of the presentation of the Petition, which appears not to have taken place till the 6th.

[3] Evidently *engage* is meant.

usefull for us. And we pray you informe yor selfe fully touchinge the season of yor departure hence; what extraordinary provision of victuall or anything els is to bee made; and all other necessaries for the voyadge. Wee are not, as yet, assured of any fitt shipp for the voyadge; but, this morninge, Mr Elbridge hath written to a frynd in Barnestable, to inquire for one there; and wee will omitt noe opportunitie to bee sped[1] here or elsewhere. Herein closed you shall receyve a letter directed to the Lord Trer,[2] the Coppie whereof wee send you herewith. Wee leaue it to yor selfe to intimate unto his Lop that, yf others did not attempt to appropriate the benifitt of the discouery, beinge once p'formed, wee, for or parts, could bee content to leaue it open for the benifitt of his Mats subjects in generall. Thus, desiringe you hasten home wth all convenient speed, [and] concluding wth or hartie salutac̃ons, wee rest,

"Yor loveinge Frynds,

"HUMFRIE HOOKE. JOHN TAYLER.
JOHN BARKER. GILES ELBRIDGE."
RICHARD LONGE.

[Bristol, ? about Feb. 10th, 1630-31.]

The copy of the King's reply preserved at Bristol is endorsed: "His Maties answere to the petic̃on, signed by the mr of Requests," and runs as follows:

"Att the Court at Whitehall, 3rd[3] Feb. 1630[-31].

"His Matie is gratiously pleased, for the furtherance of the said discouery, to graunt to the petitioners such equall Liberties and priviledges of trade as are already graunted

[1] Query, *speedy*. [2] Printed on p. cxlv.

[3] This date must be wrong if the Petition was not presented till the 6th, as appears to have been the case. I have searched in vain at the Public Record Office for the original of this document.

to the adventurers of the Citty of London for that purpose. And, yf the said discovery bee made by either of them, then the Adventurers of both the said Citties are to trade hereafter, and to haue advantadge accordinge to the proporc̃on of their seũall adventures. And his Ma[ts] Atturney generall is to prepare a graunt accordingly ready for his Ma[ties] Royall signature.

"SIDNEY MONTAGUE."

The King's reply was remitted to the Bristol merchants by Captain James in a letter which is endorsed, "Capt. James his letter to the Company of his proceedinge and intent." It runs as follows :—

"To the Woo[rll] Mr. John Barker, Alderman, in Bristoll.

"S[r],

"I send you here inclosed the Kinges answere to o[r] petition, signed by his expresse order by the Maister of the Requests for the tyme beinge, and I hope will giue you content to yo[r] desires. They will take out noe Pattent here in London till the voyadge bee effected ; and I thinke you need not, goeinge equally side by side with them, w[ch] I leave to yo[r] consideracon. I must stay here twoe or three daies, to take instruccons, and to enable my selfe by conversacon for the accon ; but, on wenesday next, I hope to bee in Bristoll, and render accompt of my ymployment ; till when, I rest,

"Yo[r] servant,

"THOMAS JAMES.

"London, Feb. 3°."[1]

On Feb. 10th, Sir Thomas Roe also wrote to the Bristol merchants to inform them that the King had

---

[1] Probably this should be either the 8th or 13th. If the former, "Wednesday next" would be the 9th of the month : if the latter, it would be the 16th.

acceded to their petition presented by Captain James. His letter, which is endorsed, "Sir Thomas Roe his l^{re} of his Resoluc̃on to sollicite on the marchants behalfe," runs as follows :—

"My worthie Fryndes,

"I haue receyved yor letter,[1] and doe much value the trust wch you repose in mee, and ẏo^{r} owne merritt, whoe soe cheerefully undertake a noble and publique worke His Matie hath gratiously signed the petic̃on, wch remaynes for a warrant in my hand ; and you may confidently rest uppon the Justice of his Matie, wch I will add all my indeavo^{r} that you shall haue equall rights and priviledges wth those or a parte that shall take out a Pattent, and when wee shall beginn to thinke uppon that (beinge it busye to p'vide the shipp), I will write to you, that you may appoint a solliciter to attend the passinge wth Mr. Atturney, and to pay the Charges. In all other things wherein I may bee usefull to you, to yor Citty, or to yor p'sons, I shall thinke it an hono^{r} to bee ymployed, and to approue my selfe

"Yo^{r} very affectionate Frynd
to serve you,

"THOMAS ROE.

"S^{tt} Martin's Lane,
"10th Feb. 1630[-31]."

This letter was sent to the merchants of Bristol by the hand of Captain James, who evidently returned home about the time he intended. In reply, the merchants wrote a letter, the copy of which is endorsed, "The Comp. letter to Sir Thomas Roe, giveinge an Accounte of their desires and doeings

---

[1] Doubtless that printed on p. cxxxix.

in the affaires of the pretended voyadge." It runs :—

"To the Hoble Sr Thomas Roe, knight, at his house in St. Martin's in the Field.

"Hoble Sr,

"Wee have receyved yor letter by Captaine James, whereby you are pleased to expresse the continuance of yor undeserved loue and favour unto this Citty in generall, and in p'ticuler unto the Company of M'chants ; whereof wee are all very ambitious, and will striue to make or selves worthie of it. And, that posteritie may take notice of yor noble goodnesse to us in this enterprize, and yor affec̃c̃on to all vertuous and laudable actions, wee haue caused both yor letters to bee transcribed in the Register of the Company of Marchants, to perpetuate the remembrance of yor Curtesie. Wee desire not to take out a Pattent till the discouery bee made, beinge till then fully satisfied wth his Mats signature to or petic̃on. And, yf God shall please soe to Crowne or indeavors as to make us the instruments thereof, wee should accompt the service done thereby to God, our Kinge, and Countrie a reward and recompence all sufficient and abundant. And, if others seeke not to appropriate it, wee shalbee willinge to leaue it open to all the Kings subjects indefinitely. Wee haue provided a shipp for Captaine James, of fowerscore tons burthen, wch is nowe in the docke under the Carpenter's hands. And, in honour of the Queene, whose Chamber this Citty is reputed, as aunciently it hath byn of her highnesse predecessors, Queenes of England, wee haue destined the shipp's name to bee the *Mary*, in expression of or humble thankefullnesse for the gratious favours wch we haue already receyved from her Matie nowe lyveinge. The shipp (god willinge) shalbee well manned, furnished wth all necessary provisions, and ready to sett saile by the end of Aprill. Thus, Sr, haueinge giuen you an accompt of or desires and doeings in theis affaires, and rendringe you most humble thanks for the honor you haue done us in genrall, and Captaine James in p'ticuler, by yor recomendac̃on of us to the Kinge, o^r^ Soveraigne, from whome wee haue receyved a message transcendently gratious, wisheing unto you the increase of all happines and honour, wee humbly kisse yor hands, and doe remaine,

"Yo^rs^ in all due respect and observance,[1]

"Bristoll, Feb. 26, A^o^ 1630[-31]."

On his return home, Captain James also reported to the merchants the services which the Earl of Danby had rendered to their cause, as related on p. cxlvi. The merchants thereupon wrote to his lordship a letter, the copy of which is endorsed:—"The Company theire letter to the Earle of Danby in thanckfullness of his favour shewed them in procureinge Capteyne James speedie successe to his Matie and recomending the service of this Citty unto him." It reads thus:—

"To the right Ho^ble^ o^r^ very good Lord, Henry Earle of Danbye.

"Right hob^le^,

"Wee understand by the report of Captaine James since his retorne home, as alsoe by the viewe of the letter wch yor Lo^p^ was pleased to write to Sr John Worstenholme,[2] yor ho^ble^ free and undeserved favo^r^ extended unto us in the p'son of the said Captaine, by procuringe him soe speedie and happie Successe to the Kinge, o^r^ Soũaigne, and for soe recomendinge to his Matie the weake and unworthie services of this Citty and the marchants here, as to obtayne that gratious Message from his Maties owne mouth, wch the said Captaine hath faithfully delivered

---

[1] The signatures to this letter have not been preserved.

[2] See p. cxlvi.

unto us, as alsoe the p'ticuler grace done unto himselfe, whereby wee are soe much comforted and incouraged as that wee conceave it a very good [[1]] to the successe of the present enterprize, And w^th^all wee acknowledge o^r^ selves infinitely obliged to yor honour for yor noble and free favour unto us, whereof wee are not a little ambitious, and wee will strive to be capeable of the continuance and increase of it. Wee haue already provided a meete shipp for Captaine James, of Fourscore tons burthen. And, in honour of the Queene, whose Chamber this Citty is reputed (as aunciently it hath byn of her hignes predecessors, Queenes of England), we haue destined the shipp's name to bee the *Mary*, in expression of o^r^ humble thankefulnesse for the great favours wch wee haue already receyved from her Matie. The ship, God willinge, shalbee well manned, furnished wth all necessary provisions, and ready to sett saile by the end of Aprill. Thus, right ho^ble^, wth a second acknowledgm^t^ of dutie and thankefulnesse from this Citty in generall to yor Lo^p^ for yor noble goodnesse unto us, wee most humbly take leave, and doe remayne

"Yo^r^ Hono^rs^ most humble devoted
in all dutie and service,[2]

"Bristoll,
"Feb. 26th, a 1630[-31]."

The Adventurers' money being at once forthcoming and placed in the hands of a treasurer, Captain James set about choosing a suitable ship. It was his opinion that the enterprise could be best carried out with a single ship, and not by two sailing together, because of the difficulty of keeping them together in icy seas (see p. 456); but in this

[1] A blank is here left in the copy at Bristol.

[2] No signatures are given in the copy of this letter.

opinion most later explorers have differed from him. He also held that a small ship was better for the purpose than a large one, and he accordingly chose "a well-conditioned, strong ship of the burthen of seventy tons" (see p. 456). He also held that the enterprise could not be effected in less than eighteen months; so the ship was provisioned for that time. He calculated, too, that the most suitable crew for his ship was one of twenty-two men, which was also the strength of Captain Foxe's crew. James says (p. 457) this was "a small number to performe such a businesse, yet double sufficient to sayle the ship." In choosing his crew, he refused all "voluntary loyterers", and would have none but "unmarried, approved, able, and healthy seamen". He soon had sufficient applicants, all strangers to one another and to himself, but well recommended "by worthy merchants for their ability and fidelity". For some incomprehensible, and certainly insufficient, reasons (which he says he thinks it unnecessary to state), he "utterly refused" all applicants who had had any previous experience of Arctic navigation, although the merchants had, as we have seen (p. cxlviii), expressly advised him to the contrary. In having no such men among his crew, James resembled Foxe, though the latter, more wisely, had sought, but failed, to obtain any. From what James does say on this point, it looks very much as though he feared that, if he had any such men among his crew, they would show to the others his own utter inexperience in Arctic navigation — not a very worthy motive, surely!

James nowhere gives a list of the names of his crew, but he incidentally refers to a good many of his men in the course of his narrative; and a list of them (so far as it is now possible to compile one) is here appended:—

CREW OF THE SHIP "MARIA", 1631-32.

Captain . . .	Thomas James.
Lieutenant . .	William Clements.
Master . . .	Arthur Price.
Master's Chief Mate .	John Warden.[1]
Master's Second Mate .	John Whittered.[2]
Cheirurgeon . .	Nathaniel Bilson.
Boatswain . .	John Palmer.
Gunner . . .	Richard Edwards.[3]
Gunner's Mate . .	John Barton.[4]
Carpenter . .	William Cole.[5]
Cooper . . .	————?
Cockswain . .	————?
Baker . . .	————?
Brewer . . .	————?
Butcher . . .	————?
Seaman? . . .	George Ugganes.
Seaman? . . .	David Hammon.
Seaman . . .	————?
Seaman . . .	————?
Seaman . . .	————?
Boy . . .	————?
Boy . . .	————?

During the time he spent in making his preparations, he says (p. 458) he endeavoured to perfect

[1] Died May 6th, 1632.

[2] Chief mate after Warden's death.

[3] Lost his leg Aug. 20th, 1631; died Nov. 22nd, 1631.

[4] Drowned Oct. 17th, 1631.

[5] Died May 18th, 1632.

his studies in Arctic exploration, and "sought after Journals, Plots, Discourses, or whatever else might help my understanding".[1] With nautical instruments he took care to provide himself most liberally; and he gives, in the form of an appendix (p. 604), a list of those he took with him, most of which seem to have been made specially for him, with the utmost care, by the best and most skilled workmen of his day. Of his numerous instruments, Captain James, in the course of his voyage, made good and frequent use. In this, of course, he was greatly helped by his extensive mathematical knowledge, of which even his rival, Captain Foxe, speaks highly. In short, from everything that James says in his "Preparations", we may gather that the merchants of Bristol placed unbounded confidence in him, and equipped him for his expedition with no niggardly hand. Everything that he thought he stood in need of seems to have been liberally provided.

James says that his preparations were complete by the 1st of April 1631. He then, he says (p. 459), made a journey to London (apparently for a second time) to know his Majesty's further pleasure; and soon after he received, through Sir Thomas Roe, the King's Instructions. Thereupon, returning to Bristol, he got his ship out of dock, and awaited a fair wind to commence his voyage. The following

---

[1] It is clear from what he says in the closing passage of his narrative (see p. 592) that he had formerly made, like Foxe, an exhaustive study of his subject, and a careful search for any charts or narratives which could throw any light upon it.

document, endorsed, "The Coppie of a warrant for paiment of the companies wages of the Henrieta Maria," is preserved in the Company's archives at Bristol, and is of interest at this point:—

"Whereas, for the setting fourth and furnisheinge of the shippe called the Henrieta Maria, of Bristoll, whereof Thomas James goeth Captaine, pretending a voyadge, by the grace of god, for the discoverie of the North-west passage to China, at the costs and charges of the societie and companie of marchants adventurers of the cittie of Bristol: It was, at a generall assemblie of the said Societie, in ample nomber, ordered and agreed that Humfrie Hooke, nowe Maister of the said Societie, Andrew Charlton, Miles Jackson, and Thomas Cole, with the assistance of the said Captaine Thomas James, being adventurers of the said Voiadge, should bee husbands and stewardes for the mannageing and furnishing of the said shippe with all prouisions and necessaries; and, forasmuch as the said husbands haue accordinglie fitted, furnished, and made readie the said shipp, not onelie with all sorts of munition, ordnance, artillerie, victualls, prouisions, and necessaries whatsoeuer requisite for the said voiadge, but alsoe with an expert and able Maister, twoe M^rs^ mates, and other marriners, and serviceable seamen and boies, whose seũall names and monethlie wages agreed uppon are menc̃oned and expressed in a schedule or portledge bill hereunto annexed.[1] Now these p'sents witnesseth that the M^r^, wardens, and Comminaltie of the arte or misterie of marchants adventurers of the citty of Bristoll, for the accomplishing of the agreement of the said husbands and rewarding of the M^r^ and companie of the said shipp for their paines and seruice to bee donne in the Voiadge, doe hereby graunt, promise, and undertake that euerie person whose name is menc̃oned in

[1] The Schedule in question has unfortunately not been preserved.

the said Scheedule serueing in the said shipp and voiadge shalbee fully satisfied and paid his seũall and respective monethlie wages, according to the rates sett downe in the said schedule; that is to say: for euerie moneth and parte of a moneth that each man shall serue and continue in the said shipp, under the commaund and obedience of the said Captaine Thomas James and his deputed officers. In witnes whereof, to one part of this writing Indented remaining with the M^{r} and companie of the said Shipp, the said M^{r}, Wardens, and Comminaltie haue caused their Common Seale to bee sett and affixed; and, to the other part hereof, remaineing with the said Maister, Wardens, and Comminaltie, the Maister and companie of the said shipp haue subscribed their names, the thirtith day of Aprill, anno dñi 1631, in the seaventh yeere of the Raigne of O$_{r}$ Soũaigne lord Charles, by the grace of god, Kinge of England, Scotland, France, and Ireland, defend^{r} of the faith, etc."

As to the precise tenor of the Instructions under which Captain James sailed, we are (as in the case of Captain Foxe) left to a large extent in the dark; for, although James tells us (p. 459) that, when his preparations for his voyage were complete, he made a special journey to London, and that he there "received his Majesty's Royal Letters, with Directions for proceeding in my Voyage, and my Discharge", he omits to tell us what those Directions were. Nor is it easy (as it is in Foxe's case) to gather, from passing remarks in his narrative, a clear idea of their purport. There can, however, be very little doubt that they were, to all intents and purposes, identical with those given to Foxe (see p. xcvi). The latter, in telling us how he received from the King his own Instructions, with a letter to the Emperor of Japan, and other documents, adds the words: "Copies of all which Captain James had." This might be taken to mean that Captain James had been provided with copies of the other's papers merely for his convenience and guidance; but it seems not improbable that the words may be taken to mean that the Instructions of the two were identical, especially as Foxe afterwards, in discussing with Captain James the question of exploring to the north-west of Nottingham Island, announces his intention of so doing, "as both hee and I were instructed" (see p. 360). In his narrative, James gives us very little information, but in the letter left at Charlton Island he says: "by my directions, I was to search especially two places—one from Digges Island to the northwards, and, failing there, to go to the Checks and Hubbert's Hope, and so to search it to the Southward." Moreover, we know that (like Foxe) he endeavoured in the first instance to explore to the northward, but without success; and that he then (also like Foxe) made for the western shore of the bay, intending to meet it in about lat. 63° N., though in this respect (unlike Foxe) he failed. After this, on reaching the western side of the bay, a little to the north of Port Churchill, he coasted eastward along its southern shore to Cape Henrietta Maria (following here exactly the route pursued by Foxe); and, in the following year, after wintering in James's Bay, he sailed northward as soon as the ice would allow him, and, lastly, again endeavoured (as Foxe had also done) to explore north-westward from Nottingham Island. On the whole, therefore, Foxe and James followed (or rather endeavoured to follow) very similar routes, except that the latter entered what has since been known as James's Bay, though he did this simply for the purpose of wintering as far south as possible. It seems, therefore, a fair assumption that the Instructions given by the King to the two navigators were identical. Moreover, we know that Sir Thomas Roe had suggested that, in drawing up the Instructions to Captain James and Captain Foxe respectively, the Bristol Adventurers should consult with the London Adventurers for mutual advantage (see p. cxli).

On the 3rd of May 1631, Captain James commenced his voyage. On the 4th of June, Greenland was sighted, and the ship received her first damage from the ice. On the 17th, Resolution Island was sighted, and, a day or two later, James entered Hudson's Strait, and at once fell into difficulties with the ice. For several days, according to his account, the ship was in extreme danger; but, by the 24th he got free, and continued his voyage westward, constantly in the utmost danger because of the ice, till the 15th of July, when he was between Digges Island and Nottingham Island. Finding, as already stated, that it was impossible, on account of the ice, to explore to the north-westward from the latter island, he sailed away W.S.W. for Mansell Island, which he reached on July 16th. Two days later, he sailed westward, intending to reach the western shore of the bay in about lat. 63° N. Ice was at once encountered, and the ship seems to have been almost constantly beset till the 11th of August, when the western shore of the bay was at last encountered, though nearly three degrees and a half to the southward of the point for which James had been making. From this point, he coasted southward. Two days later, he ran upon some rocks (probably Nelson Shoal), and, according to his own account, was for a time in extreme peril; but, escaping at last, he continued his course, passing Port Nelson (in which harbour Foxe was at the time engaged in fitting up his pinnace) on the 17th. Here he, for the first time, commenced actual discovery; but the honour of being the first to explore the southern coast of the bay, between Port Nelson and Cape Henrietta Maria, must be shared equally by Foxe, who (emerging from Port Nelson two days after James had passed) afterwards overtook and passed him. James named the country he passed the New Principality of South Wales: Foxe called it New Yorkshire—each thus naming it after his own locality. On the 20th, a small rope got foul of the cable as it was being got in by the capstan, with the result that many of his men were injured, one fatally. On the 29th, Captain Foxe in the *Charles* was sighted, and visits were exchanged, as already related by Foxe (p. 358). James endeavoured to persuade Foxe to winter in the Bay with him, but Foxe wisely declined. James then continued to coast to the eastward till the 2nd of September, when he reached and named Cape Henrietta Maria, after the Queen. Foxe reached this cape on the same day, and bestowed upon it another name, now discarded (see p. 367); but the two explorers did not meet at this point or afterwards; for Foxe sailed away northward, while James turned southward, intending to winter in the bay which now bears his name. The whole of September was spent in cruising aimlessly about among the numerous islands in James's Bay, not a few of which he named, though it is difficult now to identify them. More than once the ship ran ashore, and all the time (according to James) she was in most imminent peril. In the beginning of October, he found himself in the neighbourhood of Charlton Island, and, being overtaken by winter, he remained there. He then stripped his ship and landed most of his provisions, building three houses on shore in which to spend the winter. On the 17th, John Barton, gunner's mate, was drowned. On the 22nd, died the gunner, who had been injured in the accident with the capstan three months before. On the 29th, failing to secure the ship, he adopted the extraordinary expedient of sinking her. The account of the wintering is a harrowing, but interesting, story, which James recounts in great detail. The hardships and troubles they encountered are, as usual, very fully described by him. A great part of the winter was spent in building a pinnace in which to undertake the return journey, in the event of it being found impossible again to raise the ship, as was

m 2

expected. In April, efforts were begun in order to raise the ship, and in the end this was accomplished. On the 6th of May, died John Warden, the chief mate, and, on the 18th, William Cole, the carpenter. The month of June was chiefly occupied in refitting the ship. On the 25th, one of the crew was clumsy enough to set on fire the forest in which they had wintered, and James himself, as well as a good many of their effects, was nearly burned. The mosquitoes, too, troubled them very much. On the 2nd of July, they set sail from Charlton Island; and, after experiencing great difficulties with the ice, they at last reached Cape Henrietta Maria on the 22nd. At this point, however, James's difficulties seem only to have been commencing; for, being prevented by the ice from sailing due north, he was detained till August 7th, and then he coasted along to the westward, being almost always completely surrounded by ice. Having almost completed the circuit of the bay by the route he and Foxe had both sailed over (though in the opposite direction) in the previous year, he reached Cary's Swan's Nest on the 22nd, intending from there to search to the north-westward of Nottingham Isle, which both he and Foxe had failed to do the previous year. To accomplish this, he sailed northward in a violent storm, up the channel which (unknown, of course, to James) Foxe had explored the previous autumn. Up this Strait, James, in his turn, sailed as high, he says, as lat. 65° 30′ N. at least. He then became fast among the ice (apparently on the western side of the channel), and, after consultation with his officers, resolved to turn south and make for home. It is noticeable that James claims to have attained a point only about 1° 17′ south of that which Foxe had undoubtedly reached the year before ; but, unlike Foxe, he named no localities, and did not lay down an inch of coast-line. Resolution Island was passed on the 3rd of September, and Bristol was reached on October 22nd.[1]

Very shortly after his return, the Captain must have proceeded to London to deliver his account to the King and others ; for, among the documents at Bristol (see p. cxxxix), we find one endorsed, "The Coppie of the letters sent to the Lo: Treasurer & to the Earle of Danby uppon the arryval of Capt. James from the Northwest Passadge. The

[1] A contemporary reference to the return of Captain James is to be found in *The Court and Times of Charles I* (vol. ii, p. 189), wherein is printed a letter in which Mr. Pory, writing from London on November 1st, 1632, to Sir Thomas Lucy, says: ". . . The two Bristol men, after the expense of two summers in discovering a North-West passage, are returned back *re infectâ* ; only I heard John Tradescant tell my Lord of Ca. that they had discovered an island where were store of unicorns' horns, long and wreathed like that at Windsor, which I have heard to be nothing else but the snout of a fish, yet very precious against poison."

The reference to "*two* Bristol men" is, of course, an error. John Tradescant was a great naturalist, traveller, and antiquary, as well as gardener to Charles I. He died about 1652. His collections became the nucleus of the Ashmolean Museum. Reference has already been made to the properties formerly attributed to the horn of the unicorn (see pp. 45 and 309).

like sent to Mr. Tho: Carye of the Bedchamber." It runs :—

"Right ho^{ble},

"This bearer, Captaine James, being emploied about eighteene moneths past by the companie of marchants in this Cittie for discouerie of the Northwest passage into the South sea, with the approbaticn of our Soveraigne lord the Kinge, yor lo^{p}, and some other noble personages, and haueing escaped manifold dangers and extremities in the Voiadge through gods mercie : Hee doth now addresse him selfe to giue his Matie the first account of his industrie and endeavo^{r}, according to that royall commaund giuen before his departure, wherein wee most humblie crave the continuance of your lo^{ps} favour that his faithfull and indefatigable seruice, togeather with o^{r} purpose and designe in the Voiadge, tending principallie to the honour of his Matie, and common benifitt of o^{r} natiue countrie, may, by your wisedom and power, bee soe presented to the King that it bee gratiouslie accepted ; which, being obtained, the fruition thereof wilbee esteemed an abundant Compensation of his labour and our charge, and will more oblige him and us to pray for the encrease of yor lo^{ps} honour and happines, and to devote our selues

"Yo^{r} Lo^{ps} most humble servants,

"HUMFRYE HOOKE, RICH: LONGE,
JOHN BARKER, THO: COLSTON,
ANDREW CHARLTON.

[Bristol,]
"November the second, 1632."

Another document is endorsed, "The Coppy of the letter sent to Sr Tho: Roe upon Capt. James returne," and runs as follows :—

"Ho^{ble} S^{r},

"As our first intendment and designe for discouerie of

the Northwest passage, by yor powerfull and prudent recommendac̄on thereof to the Kinge, o^{r} Soũaigne, was gratiouslie interpreted by his Matie, as undertaken for his honour and the common benifitt of our native countrie, soe wee humblie pray that the first account of Captaine James his proceedings in the Voiadge may now by your like favour and wisedome bee soe presented to his Matie as that his resolution and industrie, as alsoe our endeavo^{r} and charge, may bee crowned by his Ma^{ts} benigne acceptac̄on thereof. This Citty in geñall, the companie of marchants especiallie, and some members of both in particular, doe most humblie and thankefullie acknowledge yor many undeserued favours, desiring to bee esteemed worthie the continuall fruition of them and to bee soe happie as to giue some reall testimonie that wee are as wee professe.

"Yo^{r} most humble and obliged servants,[1]

[Bristol,
? Nov. 2nd, 1632.]

Finally, in the same series, we find the copy of a letter to the Bristol merchants, written from London on November 19th, and endorsed, "Captain James his letter to the Comp., having delivered their letter to the Lords, and given his Matie an account of his perigrinac̄on in the voyadge."

"To the Wor^{ll} Mr Humfrie Hooke, Alderman in the Cittie of Bristoll.

"S^{r},

"Soone after my arriuall at london, I deliũed yor gen'all l^{res} to those ho^{ble} p'sonages to whom they were directed, except to the Earle of Danby, who was not in towne. They were receiued, and my selfe (by your undeserued commendac̄ons) honorablie entertained; and, as

[1] No signatures have been preserved.

by experience I haue found, they respect the worsest of yor servants, the more amplie to make to appeare how much they favo$_{r}$ you in gen'all, and any of the noble-minded of the Citty in particuler.

In convenient time, they presented mee to his Matie, whoe accepted of y$_{e}$ plot I presented him of my peregrinac̄ons, and wth a gratious patience heard mee read the briefe of my endeavo^{rs}; wth wch the conference, by way of questions and answeres, continued aboue twoe houres time. His Matie seemed to bee well contented wth my faithfull p'se-verance in the acc̄on, and commaunded mee a second time to attend him and giue him a further relation, and to bring wth mee Captaine Fox and Captaine Bruton, wch went forth this yeere,[1] the better comparatiuelie to iudge of o^{r} proceedings. My weeke seruice, by yor encouragem^{t} and bountifull accomodac̄ons in my setting forth, appeared in all humilitie, so that his Matie welcomed mee home, and was pleased to say it satisfied his expectac̄on. Hee hath commaunded mee to attend here in london, and make an abstract of my Journall, and p'fect my observations, and bring it to him; wch I am about, and brieflie intend to effecte, and know his further order. As I am but one of many, and albeit a parte to that wch was necessarie to perform the acc̄on, I was not forgetfull to make knowne to him the charge you were at, and the spetiall kindnes wch for my encouragem^{t} you haue shewen to mee: as, likewise, to manie lords at whose tables I haue beene entertained. All wch doth the more bind mee ever to remaine

Yo$_{r}$ faithfull servant,

"THOMAS JAMES.

"London,

"Nov. 19, 1632."

That Captain James was received by the King in a most gracious manner is clear from this letter. It

[1] This should, of course be *last* year.

would be most interesting to know what took place at the second audience with the King, when James was ordered to bring also "Captaine Fox, and Captaine Bruton, which went forth this year". Who the latter individual may have been I am quite unable to suggest, unless the name of Foxe's master was Bruton, instead of Dune (see p. cxii).

In reviewing generally the circumstances connected with James's voyage, the most striking feature will be found to be the extraordinary and incessant perils (chiefly arising from difficulties with the ice) which he encountered. If James has not exaggerated the dangers and the difficulties, his is certainly the most remarkable and perilous voyage ever undertaken from which any considerable number of those engaged have escaped to tell the tale. Nor is this feature noticeable only in any special portion of the voyage, or in any particular part of the region explored: it is observable all through the narrative. James had scarcely entered Hudson's Strait before he found himself, according to his own account, in frightful danger, and likely at any moment to lose his ship. Again, he had no sooner entered the Bay than we hear of almost hourly hair's-breadth escapes from the ever-present ice. Scarcely had he got free before he ran upon some rocks; and scarcely had he escaped this danger before an accident happened with the capstan, causing a number of his men to be injured, one of them fatally. On getting down into the Bay in which he afterwards wintered, and which has been called after him, his narrative becomes little more than an account of how he escaped from incessant dangers, arising from the shoals which, we are told, surrounded him in every direction, and on which he struck more than once. Time after time, we read how the ship survived some danger which had for hours threatened to instantly destroy her, and how the crew then fell to prayers and thanked God matters were no worse. To quote instances would be to quote a very large portion of the narrative. The wintering must undoubtedly have been a time of great hardship; but it is difficult to understand how it can have been really necessary to adopt the extraordinary expedient of sinking the ship in order to preserve her. On the return voyage, matters were much the same. For weeks together, the ship was almost immovably beset by the ice on the southern coast of Hudson's Bay. On at last escaping, he says: "We then went to prayer and gave God hearty thanks, that had delivered us out of it. For we were hourly, for the space of six weeks, as it were, in the jaws of Death."

It would be difficult to sum-up the foregoing remarks better than in the words of Sir John Barrow, who humorously says (*Chronological History of Voyages in the Arctic Regions*, p. 250): "Captain James's history of his voyage may be called a book of 'lamentation and weeping, and great mourning'. It is one continued strain of difficulties and dangers and complainings, from the first making of the ice off Cape Farewell, till his return to the same point." On the same subject, Ellis, in his *Voyage to Hudson's Bay* (London, 8vo, 1748, p. 64), says: "In the long account which Captain James has given us of his wintering, there is such a detail of miseries and hardships as might have been sufficient to have deterred any from venturing again into this Bay." Mr. Ivor James has expressed the same idea in his *Source of "the Ancient Mariner"*, wherein he says: "The most superficial examination of the narrative would excite interest. Once taken up, the book is not to be laid aside. . . . I know of nothing like it in the whole range of the early literature of the sea. . . . The narrative is one long thrilling description of a fearful struggle with the forces of Nature on unknown shores; of frightful storms in blinding mists and snows, in raging seas, among mountains and islands of ice, resistlessly driving and crushing ship and all against other mountains and islands of ice. . . . The story throughout is one of terrors piled on terrors, until they weary with their endless monotony. . . . The account of James's voyage is, in many respects, I will venture to say, one of the most remarkable productions in the English language."

It would, I think, be difficult for anyone to review all the circumstances of the case without coming to the conclusion that the difficulties and dangers encountered were largely due to the captain's incompetence, if not as a seaman, at least as a navigator of ice-covered seas, which is not necessarily the same thing. This may, I think, be easily inferred from the general tenour of his narrative; but it is, I think, proved by the fact that Foxe, in sailing over almost the whole of the same ground, and in the same year, scarcely experienced any serious difficulties at all. When we compare Foxe's brief but pertinent comment when he had injured his cutwater—"the first harme, and all I received"—with the pages and pages of hair's-breadth escapes and accounts of appalling accidents given by James, we shall at once see the extraordinary difference that exists between the two men and their experiences. Barrow, a high authority, says (*Arctic Voyages*, p. 250) that "the sufferings of his crew [from the intense cold] . . . . could only have arisen from mismanagement". The author of the "History of Maritime and Inland Discovery" (Lardner's *Cabinet Cyclopædia*, vol. ii, p. 180) also remarks: "James was evidently an unskilful navigator; and, if not actually timid, was at least well disposed to magnify difficulties. He was one of the few who maintained the improbability of a North-West Passage; and his opinion had less weight, as it was contradicted by those who displayed far more sagacity and skill in the conduct of a similar enterprise."

The same view has been taken by most other writers of authority, who fully uphold the estimate which Foxe had formed (see p. 359) of James's incompetence as an Arctic navigator.

The next most striking feature which the student of Captain James's narrative will observe is the very small amount of useful geographical discovery he accomplished. The credit of having first explored the southern coast of Hudson's Bay belongs, as has been said, equally to both Foxe and James. The only other direction in which both made any new discoveries was in Foxe Channel; and here James (who was a year later than Foxe, and did not sail nearly so far north) did not name a single feature nor locate a single inch of coast-line, while Foxe did both and much more. James had one opportunity which Foxe had not, in that he spent a considerable amount of time in James's Bay; yet the very poor account to which he turned this advantage is shown by the fact that he ascertained little about it except that it was a large bay containing many islands. But this much had been ascertained by Hudson just twenty years before. James's map of it is so unsatisfactory, that it is quite impossible, even with the help of the narrative, to follow his track with precision, or to identify more than a few of the islands he visited. His map of James's Bay, in fact, was of very little more value than that brought home by the survivors of Hudson's mutinous crew in 1610.[1] Clearly, then, Barrow

[1] Sabin's statement on this point (*Dictionary of Books relating to America*, vol. ix, 1877, p. 214) is quite misleading. He says of James that he "has been censured for the lack of geographical information contained in his book, but the map is assuredly free from any such blame, for it contains a singularly-correct delineation of the high latitudes in which Hudson, Baffin, James, and Foxe pursued their search for the North-West Passage."

was correct in saying (*Arctic Voyages*, p. 251) that, "with regard to discovery, James contributed nothing to what former navigators had effected". Captain James was absent from England over three times as long as Foxe, but his discoveries, such as they were, were certainly not equal in value, from a geographical point of view, to Foxe's.

Another feature in James's voyage which will strike the student is the fact that he wintered in the Bay, which Foxe did not do. The account he has given us of the way in which he spent this winter is certainly the most generally interesting portion of his book. It forms one of the earliest and fullest accounts we have of the severe cold and the hardships of an Arctic wintering, though it must not be forgotten that his winter home was devoid of many essentially-Arctic characteristics; for it lay about fourteen degrees south of the Arctic Circle, and (being far to the south of the limit of the growth of trees) was surrounded by a considerable pine-forest. Nor can the fact be overlooked that his wintering at all was a gigantic mistake, as Foxe had endeavoured to show him it would be, when they met at the end of August (see p. 358; also pp. 407 and 415). Had he returned to England (as Foxe did) when the season for navigation closed, he would have avoided all the hardships he and his crew suffered, and would not have lost the four men who succumbed. Moreover, he might easily have returned to the same spot, with a sound ship and a refreshed crew, at a period of the year quite as early as that at which he actually found himself free; for, whereas he was unable to recommence exploration in his second year (1632) before the 3rd of July, both Foxe and he, when sailing out from England in the previous year, had entered the Bay at an earlier date. Foxe, as we have seen (pp. civ and 407), was severely blamed on his return for not having wintered; but it is impossible to deny that his was the wiser course, as the experience of the two proves. Perhaps, however, as Seyer suggests (*Memoirs of Bristol*, vol. ii, p. 286), James "suspected that he should not be able to persuade his employers to a second attempt, and was therefore resolved to use to the utmost the golden opportunity which then presented itself."

Still, although James's expedition was not a great success geographically, we should remember that he was an accomplished gentleman, of good family and education, devout, considerate of others, and personally in all respects an interesting and estimable character. Nowhere in his history do we meet with unworthy conduct on his part, either in word or action. One puts down his book feeling nothing but respect for his endeavours, however much we may recognise his shortcomings as a sea-captain and an explorer.

In the course of his voyage, Captain James named various capes, islands, and other features, though fewer than Foxe had named, and nearly all of them situated in James' Bay. A list of these names here follows, those that have survived to the present day (which are very few) being printed in *italics*. Remarks upon the origin of these names, and on other points connected with them, will be found in the shape of foot-notes appended to each as they appear in the narrative.

A List of Islands, Capes, etc., named by Capt. Thomas James, 1631-32.

Name.	*Latitude, N.*	*Longitude, W.*	*Date of Naming.*	*Page.*
Harbour of God's Providence.	61° 20′	65° 00′	June 22, 1631	470
Price's Cove . . .	61° 20′	65° 00′	„ 23 „	471
Button's Bay . . .	—	—	—— ? „	—[1]
Briggs his Bay . .	59° 00′	94° 00′	—— ? „	—[1]
New Principality of South Wales.	57° 00′	—	Aug. 20 „	485
New *Severn River* . .	56° 00′	89° 10′	„ 26 „	—[1]
Lord *Weston's Island* .	52° 50′-53° 35′	81° 00′	Sept. 10 „	493
Earl of Bristol's Island .	53° 10′	80° 00′ ?	„ 19 „	500
Sir Thomas Roe's Island	52° 15′	79° 10′ ?	Sept. 22 „	503
Earl of *Danby's Island* .	51° 55′	79° 10′	Oct. 2 „	506
Sir John Winter's Forest .	52°	79°	Dec. 24 „	529
Brandon Hill . . .	52°	79°	May 6, 1632	547
Charlton Island . .	52°	79°	„ 29 „	552
Charles' Town . . .	52°	79°	„ 29 „	552
Cary's Island . . .	52°	79° 10′	—— ? „	569
Isle of God's Favour .	— ?	— ?	—— ? „	598
James his Bay . . .	51°-55°	79°-83°	—— ?	—[1]

James reached Bristol on the 22nd of October 1632, and he must at once have set about preparing his narrative, which he completed and issued with commendable promptitude; for, within five months of his return, his book was ready for issue. In Prof. Arber's *Transcript of the Registers of the Stationers' Company* (vol. iv, p. 267), we meet with the following entry:

[1] See Chart.

19° *Martij* [1633.]

Master Partridge Entred for his Copy[right] vnder the hands of master WECKERLYN and master WEAVER, warden, a booke called *the strange and miraculous voiage of Captayne* THOMAS JAMES *to the South sea for the discouery of the North west passages, begunn in the yeare* 1631 *and ended in* 1632, by the said Captain THOMAS JAMES . . . vj^d^.

It is true that James's volume is a much smaller one than that of Foxe (notwithstanding that the former was absent from England nearly three times as long as the latter); but its prompt publication was no doubt mainly due to the fact that (unlike Foxe) James was a well-educated gentleman, who had had considerable literary experience. In his narrative, he expresses himself so clearly, his diction is so good, and what he has to say is so interesting, that it is a pleasure to read his book; while to peruse that of Foxe (which took a much longer time to prepare) is a tedious and difficult occupation.[1] It is undeniable that to this great merit in James's book is due the fact that it has been very frequently reprinted (as mentioned hereafter), and that it has made a distinct position for itself in literature, which Foxe's work has utterly failed to do, although it is unquestionably of the greater geographical value.

James's work, like Foxe's, was printed by Royal command. It was issued early in 1633 as a pott

[1] This was, no doubt, partly due to the attention bestowed upon the copy by the Rev. William Watts (see p. 626).

quarto volume, of about one hundred and fifty-two pages. It was "printed by John Legatt[1] for John Partridge". The title-page has herein (see p. 449) been reproduced as nearly as possible in facsimile. The pecuniary value of a copy is now very considerable. Messrs. Hy. Stevens and Sons, the well-known American booksellers, recently advertised a copy, without the original map, at £15, describing it as "a volume of the greatest rarity, and seldom found with the original map"; adding that "copies with the original map have been recently sold at from £35 to £40". Mr. Quaritch has also been good enough to inform me that he considers the sum named to be about the value of a perfect copy at the present time.[2]

---

[1] We learn from the Addendum to Arber's *Transcript of the Registers of the Stationers' Company* (vol. iii, pp. 699-704), that John Legatt, or Legate, was an "authorised" London printer, and was licensed to have two presses. He succeeded his father, formerly of Cambridge, and of the same name, about the year 1620. Lowndes carelessly describes Watts (the author of the "Advise") as the publisher.

[2] The collation of the work seems to be as follows: Title, with reverse blank [2 pp.]; Advertisement of nautical instruments for sale by Andrew Wakely, on Redriff Wall, near Cherry Garden Stairs, with back blank [2 pp.] (this leaf blank in most copies); Address "To the King's Most Sacred Majesty" [2 pp.]; Letter signed "Thomas Nash", and addressed "To my worthy friend and fellow-Templar, Captaine James" [1 p.]; the Printer's Apology to the Reader [1 p.]; "The Preparations to the Voyage," pp. 1-4; The Narrative of the Outward Voyage, pp. 5-53; The Narrative of the Wintering, pp. 54-89; The Narrative of the Homeward Voyage, pp. 90-111; Copy of the Letter left at Charlton Island, pp. 112-120; List of Nautical Instruments used upon the Voyage

The map (which is very often absent from otherwise perfect copies) is of far less historical interest than that which accompanies Foxe's book, because, instead of depicting (like Foxe's) nearly the whole of the Arctic regions, as known at that day, it shows little more than the Bays of Hudson, Baffin, and James, and Hudson's Strait. It is entitled, "The Platt of Sayling for the Discovery of a Passage into the South Sea, 1631-1632." In the left-hand upper corner is, "The trve portraict of Cap. Thomas Iames. Ætatis suæ 40." He wears the pointed beard, long hair, and large embroidered collar of his time. Beneath the quarter-length figure are the words, "Some has a time." This reads much like a family motto; but I cannot

---

[2 pp.]; The Manner of taking the Variation of the Compass [4 pp. and blank leaf]; "An Appendix touching Longitude," signed "H. Gellibrand" [6 pp.]; An Address to the Divinity Students at Cambridge, signed "X. Z." in most copies [10 pp.]; Folding map, with portrait of author in corner (variously placed).

There are in the British Museum three copies of James's work, which differ somewhat from one another in various details. One copy (C. 32. d. 9), though not in very good condition, is quite perfect, with the exception of the Advertisement. It is remarkable (and, so far as I know, unique) as having the signature "W. Watts", instead of the usual "X. Z." The Grenville copy (G. 7166) is in superb condition and perfect, even to the Advertisement. It contains a few MS. notes signed "J. B. [? Barrow], 1791." The copy in the King's Library (♔ 362, c. 3), though erroneously marked perfect, wants both the map and the Advertisement, while the Letter from Nash precedes the Address to the King, instead of following it, as usual. There is also a separate copy of the map (Maps 70095-3), which is perfect, with the exception of the portrait, which has been cut out.

discover that it was ever borne either by James or any other family. The fact of the portrait appearing upon the map has, as one writer observes, "occasioned the destruction [or, rather, mutilation] of many copies of the book by the *Grangerites*".[1] The portrait was re-engraved, on a considerably enlarged scale, for the Supplement to Richardson's *Collection of Portraits illustrating Granger* (vol. ii, No. 13, London, 4to., 1822).[2] The plate is now in the possession of Mr. W. V. Daniell, of 53, Mortimer Street, Cavendish Square, and from it has been printed the portrait of James which appears as the frontispiece of the second volume of this work. The facsimile of James's autograph, which appears below the portrait, has been copied from a document in the Record Office. On the opposite, or right-hand, upper corner of James's map, James's Bay is reproduced on an enlarged

---

[1] The "Grangerites" were the followers of the Rev. James Granger, author of *The Biographical History of England* (London, 2 vols., 4to., 1769), whose collection of engraved portraits illustrating his great work, amounting to 14,000, was sold in 1778. Several supplements to his work have been published. He was born in Dorsetshire in 1723, and died at Basingstoke in 1810.

[2] Seyer says (*Memoirs of Bristol*, vol. ii, p. 283) he found this note in a copy of James's work lent him by C. J. Harford, Esq.: "Three guineas, Harding assured me, have been given for it [*i.e.*, the portrait]. He re-engraved it [? for Richardson], and that cannot now be had. I have in vain searched for a portrait of this certainly great seaman in the Council-House and Merchants' Hall [at Bristol]."

scale, Charlton and the neighbouring islands being named.[1]

James's narrative, unlike Foxe's, has been frequently reprinted in a more or less incomplete form. This certainly cannot be ascribed to its having possessed any great geographical value, for it has but little such—far less, in fact, than Foxe's. But it had what Foxe's book lacked: it had considerable general interest as a narrative of adventure by sea in a then-almost-unknown portion of the world. It had also been well and clearly written, which feature Foxe's book conspicuously lacked.

The "Second Edition, revised and corrected", appeared in 1740 as a demy octavo volume of [10]-142 pp. and a map.[2] This edition, however, is immeasurably inferior to the original. The punc-

---

[1] Mr. Coote has kindly drawn my attention to the fact that there is in the British Museum (*Add. MS.* 5415. G. 1) a curious old MS. map which is identical in all particulars with that of Capt. James. It may be the original from which the engraver worked, or a copy of later date; but the appearance of the paper proves its age.

[2] The title-page bears the following imprint: "London: Printed in 1633, and now Reprinted for O[LIVE] PAYNE, at *Horace's* Head, in *Pope's* Head Alley, *Cornhill*, over-against the Royal *Exchange*, MDCCXL. Price bound Two Shillings and Six Pence." An "advertisement concerning this Edition", signed by the publisher, says: "The universal good Character this Voyage has among the Judicious for its Integrity and Simplicity, and the great scarcity of it (having been sold for 15*s*. and a Guinea in several Auctions), are sufficient Motives for the Reprinting of it; likewise, we hope it will prove useful and agreeable to the Publick."

tuation and orthography were brought up to date throughout; but the matter does not seem to have been in any other respect tampered with, except by the omission of James's two poetic effusions,[1] the interesting letter left at Charlton Island, and the irrelevant "Advise" at the end. The map was re-engraved entirely, but without any alteration whatever, except in the spelling of certain names and the omission of the portrait of James. That there should have been no alteration in the map, especially in that part of it depicting James's Bay, was unpardonable; for, in 1740, the configuration of that bay was tolerably well known, the vessels belonging to the Hudson's Bay Company having made many voyages to the Company's posts or "factories" around its shore.[2] The map is often missing from this edition.[3]

The next and last separate edition of James's book appeared in 1807, as a tiny pott octavo volume of 60 pp., entitled *The Voyages and Distresses of Captain T. James and Mr. H. Ellis for the Discovery of a North-West Passage to the South Seas.* There is an engraved frontispiece representing the

---

[1] See pp. 505 and 564.

[2] Moreover, only eight years later (in 1748), there was published a map of Hudson's Bay, which (except as regards minor details) shows the Bay almost as correctly as maps of the present day, namely, that in Ellis's *Voyage to Hudson's Bay* (London, 8vo., 1748).

[3] There is one perfect copy in the British Museum (G. 15982), which seems to have served as the copy for some later reprint, as several passages are marked for omission.

incident which occurred on Charlton Island on June 25th, 1632, when Captain James, who had climbed a tree, barely escaped with his life from a forest-fire which he had ordered one of his men to light. James's narrative, which occupies the first thirty-two pages, has been greatly abridged. There is nothing to show by whom this condensation was effected, but it seems to have been done judiciously. The work was printed by T. Maiden for Ann Lemoine and J. Roe.

Beside the above editions in separate form, Captain James's work has been frequently reprinted, more or less abridged, in various collections of voyages and travels. The following is a chronological list of the chief of these reprints :—

1705. In Harris's *Navigantium atque Itinerantium Bibliotheca, etc.*, vol. i, pp. 593-608 (London, 2 vols., fo.). The map and most of the preliminary matter and appendices are omitted; the narrative is in part an almost verbatim, and in parts an abridged, reprint.

1704. In A. and J. Churchill's *Collection of Voyages and Travels*, vol. ii, pp. 479-544 (London, 6 vols., fo., 1704-1732). A very perfect reprint; nothing omitted except the map, the Address to the King, and the letter from Nash; even the "Advise" is inserted; the orthography is to some extent brought up to date.

1741. In the first of the three parts of Daniel Coxe's *Collection of Voyages and Travels* (London, dy. 8vo.). Identical with the second separate edition of 1740 (see p. clxxxi), and published, like it, by Olive Payne, who clearly sought to dispose of his "remainder stock" of this and two other works he had previously published by binding up in one volume with a general title-page. Coxe, although his name appears on the title-page, seems only to have written the last of the three parts, namely, *A Description of the Province of Carolana*, which had appeared in 1722.

1744. In Churchill's *Voyages ana Travels* (another edition), vol. ii, pp. 407-466. Apparently identical with the reprint of 1704.

1748. In Harris's *Navigantium atque Itinerantium Bibliotheca*, 2nd ed., vol. ii, pp. 406-436. A much more complete reprint than that of 1705 in the same work; it is entitled, "The Accurate and Admirable Voyage of Captain Thomas James, . . .; extracted from his own account and delivered in his own words"; a Table of Contents, which nearly fills a page, is provided.[1]

1752. In Churchill's *Voyages and Travels* (another ed.), vol. ii, pp. 407-466. Apparently identical with the edition of 1744.

[1] Five reasons are given at great length for reprinting the narrative almost verbatim, chief among which was the fact that "it is very justly looked upon as the very best work of its kind that ever was published, and this in every respect." Other reasons were that the original edition of 1633 was "become exceeding scarce, and, indeed, hardly to be met with"; that the narrative gave such a graphic account of the methods of conducting an arctic voyage, and of the dangers, difficulties, and climatic rigours to be met with, that its insertion entire would render unnecessary the insertion of any other narratives of a like kind; and that it shows the grounds on which later navigators believed in the existence of, and sought for, a North-west Passage. Certainly this is placing too high a geographical and general value on James's narrative. Some other remarks summarising the circumstances connected with James's preparations for his voyage are decidedly incorrect as to fact; but the narrative of the voyage is reprinted almost verbatim. It is preceded by the Address to the King, which is omitted from most other reprints, and concludes with the "Advise", but lacks the map, the letter left at Charlton Island, and the other appendices. In their place, we have a long and not-very-valuable discourse on the soundness of James's views as to the non-existence of a navigable passage.

1764. In Harris's *Navigantium, etc.* (another ed.), vol. ii, pp. 406-436. Apparently identical with the edition of 1748.

1768. In E. C. Drake's *Universal Collection of Authentic and Entertaining Voyages and Travels*, pp. 319-331 (London, fo.). A fairly-good abridgment; there are also later editions.[1]

1773. In *The World Displayed, or a Curious Collection of Voyages and Travels*, 3rd ed., vol. x, pp. 151-202 (London, 8vo.). A mere abstract; from it, apparently, the separate edition of 1807 was still further abridged, as both are provided with very similar engravings of James's adventure with the fire, and have appended an abstract of Ellis's voyage of 1746.

1784. In J. R. Forster's *Geschicte der Entdeckungen und Schifffahrten im Norden*, pp. 423-432 (Frankfort, 8vo.). A good, though brief, abstract.

1786. In Forster's *History of the Voyages and Discoveries made in the North*, pp. 367-376 (London, 4to.). A translation of the above.

1789. In M. D.'s *Histoire des Naufrages*, vol. i, pp. 100-159 (Paris, 3 vols., 8vo.). An admirable summary, both of the voyage and the events leading up to it, by an anonymous writer.

1796. In Mavor's *Historical Account oj the Most Celebrated Voyages and Travels*, vol. iii, pp. 1-28 (London, 8vo.). A brief, though fairly-good, abridgment.

1810. In Mavor's *General Collection of Voyages and Travels*, vol. i, pp. 261-283 (London, 28 vols., 8vo.). Identical with the above.

1818. In Barrow's *Chronological History of Voyages into the Arctic Regions*, pp. 243-252 (London, 8vo.). A good, though brief, abstract and criticism.

1818. In Maria Hack's *Winter Evenings*, vol. iii, pp. 126-

---

[1] This work contains the erroneous statement (see p. cxxxiii) that "Captain James had already gone several voyages to the North".

211 (London, 4 vols., 16mo.). There are later editions; the narrative forms one of her "Tales of Travellers", and is put into simple language for the proverbial "Harry and Lucy".

1849. In Rundall's *Narratives of Voyages towards the North-West*, pp. 186-224 (London, Hakluyt Soc., 8vo.). A considerable extract, though not correct in all details.

1870. In J. F. Nicholls's *Bristol Biographies*, No. ii, pp. 66-88 (Bristol, 8vo.). A good abstract.

In spite of the number of times James's work has thus been reprinted in more or less incomplete form, it has never, until now, been reprinted verbatim, fully edited and annotated.

James's work may be said to have taken a recognised position in English literature; for, in addition to having been frequently reprinted, it has been several times drawn upon or referred to by writers of more or less eminence.

A contemporary reference to James's voyage is to be found in a scarce little volume of verses by Thomas Beedome, entitled *Poems, Divine and Humane*, published posthumously in 1641.[1] The reference to James consists in two Epigrams, which, as they are short, are here inserted, though they have no poetic merit.

---

[1] This book, which was edited by Henry Glapthorne, the dramatist and poet, contains also verses by several other writers in commendation of the author. Beedome's poems are of no value, and nothing is known of his life, except that he died young, apparently in, or shortly before, 1641. He was probably a friend of Captain James, and it may be surmised that he was a Bristol man.

To the Heroicall Captaine THOMAS JAMES, *of his discovery made by the Northwest Passage towards the South Sea,* 1631.

Heroicke Soule, thy memory must live
Beyond those stone-built structures, that can give
Their earth an Ages talke ; or can assure
The effigies of some mony Gull shall dure,
Till spiders eate his memory : Oh poore glory
T' inscribe a Marble with the tedious story
Of some stout Sir, ...
But thou (great James) hath by thy actions fram'd
A trophie that, hereafter, thou being nam'd,
Men shall rise up with reverence, and keepe
Thy fame from freezing, when thy Ashes sleepe.

To the same Captaine on his Couragious and pious behaviour in the said voyage.

Matchlesse Commander; when fierce winds did hurle
Water to aire, and made the old waves curle
To mounts of solid liquor ; when strong streames
Of moving marble did assault thee, James,
Did not thy conquer'd courage, like the rest,
Flag, and sit heavie on thy hopelesse breast ?
Didst thou not faint to heare the Thunder roare,
And furious seas rebell against the shoare ?
Didst thou not quake at this ? why then I see
Thy soule (though prison'd in thy flesh) was free ;
Thou wert above a man ; thy zeale like fire
Dissolv'd th'opposing Ice, and did aspire,
Through all the storms of dark-condensed ayre,
Wrapt in a sheete of storme-contemning prayer ;
These were prevailing blowes, and broke more Ice
At once then all your hands at ten times twice,
This man'd your ship securely through the maine,
And stered you safely to your home againe.

The Hon. Robert Boyle,[1] in the latter part of his *New Experiments and Observations touching Cold* (London, 8vo., 1665), makes many references to Captain James and extracts from his work, while the following passage occurs in the author's preface:—

"Captain *James*, [is] a person from whose journal I have borrowed more observations then from those [? *that*] of any other seaman. . . . . . This gentleman was much commended to me, both by some friends of mine who were well acquainted with him, and by the esteem that competent Judges appear to have made of him. For, having been not only imployed by the Inquisitive Merchants of *Bristol* to discover a *Northwest Passage* into the *South Sea*, but designed for so difficult a work by so judicious a Prince as the late King; and, having at his return published his Voyages by his Majesties command; as by these circumstances (though not by these only) this Gentleman's Relations may well be represented to us as likely to deserve our consideration and credit: so, by his breeding in the University, and [by] his acquaintance with the Mathematicks, he was enabled to make far better use then an ordinary sea-man would have done of the opportunity he had to observe the *Phaenomena* of Cold, by being forced to winter in a place where he

[1] The Hon. Robert Boyle, an eminent chemist, theologian, and natural philosopher, was the seventh son and fourteenth child of the Earl of Cork, and was born at Lismore Castle, Ireland, in 1627. He was highly educated, both at home and abroad, and early showed a great aptitude for scientific studies. He was one of the founders of the Royal Society, was the author of many learned works, and was regarded as the leading philosopher of his own day. During the latter part of his life, he resided in London, where he died in December 1691.

endured little (if at all less) extremity of cold then that of *Nova Zembla*."[1]

Southey, who was (like Captain James himself) a native of Bristol, noticed the poetic efforts contained in James's narrative; for in his *Omniana* (vol. ii, p. 118; London 12mo., 1812) he quotes the lines appearing on p. 505 with the remark that "The circumstances under which they were written would alone render them curious, even to those who cannot pardon the mannerism of that age. But it is hoped there are many readers who are capable of understanding the strain of fine and manly feeling which is breathed in them." He then refers to the lines occurring on p. 564 as "the other and far finer poem". Such a tribute of praise from Southey is of much interest. These lines are, moreover, considered by Mr. Ivor James (*Source of "The Ancient Mariner"*, p. 79) to be "remarkable as containing one of the best of the conceits which distinguish the poetic literature of the early seventeenth century."

Without doubt, however, the most interesting connection between Captain James's work and more recent English literature lies in the fact that James's narrative seems to have been a source—to some extent, at least—of Coleridge's *Ancient Mariner*. This fact was, I believe, first pointed out by the late Mr. J. F. Nicholls, Librarian to the

[1] As the editor of *l'Histoire des Naufrages* remarks (vol. i, p 153):—"L'usage que le savant Boyle en a fait dans ses ouvrages a donné un grand relief au Journal de James."

City of Bristol, who, in his account of Captain James (*Bristol Biographies*, No. 2, p. 76), says: "It is very likely indeed that S. T. Coleridge, who was a regular and daily frequenter of our old City Library, derived his marrow-chilling scenes, depicted in that unique and immortal poem, *The Ancient Mariner*, from Captain James's *Strange and Dangerous Voyage*."

The credit, however, of having more fully investigated and drawn attention to the matter belongs undoubtedly to Mr. Ivor James, of the University College of South Wales and Monmouthshire, at Cardiff, who, in an interesting and scholarly pamphlet, entitled *The Source of "The Ancient Mariner"* (Cardiff, 88 pp., 16mo., 1890), discusses the question very minutely. After a careful perusal of Mr. James's arguments, I have come to the conclusion that (although he has endeavoured to prove too much in attempting to show that Captain James's *Strange and Dangerous Voyage* was THE source of the poem) he has amply made out a good case, and has proved that the voyage in question undoubtedly formed an important source of that rugged and gruesome poem, *The Rime of the Ancient Mariner*.

That Coleridge should have borrowed the design and moral of his poem from an earlier source was in no way remarkable, for Shakespeare and many (if not most) other writers of higher standing than he have done the same.

Nor was there anything at all remarkable in the fact that Captain James's narrative should have afforded to the poet many of the main ideas of his now celebrated poem; for the very striking, and indeed unique, character of that narrative has been already alluded to (*ante*, p. clxxi). Moreover, James's work contains many things likely to interest a resident in Bristol, which Coleridge was for some years previous to the publication of the *Lyrical Ballads* (in which *The Ancient Mariner* first appeared), having spent the greater part of his time, from the autumn of 1794 to the close of 1798, in or near that ancient seaport.[1]

After stating that the poem was first published in 1798, but without the marginal gloss that was afterwards added, Mr. James goes on to show that, although the poem was substantially Coleridge's, Wordsworth was responsible for some of the passages, and probably suggested some of the incidents, including that of the shooting of the albatross, which he had borrowed from Captain Shelvocke's *Voyage round the World by Way of the Great South Sea* (London, 8vo., 1726, p. 72). Mr. James also alludes

[1] Mr. James devotes a long note to proving that Coleridge had, in all probability, seen a certain copy of Captain James's work which is preserved in the Old Bristol Library. It was, however, afterwards pointed out (*Athenæum*, March 8th, 1890) that the copy in question was not in the Library in Coleridge's time, so that he could not have seen it. That he should have done so is in no way material to the argument; for he might easily have seen another copy, or he might have read the reprints in the collections of Churchill or Harris. At any rate, we may assume with confidence that Coleridge knew James's work, for we know that Southey (who was his inseparable companion) knew of it and quoted from it.

to the statements of Wordsworth, De Quincey, and others as to the sources whence Coleridge obtained the main idea of his poem, and he then states his own contention, which he summarises as follows (p. 18): "In the captain's journal, I am strongly inclined to believe, Coleridge did find what ultimately proved to be the leading incidents of the 'Ancient Mariner' of the poem. The story, in truth, must have haunted him. . . . . . Between the poem and the narrative there are a great many coincidences which cannot be explained, except on the assumption that the poet's source is the captain's journal of his voyage to the North-West."

Mr. Ivor James endeavours to prove this contention by adducing similar passages or ideas taken from both works. This he does under various headings, the first of which relates to mentions of ice, snow, and cold. This is not the place to enter upon a purely literary discussion of the matter, but I propose hereafter to enter upon such a discussion elsewhere. It will suffice, therefore, to say that Mr. Ivor James brings forward so many small points of resemblance between the poem and the narrative that it is difficult to explain them except on the assumption that the former is, to some extent at least, derived from the latter. A few such points of resemblance might be regarded as mere coincidences, but the number of them in this case is too great to permit of this supposition, though the evidence afforded by individual instances is very small. Thus, the very name of the poem—

The Rime of the Ancient Mariner—is of interest; for, in 1798, when Coleridge wrote, James (who had sailed one hundred and seventy years earlier) might have been well described as "an Ancient Mariner". The use by Coleridge of the word *Rime*—a sixteenth-century form of the word—instead of the more modern *Rhyme*, is not without significance, for Captain James describes some of his own verses as "ragged and teared Rimes" (see p. 505).

On the whole, as I have already said, I think it has been clearly shown that James's *Strange and Dangerous Voyage* was at least one of the sources —perhaps the *chief* source, but certainly only one among several—of Coleridge's *Rime of the Ancient Mariner*.

Of Captain James's short but useful career, between his return from the North-West in October 1632, and his death early in 1635, we have a tolerably clear outline in the State Papers of the period, and *The Earl of Strafforde's Letters and Dispatches* (2 vols., fo., London, 1739), wherein his name frequently appears. His deserts seem to have been much more fully recognised by the authorities than were those of Button, Foxe, and some others who had previously distinguished themselves in the search for a North-West Passage; and, during the period abovenamed, we hear of him as the trusted commander of one of the King's ships in the Irish Seas.

It is not quite clear why James should have been thus favoured; for, although whilst upon the voyage he had certainly shown much perseverance, he had by no means distinguished himself as an explorer, and his results had fallen far short of what they should have been. Nevertheless, it is quite obvious that his proceedings met with the almost universal approval of his contemporaries. The King, the Court, the Bristol merchants, and the general public all seem to have been satisfied. Probably sympathy with him in the extreme hardships he had undoubtedly undergone had a good deal to do with the matter (his admirers being unable to see, as we can now, that his innumerable misfortunes had been largely due to his own bad management), especially as he was in himself an estimable gentleman, and, under ordinary circumstances, was, without doubt, a skilful sea-captain. Still, the honours James received must have proved very galling to Foxe, who met with nothing but neglect, although he had, under similar circumstances, accomplished much more than James, and whose skill and good judgment had enabled him to avoid those innumerable perils which seem to have gained for James so much misplaced sympathy.

On April 6th 1633, about six months after James's return from the North-West, and within a month of the publication of his book, we find him appointed by warrant of the Lords of the Admiralty to the command of the *Ninth Whelp of the Lion* (*State Papers, Domestic*, vol. ccxxviii, fo. 36), in the place of a certain Captain Cooper, who had quarrelled with some of his chief officers. The *Ninth Whelp* was a pinnace then undergoing extensive repairs at

Bristol, and was intended for the guard of the Irish coast and the Bristol Channel.[1] Ten days later, James writes to their Lordships reporting his arrival at Bristol from London on the 13th, and stating that, having received an order for money, he had arranged for the immediate preparation of the ship for sea, which he thought would occupy six weeks[2] (*ibid.*, vol. ccxxxvii, no. 2). At the same time, he wrote to Edward Nicholas, the Secretary to the Lords of the Admiralty (see p. 397), to much the same effect, and asking for instructions as to what he was to do with reference to the differences existing among his crew, as already mentioned. "I finde [he says] much dissention and mallice amongst the companie. I would knowe the lordes pleasure, whether I maye not dismiss such as have bene faccious and opposit to the Mayster, Mr. Brooke," etc., etc. (*l. c.*, no. 3). Brooke also wrote to Nicholas at the same time, reporting the captain's arrival and other matters (*l. c.*, no. 4). On April 22nd, James wrote to Nicholas respecting some stores he wanted (*l. c.*, no. 22), and on the 28th he reports to him that so

[1] From January 5th, 1629-30, to July 31st, 1631, she had been engaged in the same service under the command of Sir Thomas Button (*State Papers, Domestic*, vol. ccciii, no. 10).

[2] This and all the other letters preserved among the State Papers are wholly in James's handwriting, and are sealed (like his will) with a seal bearing: Quarterly, 1st and 4th, an elephant's head erased; 2nd and 3rd, three millrinds (?) or castles (?), two and one. Crest: out of a coronet, a hand (?) holding a spear (?) or sceptre (?). I can find no such coat in Burke's *Armoury* or Papworth's *Dictionary*.

much diligence had been used that he expected the ship to be ready for sea by May 18th. He begs Nicholas to send down the gunner's stores, and the instructions from the Lords of the Admiralty; and (as though to hasten him) he adds that pirates were about, and that ships entering Bristol reported having been chased by them (*l. c.*, no. 57). On May 13th (the stores and instructions not having arrived), James wrote a still more urgent letter to Nicholas, begging him to hasten their dispatch. He adds that he expected to have his ship in the Roads by the 15th, and that the merchants were beseeching him to get to sea for their protection against the pirates then infesting the Bristol Channel (*ibid.*, vol. ccxxxviii, no. 66). Three days later, Robert Kitchen, the agent of the Admiralty in Bristol, wrote to the Admiralty from on board the *Ninth Whelp*, reporting that the ship had been got into the King's Road on the 15th; that the gunner's stores were expected on the 18th; and that, immediately after their arrival, the ship would put to sea. Kitchen also reported that an agent of Captain Cooper had arrived from London with a warrant, and had arrested several of the officers, who were charged by Cooper with irregularities whilst under his command; but that James had taken the part of his officers, whom he found to be good and competent men; and that, having been before the Mayor of Bristol, James had obtained their release, saying that their arrest had been malicious and would prevent his departure (*l. c.*, no. 81). At the same time, James reported these facts to Nicholas, adding that he intended to sail on the 21st (*l. c.*, no. 82). On the same day, too, the Mayor also wrote reporting his action in the matter to Nicholas (*l. c.*, no. 90). James set sail as intended on May 21st, as he reports in a letter to Nicholas penned on that day, in which he again repeats his statement that pirates were in the vicinity (*ibid.*, vol. ccxxxix, no. 18). Nor was this statement baseless, for, on July 2nd, Admiral Sir Richard Plumleigh, under whom James was serving, reported to Nicholas that he had sent the *Ninth Whelp* to search the Severn, and that she had lighted upon a pirate in Milford Haven, commanded by an "arch-rogue" named Gosman, who, having previously seized a Plymouth bark in Youghal Harbour, was trading with some knaves on shore for powder and muskets, but that his market was spoiled, and four of the principals delivered to justice, while the rest fled inland (*ibid.*, vol. ccxlii, no. 13). On the 4th, Captain Charles Keane of the *Antelope* (the consort of the *Ninth Whelp*) also informed Nicholas that Captain James had "latelie crushed a small piratt in the egge" (*l. c.*, no. 27). This fact, Nicholas duly reported to the Lords of the Admiralty on the 6th (*l. c.*, no. 36). A month later, on August 3rd, their Lordships instructed him to continue still upon the Irish coast, and to apply himself to guarding the same and the mouth of the Severn (*ibid.*, vol. ccxliv, no. 14). On the 15th, James was back in Bristol with his ship, replacing his mainmast, which had been sprung. He then made application for four more brass guns, of which he stood in need (*l. c.*, no. 65). After this he was probably engaged for some time at sea, for we do not hear of him again until October 4th, when he was off the Isle of Man, interviewing the Governor, Captain Christian, to obtain information concerning pirates, by order of the Lord Deputy of Ireland, the Earl of Strafford (see p. 352). Captain Christian seems to have had more sympathy with the pirates than with James; but, after some trouble, James got him to write a report to the Lord Deputy of what had occurred (*Letters and Dispatches*, vol. i, p. 118). This report, James enclosed in a dispatch of his own to the Lord Deputy, dated October 16th, in which he related the events of his visit to the Isle of Man (*l. c.*, vol. i, p. 126). The Lord Deputy enclosed both of these letters in a dispatch of his own, dated October 23rd, to the Lords of the Admiralty, in which he reports James's arrival at Dublin from the north, which part of the Channel was free from pirates. The Lord Deputy adds that, as three valuable laden merchant ships had for some time lain in Dublin Harbour, not daring to sail for fear of pirates to the southward, he had appointed Captain James to convoy them as far as Scilly, and then to fall off towards Kinsale. That Captain James's services had given great satisfaction to the Lord Deputy may be inferred from a passage in the same letter, in which his lordship says: "I must give the Captain the testimony of [being] a very diligent attendant upon his charge; of [being] a very civil man in his conversation, and [an] able man in his profession and so [I] recommend him to your Lordship's favour; for, in truth, he deserves to be esteemed and remembered as occasion serves" (*l. c.*, vol. i, p. 136). If Captain James had given satisfaction, however, it is clear that he had himself much cause for dissatisfaction, to judge from a letter written by the Lord Deputy to Secretary Coke on the same day as the foregoing, in which appears the following passage:— "Now comes in Captain James, [who] tells me again he hath not eat one sweet piece of beef since he went last to sea; that his men are almost poisoned with it, their lips broke forth, distempered in their health by it, and all ready to run away from him, which is the greatest shame and dishonour in the world." Referring to those who were responsible for this disgraceful state of things, the Lord Deputy says: "And, considering this villainy was done at land, hang them if there be law for it; or, at least, so pillory and slit their ears that others shall take little pleasure to serve the King so hereafter" (*l. c.*, vol. i, p. 135). The perfect satisfaction which James's conduct had given to the Lord Deputy may be further seen from a letter which the latter wrote from Dublin to Secretary Coke on November 8th, in which he says:—"If it may please my Lords of the Admiralty, [let them appoint, for next year's service in the Irish Channel], no other Admiral or Captain than Sir Richard Plumleigh and Captain James, who have acquit themselves passing well in this year's service, which I desire may be represented to my Lords the Commissioners of the Admiralty" (*l. c.*, vol. i, p. 152). Writing on the same day to the Lords of the Admiralty, the Lord Deputy reports James as being then at Kinsale, but adds that, by the 14th of December, he would return for repairs to Bristol, there to remain until the following March, by the 15th of which month he wished him to be back at Kinsale (*l. c.*, vol. i, p. 154). Accordingly, on December 2nd, orders were issued from the Admiralty that the *Ninth Whelp* was to return to Bristol by the end of the month, that her crew was to be discharged, and that she was to be got ready for service again by the 1st of March following (1633-34), when she was to have a complement of sixty men (*State Papers, Domestic*, vol. ccxxviii, fo. 98*a*). These orders were doubtless carried out, for, on December 31st, we find James writing to their Lordships desiring to know whether or not he was to come up and attend them in person or no (*ibid.*, vol. ccliv, no. 19, and vol. cclviii, no. 15). No reply is recorded, but there can be no doubt that James did come up to London at this time; for, on January 16th, 1633-34, Sir Richard Plumleigh, writing to the Lord Deputy in Dublin, says: "Captain James and myself give our daily attendance on the Lords [of the Admiralty], and are warned to be ready by the 1st of March to go on shipboard, but I fear we shall (except your Lordship assist us) fall short of that reckoning" (*Letters and Dispatches*, vol. i, p. 180). On January 11th (probably because James was in London), his petition of four months earlier for four additional guns was considered by the Lords (*State Papers, Domestic*, vol. cclviii, no 51), but no decision is recorded. Three days later, however, the matter was again considered, probably because in a long letter written on the 10th (*l. c.*, no. 43) James had again asked urgently for the four guns, together with many smaller pieces of ordnance, pistols, and pikes, those of the latter in use on his ship having their heads eaten up with rust and their handles rotten. This time the curt reply was, "There are none to spare"! (*l. c.*, no. 65). Meanwhile, the *Ninth Whelp* was refitting at Bristol, and Kitchen reports that she would not be ready for sea again before the 10th or 15th of March. Captain Cooper also renewed his attempt to arrest some of the officers of the ship; but, on the matter being officially investigated, they were discharged (*ibid.*, vol. cclviii, no. 60; vol. cclix, nos. 9, 24, and 25; and vol. cclx, no. 34). On January 28th, one Lewis Gwillim wrote to Captain James informing him that two pirates were frequenting Lundy for the purpose of robbing the vessels passing between Bristol and Ireland, and begging him to get to sea again as soon as possible (*ibid.*, vol. cclix, no. 52). It was probably this letter which was considered by the Lords of the Admiralty on February 8th (*ibid.*, vol. cclx, no. 34); but nothing could be done then, for the ship was still not expected to be ready for sea until the following month (*l. c.*, no. 73), and the warrant reappointing James to the command of her was not issued until February 19th (*ibid.*, vol. ccxxviii, fo. 116b.) On the 24th of February, James writes to

Nicholas reporting his arrival in Bristol, where he found his ship likely to be ready by the 8th or 9th of March, and he asked for his instructions from the Lords. He adds that the Bristol merchants were in great fear of the pirates then infesting the Channel (*ibid.*, vol. cclx, no. 101). On the 15th of March (foul weather having prevented an earlier start), James wrote to the Lords of the Admiralty reporting that, with the first fair wind, he should put to sea; and, after seeing that the Channel was clear of pirates and touching at Milford, he would make for the Irish coast, where, after informing the Lord Deputy of his arrival, he should range about Cape Clear (*ibid.*, vol. cclxii, no. 79). On the same day he also wrote to Nicholas to the same effect (*l. c.*, no. 81), and Kitchen reported the same facts (*l. c.*, no. 89). During the following month, in a kindly letter to Nicholas,[1] he strongly recommended the master of the *Ninth Whelp*, William Brooke, who was one of those arrested, but of whom James held a high opinion (*l. c.*, cclxvi, no. 77). After this, for some totally unexplained reason, we hear almost nothing of James during the rest of the year 1634. There are no letters from him to the Lords of the Admiralty or to Nicholas, and there is only a passing mention of him in *The Earl of Strafforde's Letters and Dispatches.* One cannot even trace the movements of his ship.

---

[1] This letter is undated, but it may be safely assigned to April or May 1634, as it contains a reference to the recent death of Sir Thomas Button.

There is little doubt that she was still employed in the service to which she was appointed; for the Lord Deputy, writing to Secretary Coke on August 18th, speaks of sending back some unserviceable powder by Captain James (*Letters and Dispatches*, vol. i, p. 280. Further, the Lord Deputy, writing to the same correspondent on June 24th (*l. c.*, p. 269), says that the merchants had been "much comforted, finding these rovers [*i.e.*, the pirates] in a manner quite driven off this coast by the guard of His Majesty's ships, who have indeed so well attended their charge as I hear not of any complaint at all this summer." Either James performed his duties so satisfactorily that there was nothing to report concerning him, or else he became ill and had to resign his command. The latter seems a not unlikely supposition in the light of a letter which he wrote from Bristol to Nicholas, on January 29th, 1634-35, in which he says: "It has pleased God to visitt me with sickenesse, that for this year I am vtterly disabled for any Imployment; yet my Hart is sound and strong, and full of desire to doe his Mat[ie] service, which I will be forward to tender as soone as I can recouer my strength and former sufficiency." James then proceeds very strongly to recommend to Nicholas his master's mate, William Purser by name, who had served under him two years, and who, although he had lost one hand, James considered one of the very best mariners and navigators, both practical and theoretical, he had ever known (*State Papers, Domestic*, vol. cclxxxii, no. 104). This frequent recommendation of his officers for promotion shows James in a very favourable light.

But if James's heart was still sound, his illness was to prove his last. After writing to Nicholas he probably grew rapidly worse, and a month later he made his will as follows :—

In the name of God, Amen. The eight and twentieth day of ffebruary Anno Dñi one thousand six hundred thirtie foure [1634-35], Annoq̃. Regis Caroli rex decimo. I, Thomas James, of the Citty of Bristoll, Gent., being sick in body but of sound and p'fect memory, doe make my last Will and Testament in forme following: *viz.*, first and principally, I coñend my Soule into the handes of Almighty God my Maker, hopeing to be saved onlie by the death of Jesus Christ my Savio^r^, and my body to the earth from whence it came, to be layed in Christian buriall. Item: As touchinge my worldly estate, which God of his great mercy hath lent me (haveing already disposed of my lands by deede), I doe forgive to mine elder brother, John James, Esquier, all such debts and soñes of money wch hee oweth mee, wch is twoe hundred and fyftie poundes and upwards. All the rest of my goods, cattels, chattels, debts, jewells, money, and other estate whatsoever (my debts payed and funeralle discharged) I give and bequeathe to my loveinge sister Katherine Lacie, widdowe, whome I make and appoint to be full and sole Executrix of this my last Will and Testament. In witness whereof, I have put to myne hand and Seale the day and yeare above written. THOMAS JAMES.[1]

Signed, sealed, and acknowledged the day and yeare above written in the p'sence of Richard Pownall, Willm Yeamans, George Bowcher, Richard Henry, ffra. Yeamans.

---

[1] This will (of which the signature only is in James's hand-

In the instructions dated March 3rd, 1635, from the Lords of the Admiralty to Sir Richard Plumleigh, as Admiral of the Fleet appointed for the guard of the Irish coast, we find a reference to the appointment of Sir Beverley Newcomen as Captain of the *Ninth Whelp* "during her present employment", and in place of Captain James (*State Papers, Domestic*, vol. cclxiv, fo. 85*a*).[1] This may mean that the appointment was only temporary, and that James was still expected to recover; but, if he was alive on March 3rd, he must have died soon after, for his will was proved on May 4th, 1635. It may very well be that his health had been undermined by the hardships endured upon his voyage, but of this we have no proof.

It is strange that Foxe and James should have died within a few weeks of one another, but such was the case.

James's place of burial is unknown; but it was probably in St. Mark's Church, otherwise known as the Mayor's Chapel, at Bristol.[2] On the monument of Alderman Thomas James, in the south end of

---

writing) is sealed with the same seal as his letters preserved in the Public Record Office (see p. cxcv, *note*).

[1] There are, among the State Papers, many later references to the *Ninth Whelp*, as being either under repairs at Bristol or serving in the Irish Channel during the years 1635 and 1636, with Newcomen as Captain and Brooke as Master.

[2] It is not now possible to substantiate this probability, as the rector, the Rev. J. H. Bright, has been good enough to inform me that the registers were unfortunately destroyed at the time of the Bristol Riots in 1832, when the then Council House, wherein they were kept, was burned.

the aisle of the said church, is the following inscription :—

"This monument was erected for Thomas James, merchant, twice Mayor of this City, and Parliament-man for the same in the Reigns of Queen Elizabeth and King James First. He died in the year 1613. Here also lieth the body of Thomas James, Esq., of Bristol, Barrister-at-Law and son of the said Thomas James. He died in the year 1665. Here also lieth the body of Alexander James, of Tydenham, in the County of Gloucester, son of the said Thomas James, Jr., Esq. He died in the year 1713."

As Nicholls has shown (*Bristol Biographies*, no. 2, p. 59), this inscription (which is of the eighteenth century, and replaces an earlier one now effaced) is full of errors. Thus, King James would not have been described as "the first" twelve years before his death and nineteen years before the second of that name was born; while Alderman James died in January 1618-19, not 1613. Further, if the Thomas James who is described as a barrister is identical (as seems probable) with Captain Thomas James, he died in 1635, not 1665. The corrosion of the stone would account for the error in both these dates. Moreover, Captain Thomas James was, almost certainly, not a son of Alderman Thomas James, as stated. Again, as Alderman James had a son Alexander (as we learn from his will at Somerset House), the Alexander mentioned above was probably a son of the Alderman, not of the Captain. The absence from the will of the latter of any mention of wife or children, and the fact that the property was left to a sister, entitles us to assume that he died unmarried.

As to whether or not Captain Thomas James was a barrister and a member of the Inner Temple, I have made considerable research. Nash,[1] writing from the Inner Temple, addresses him (see p. 451) as "my worthy friend and fellow Templar"; Boyle speaks of "his breeding in the University" (see p. clxxxviii), which probably means much the same thing; while, on the monument at Bristol, we find a Thomas James (probably the Captain) also spoken of as a barrister. I have, by the kind permission of the Treasurer of the Inner Temple, searched the ancient Register of Admissions to that body, and therein I find the following entry in 1612:—

James.	Thomas James de Gwerne y combe in com. Monmouth Generosus admissus est in sociatatem istius coitivae in consid. sexaginta sex
iii[l] vi[s] viii[d]	solidorm̃ octo denariorm̃ premanibus solut xxviij die Aprilis annoq. Jacobi rex decimo.
	CHARLES JAMES.
	plegi JOHN JAMES, jun.

As the evidence seems pretty clear that Captain Thomas James had entered at the Inner Temple, and as the records of that body contain no mention of another Thomas James, we may, I think, fairly assume that the above record relates to the Captain,

---

[1] According to the *Members admitted to the Inner Temple, 1547-1660*, p. 180, Nash (see p. 452, *note*) was "an eminent linguist and jurist", who was born at Worcester, was admitted in 1607, was called to the Bar in 1616, died in August 1648, and was buried in the Temple Church.

who would be nineteen years of age at the time, if he was born in the year 1593. In the printed list of *Members admitted to the Inner Temple, 1547-1660* (London, dy. 8vo., 1877, p. 197), we find the additional information that he was "a younger son of James ap John ap Richard Herbert", but whence this information was obtained I know not.

It is difficult to avoid a comparison between the two explorers who set forth in the same year on the same errand—Foxe, the uncouth, self-reliant sailor, of limited education and unbounded conceit, and James, the polished and well-educated gentleman, who was modest almost to a fault. It would have been hard to have found at the same time two men in any way competent to undertake the work who would have differed much more widely than these in their origin, bringing up, and personal characteristics; yet each undoubtedly accomplished useful work in his own way.

If a comparison be made between the results obtained by each, it will, I think, have to be admitted that Foxe, with his wide practical experience as a seaman, was better fitted for the post, and accomplished more, than the more-highly-educated James; for, whereas the former went through the voyage without loss of life or any mishap to his ship, the latter was in constant difficulties and dangers, losing several of his men, and rendering his narrative a veritable "Book of Lamentations". The conclusion (which is, I think, supported by the results of the two voyages) is, that it is a pity a single voyage only was not undertaken under the command of some man who combined within himself the main characteristics of both of those who actually went; for each undoubtedly possessed characteristics which were lacked by, and would have been of the greatest possible value to, the other. Foxe lacked James's literary skill and general knowledge, and his journal makes in consequence a somewhat bald and ill-digested narrative[1]; while James lacked Foxe's sound judgment and ready resource in time of difficulty, which, had he possessed them, would have enabled him to accomplish much more than he did, and would have saved him from many hardships.

In conclusion, it will be well briefly to notice those later voyages by means of which the discovery and exploration of Hudson's Bay were completed. The routes of all the voyages thus far described are shown in detail on the large chart at the end of Volume I.

After the return of Captain James in October 1632, public opinion seems, for nearly forty years, to have remained satisfied that further search for a north-westerly passage to China through Hudson's Bay was useless, and there is no reliable record that any vessel again entered the Bay until the year 1668.

---

[1] Ellis has well observed (*Voyage to Hudson's Bay*, London, 1748, 8vo, p. xx) that "Capt. Luke Fox has been pretty much censured; but, notwithstanding this, he was certainly a very good seaman, though a very bad writer."

The voyage undertaken in that year was, moreover, essentially a *trading* voyage, and was made without any serious thought of searching further for a passage. It was commanded by a Captain Zachariah Gillam, and its sole object was the establishment of a trade in furs with the Indians. The idea of it originated with two French-Canadian fur-traders, Pierre Esprit Radisson and Medart Chouart (better known as Sieur des Groseilliers), to whose joint exertions the formation of the Hudson's Bay Company may be ascribed.

Radisson, though born in France, had early emigrated to Canada, where he grew rich as the result of many inland fur-trading journeys. Des Groseilliers was also French, having been born at Ferte-sous-Jouarre, about the year 1626. About 1642, he went to Canada and became associated with the fur trade. In 1653, he married, as second wife, a widowed sister of Radisson, and the two men thus became associated in their trading expeditions.[1]

About the year 1658, these two men heard, from the natives with whom they were trading, of the existence to the northward of Hudson's Bay, and of the trade in furs which might be done with the

[1] The narratives of some of their expeditions were edited by Mr. G. D. Scull, and published by the Prince Society of Boston in 1885, under the title, *Voyages of Peter Esprit Radisson; being an Account of his Travels and Experiences among the North American Indians from 1652 to 1684, transcribed from original Manuscripts in the Bodleian Library and the British Museum, with Historical Illustrations and an Introduction* (Boston, viii-385 pp., 4to).

Indians living on its shores. On their third recorded journey, which terminated in June 1660, they made careful inquiries with a view to a journey overland to the southern shores of the Bay. During the rest of that year they stayed at home perfecting their plans. They decided not to disclose the information they had obtained, as they knew nothing except by hearsay, and wished first to investigate for themselves the possibility of reaching the Bay overland. However, the wife of one of them having revealed their secret, others sought to take advantage of it. Thereupon they approached the Governor for his licence to undertake an expedition. This being refused, apparently through jealousy, they started without leave in August 1661. On their return to Canada, about the year 1663,[1] they were received with disfavour by the Governor, who fined them heavily for having proceeded upon their expedition without his licence. This treatment so incensed them, that Des Groseilliers decided to go and demand justice in France, which he did. Failing to obtain restitution there, he returned. Subsequently, the two proceeded to Port Royal,[2] Nova Scotia, where some English merchants promised them a ship for a sea-voyage thence to

[1] See p. 321, *note*. Oldmixon says (*British Empire in America*, 1708, vol. i, p. 385) that they actually reached the Bay; but, if they did so, their own narrative does not state the fact. Moreover, their own account of their proceedings by no means agrees with those given by Mons. de Bacqueville de la Potherie in his *Histoire de l'Amérique Septentrionale* (4 vols., Paris, 1722, 12mo), and by Mons. Jérémie in Bernard's *Recueil de Voiages au Nord* (8 vols., Amsterdam, 1724).

[2] Now known as Annapolis.

Hudson's Bay in the following year (1665). This expedition duly started, but got no further than the entrance to Hudson's Straits, when the ignorant fears of the captain led to the return of the ship. In the following year, apparently, the same merchants promised them two ships for another expedition, but this apparently fell through. Meanwhile the two fur traders had met some Commissioners from the Court of Charles II, who had been in Boston on the King's service, and who persuaded the traders to accompany them to London. The party duly sailed for England on August 1st, 1666, but their ship fell into the hands of some Dutch marauders. In the end, however, they landed in England about October, and were introduced to the presence of the King, who ordered them to be entertained, and gave them hopes of a ship for an expedition in the following year. In 1667, accordingly, a ship was prepared, but the visit of the Dutch fleet to the Thames in that year prevented the departure of the expedition. In 1668, however, largely through the patronage of Prince Rupert, two ships, the *Eagle* (Capt. Stannard) and the *Nonsuch* (Capt. Z. Gillam), actually started; but, before they had proceeded far, a storm separated them, and the *Eagle*, with Radisson on board, returned home. The *Nonsuch*, however, with Des Groseilliers on board, continued the voyage, passed through the Strait, entered the Bay, sailed southward, and eventually wintered in the mouth of the Rupert river, near where Hudson had wintered nearly sixty years before. Here Des

Groseilliers established friendly relations with the natives, and built Fort Charles, the first trading-post established on the shores of Hudson's Bay.

On the return of Captain Gillam to England in 1669, through the active interest of Prince Rupert, the still-existing Hudson's Bay Company was formed, and was incorporated by Royal Charter, dated May 2nd, 1670, as "The Company of Adventurers of England trading into Hudson's Bay". Prince Rupert was nominated Governor, and a number of noblemen, knights, and gentlemen formed the members. The comprehensive and much-discussed Charter conferred upon the Company the sole and exclusive right to trade within the entrance to Hudson's Strait, together with territorial rights and absolute jurisdiction over an enormous (and then for the most part unknown) area around Hudson's Bay and Straits, which was thenceforth to be known as Rupert's Land.

Although one of the expressed objects of the incorporation of the Company was that the search for a North-West Passage might be continued, it must be admitted that, at first, at any rate, little or no effort was made to carry on the search. For thirty years after its formation, the Company was too closely occupied, either in raking in its enormous profits, or in contests with the French, to make any serious attempt at further search. After the Peace of Utrecht, however, in 1713, all the Company's posts which had been taken by the French having been restored, and peace having been established, the time for further search seemed to have come; and, in 1719, an expedition was at last despatched with the express object of continuing the search for a passage. The idea of the expedition was due to Captain James Knight, a very old man long in the Company's service, and a former governor of the factory on Nelson river. He is said to have continued to importune the Company until he got what he wanted. He was ultimately appointed to the supreme command of the expedition, which consisted of two vessels, the frigate *Albany* (Captain George Berley or Barlow) and the sloop *Discovery* (Captain David Vaughan), and which left London on June 5th, 1719. It appears probable that Knight's eagerness to proceed upon the expedition was mainly due to a hope he had of discovering certain mines of gold and copper of which he had heard, but he wisely gave out that he hoped to discover a North-West Passage; and in his Instructions[1] he was charged to find out the

[1] Printed in the *Report from the Committee of the House of Commons appointed* [*in 1749*] *to enquire into the State and Condition of the Countries adjoining to Hudson's Bay, and of the Trade carried on there.* In the same *Report* there is mention of two other vessels "fitted out by the Hudson's Bay Company on discovery of a North-West Passage" in the year 1719; namely, the *Prosperous* (Captain Henry Kelsey), which sailed from York Fort on June 19th and returned on August 10th, and the *Success* (Captain John Hancock), which sailed from Prince of Wales Fort on July 2nd and returned on August 10th. In the year 1721, the same two vessels are again reported to have been engaged in the same search. They both left York Fort on June 26th, but the *Success*, under Captain James Napper, was lost four days later, while the *Prosperous*, which was again commanded by Kelsey, returned on

mythical Straits of Anian. As he was also instructed to proceed "to the northward and westward of 64° in Hudson's Bay", it seems probable he expected to discover the entrance to the Straits of Anian, in the channel known as Sir Thomas Roe's Welcome (see p. 321). His hopes, however, ended disastrously, for neither of the vessels ever returned, and every member of the expedition perished ;. nor was it until nearly fifty years later that it became known what had become of the ships. In 1767, a trading sloop from Fort Prince of Wales, on Churchill River, discovered the remains of the expedition on Brook Cobham, or Marble, Island. The account given by Hearne,[1] who gathered it from the Esquimaux on the spot, is extremely tragic. The expedition seems to have arrived on the island late in the autumn of 1719; by the summer of 1720, the number of the men had been much reduced, and the survivors were stricken with scurvy; by the summer of 1721, there were only five survivors, and these were afterwards reduced to two, who, after watching anxiously for relief, died also, and Hearne says he himself saw their remains. The two ships, he tells us, were found sunk in the harbour.

As neither of Knight's ships returned in the year

September 2nd. It seems probable that these vessels (of the voyages of which we have no information) were engaged rather in general discovery and in extending trade, than in a serious search for a north-westerly passage.

[1] *A Journey from Fort Prince of Wales in Hudson's Bay to the Northern Ocean* (London, 1795, 4to), pp. xxviii-xxxii.

in which they sailed, many persons were inclined to believe that they had found a passage; but, when the following year (1720) had also passed, and they had not been heard of, the Company became alarmed, and set about preparing an expedition to go in search of Knight and his companions. A sloop, named the *Whalebone*, and commanded by a Captain John Scroggs, was selected and despatched in the summer of 1721; but she arrived at Churchill so late in the autumn that she was obliged to winter there, and she did not leave that port to commence the search until June 22nd, 1722. Scroggs (who does not appear to have been a man at all suitable for the command) sailed northwards at least as far as Whale Point, in Sir Thomas Roe's Welcome (which he named Whale-bone Point, after his ship), and returned on July 25th, without any tidings of the missing expedition, though he must have passed close to Marble Island.

Fifteen years later, on July 7th, 1737, two vessels, the sloop *Churchill* (Captain James Napper) and the sloop *Musquash* (Captain Robert Crow), were sent from Churchill to the northward; but the object seems to have been rather to make general explorations, to search for mines, and to open trading relations with the Indians, than to search further for a North-West Passage; and the expedition proved fruitless, for the *Churchill* returned on August 8th, Napper having died, while the *Musquash* also returned before the end of the same month.[1]

---

[1] *Report of the Committee of the House of Commons on Hudson's Bay*, 1749.

The resumption of the search after this date was due to a gentleman named Arthur Dobbs, who, from a careful study of the subject, had become convinced that all the available evidence pointed to the existence of a passage. In the end, by his persistence, he prevailed upon the Lords of the Admiralty to devote to the continuation of the search one of His Majesty's ships. The *Furnace* was selected for this service, and was placed under the command of Captain Christopher Middleton, an old servant of the Hudson's Bay Company, while the *Discovery*, a small vessel, of which Captain William Moor was master, was also placed under his command. Middleton left England late in 1741 and proceeded direct to Churchill, where he wintered, and lost a number of his men through scurvy. There he remained until July 1st, 1742, when he sailed to the northward, past Marble Island, and up Roe's Welcome, which he explored to a more northerly latitude than had been previously attained. On July 13th, he reached and named Wager River. After exploring this, in the hope of finding a passage, he proceeded further up the western shore of Roe's Welcome, and eventually reached Beach Point, and was again disappointed in finding a passage through Repulse Bay. Then, having seen and named the Frozen Strait, to the north of Southampton Island, he decided to return, and he was back at Woolwich by October 18th.

Middleton had, upon the whole, satisfactorily discharged the duties entrusted to him, and had made a successful voyage; but this was not the view taken of his efforts after his return. Arthur Dobbs became convinced that his account of his proceedings was not trustworthy; that he, as an old servant of the Hudson's Bay Company, had subordinated public interests to the interests of his old employers; and that he had in reality discovered a passage but was concealing the fact. A long and exceedingly acrimonious pamphlet-war between the two men then ensued. Middleton's defence was, as we now see, satisfactory; but public opinion then was not satisfied; and the indefatigable Arthur Dobbs at once set about organising another expedition. Failing to obtain further support from the Lords of the Admiralty or the Hudson's Bay Company, he endeavoured to raise the necessary funds by public subscription, which he succeeded in doing, as public interest had become very strongly centred upon the question, and Parliament had, in the year 1745, passed an Act (18 Geo. II, c. 17), offering a reward of £20,000 to any of His Majesty's subjects who should discover a North-West Passage through Hudson's Straits.

For the purposes of the new expedition, a small vessel named the *Dobbs*, after the promoter, was chosen, and the command of her was entrusted to Captain Francis Moor, while her consort, the *California*, was commanded by Captain Francis Smith. The two ships left Gravesend on May 20th, 1746. On the 11th of August, they made the land on the west side of Roe's Welcome in 64° N.; but, sailing to the southward, they made for York Factory, near where they wintered. On getting free in the following year (1747), they proceeded to the entrance of the inlet known as Wager River, which they explored carefully, thereby fully establishing the truth of Middleton's statement that no passage was to be found there. After this, nothing was accomplished, and the expedition returned home, leaving matters much as they were before.

The expedition under Captains Moor and Smith, in the years 1746-7, may be regarded as the last of the many fruitless attempts to discover a North-West Passage through Hudson's Bay. There were still to be found not a few persons sanguine enough to maintain that the non-existence of a passage running westward from Roe's Welcome still remained to be proved; but public opinion seems to have been satisfied, and interest in the question largely died out, as it had done after the return of the expeditions of Foxe and James, rather more than a hundred years earlier. For more than a century after 1747, many expeditions continued to seek a passage among the numerous channels and waterways further to the north; but these have no connection with the present subject. Nor is it necessary here to refer to the various overland expeditions, commanded by Hearne, Mackenzie, Franklin, Back, Richardson, Rae, Simpson, and others, which, during the same period, left the shores of Hudson's Bay, and proceeded in a north-westerly direction, either with a view to co-operate with, or succour, expeditions that had gone by sea, or to complete the discovery of the northern coast-line of North America. To all of these latter, the Hudson's Bay Company (dropping the jealous attitude it had maintained whilst there still remained the possibility of a North-West Passage *viâ* Hudson's Bay being discovered, and their monopoly thus threatened) gave their warmest support.

By the year 1747, the entire coast-line of the Bay may be said to have been made known—not, it is true, with the absolute precision now considered necessary, but still with a sufficiently near approach to accuracy for practical purposes then. Indeed, this position may be said to have been reached more than half a century earlier; for, from the time of the incorporation of the Hudson's Bay Company in 1670, trading vessels belonging to the Company began to visit the Bay annually and with unvarying regularity. If we compare James's chart of 1633 with Thornton's beautifully-drawn chart of 1685 (see p. 166), we shall see the extraordinarily-rapid progress which the vessels of the Company had made in the exploration of almost every part of the Bay between the year 1668 (when exploration, after James's time, recommenced) and the year 1685, when Thornton's chart was drawn. We have a clear portrayal of the state of knowledge of the geography of the Bay in the year 1751, in the interesting document describing the Bay, which was drawn up by Captain W. Coats as a result of his many voyages to the Bay between the years 1727 and 1751, and which was edited by Mr. John

Barrow and printed by the Hakluyt Society in 1852.

From the year 1670 to the present time, there has never been a year during which at least one vessel belonging to the Hudson's Bay Company has not entered the Bay, while in most years two, and in some years three, vessels have been so employed. Between 1670 and 1713, the Bay was also entered by various French trading-vessels and war-ships. During the present century, vessels belonging to the British navy have on various occasions convoyed the Company's ships; and, in 1848, and on other occasions, troops intended to maintain order at the Red River Settlement were landed at York Factory. Voyages into the Bay have also of late been made almost annually by American whalers, who have often wintered at Marble Island. The latest exploring expeditions of consequence have been those of Lieutenant A. R. Gordon, who was sent out by the Canadian Government in H.M.S. *Alert* to make accurate scientific and general observations in the years 1884, 1885, and 1886, when much interest was felt in Canada in a proposal to establish a trading-route to and from Manitoba and the North-West Territories *viâ* York Factory, a project which the future will probably, to some extent, see realised. It appears probable that a steam-vessel first entered Hudson's Bay on the occasion of these expeditions.

POSTSCRIPT.

WHILST this work has been passing through the press, a few facts have come to my knowledge too late for insertion in their proper place, and these may be conveniently referred to here.

Voyages of the Zeni (pp. 19-28). I find that the more recent authorities are inclined to doubt the historic value of these narratives.

Weymouth's Voyage, 1602 (pp. 80-86). It should have been stated in the long foot-note on p. 80, that the whole of the circumstances which immediately led up to Weymouth's voyage, as well as those attendant upon his return, are shown in great detail in the first volume of the Court Minute Books of the East India Company, now preserved at the India Office, which volume has been printed verbatim in Mr. Henry Stevens's *Dawn of British Trade to the East Indies* (London, cr. 8vo, 1886). In this interesting volume, there does not appear to be any reference to the man whose name seems to have been variously spelled Cobreth, Colbert, Coleburne, and Coolbrand (see p. 115, *note*), and who accompanied Weymouth as second in command. A copy of the agreement for the voyage between him and the East India Company is, however, to be found in the first volume of letters, or "Original Correspondence", of the Company, which is also preserved at the India

Office. This volume has been printed verbatim in Messrs. Birdwood & Foster's *Register of Letters, etc., of the Governor and Company of Merchants of London trading into the East Indies, 1600-1619* (London, dy. 8vo, 1893). The agreement with Cobreth is printed on pp. 190-191 of this work; that with Weymouth himself appears on pp. 21-25.

Button's Voyage, 1612-13 (pp. 162-200). Mr. C. H. Coote has kindly drawn my attention to the fact that there is in the British Museum (Titus, B. 8. 318) an interesting document in the handwriting of Edward Wright (see p. xxxviii) which goes to show that Wright was one of the advisers of Henry Prince of Wales in making the preparations for Button's voyage. The document, which is a statement of services rendered to the Prince, is undated; but a reference contained in it to Wright's *Errors in Navigation*, which was published in 1611, shows that it was drawn up subsequent to that year. It commences as follows:—

"*Busines done for his Highnes by Edward Wright.*

"A Module of an Instrument to make a Plat smaller or greater in any proportion, newely devised.

"A Module of a large generall Astrolabe of more manifould and easyer use then others before invented.

"A Sea-Chart for the North-West Passage.

"A Paradoxall Sea-Chart (as they call it) of all parts of the worlde from 30 Degrees of Latitude Northwardes."

The rest of the document is filled with a similar enumeration of charts drawn and instruments devised, together with other services to the Prince of a more personal nature, such as "reading Mathematicks vnto him since Christmas Twelvemonth", and arranging and cataloguing the Royal Library. There can be very little doubt that the two northern sea-charts above described were drawn for the Prince by Edward Wright in connection with the preparations for Button's voyage. This affords grounds for the surmise that Button's "Instructions" (see pp. xxvii and 636) were drawn up, largely, at least, by Wright.

Mr. William Foster, of the India Office, has kindly informed me of a reference to certain pieces of cloth brought home by Button, which is contained in a fragment of a hitherto unknown volume of minutes of the East India Company he has just discovered among the Records of the Company. Among the Minutes of a "Court of Committees", or Directors, held on December 18th, 1613 (three months after Button's return), occurs the following:—

"The Company being enformed that there is some cloth come home y^t^ was sent forth in [the voyage for discovering] the North-West passage and [is] supposed to be seruiceable for the Comp^a.^ seruice, they therefore desired Mr Dep^tie^ and Mr Stone to p'use them and see howe many of them will serue for there vse."

Probably this cloth was taken out by Button's expedition for the purposes of sale or barter when the vessels arrived, as was hoped, in Japan. We know (see p. lxxxiv) that pieces of cloth were chosen with the same object and sent out by Luke Foxe's expedition in 1631.

Gibbons' Voyage, 1614 (pp. 201-202). With reference to the statement on p. 201 (*note*), that Gibbons' voyage was promoted by the Company of Discoverers of the North-West Passage, Mr. William Foster has communicated to me an extract from the fragment of the hitherto unknown Minute Book of the East India Company he has recently discovered (see p. ccxxiv), which shows that the East India Company also contributed to the cost of the expedition. Among the minutes of a General Court of the Company, held on December 22nd, 1613, Mr. Foster finds the following entry:—

"300^{li} aduentured "out of the Joind "Stock to the Nor-"west discouerye."

"Sr Dudley Digges put this Courte "in remembraunce of an order former-"lie made for aduenturinge 300^{li} p' "annũ vnto the Norwest discou^{r}y for "3 yeares together; but, being nowe "entred into a new course by a Joynd Stocke, hee desired "to vnderstand their resolucõns, seeinge that himselfe and "diu^{r}s others that hád aduentured there 100^{li} a peece in-"tended to p'secute the same againe, wth hope of better "successe, by Godes assistaunce. And this Courte, rightlie "weighinge the hopefull euent that is expected, resolud by "erecõn of handes to have 300^{li} Aduentured out of the "Joynd Stocke: But, whereas lib'tie was giuen vnto those "of the East India Comp: to bee admitted into the said "Comp^{a}: [of Discoverers of the North-West Passage] for a "fine of 10^{li} before a c^{r}teine tyme were expired; and a "mocõn being nowe made alsoe in the behalf of such as "are of this Company and shalbe desirous of that freedome "that they might be accepted for the like sume of 10^{li}, it "was yealded vnto by Sr Dudley Digges and others in the "name of the said Company [of Discoverers of the North-"West Passage] and leave [was] given for any of the East

" India Company to come in betwixt this [time] and the " 25th of March next for their seu^{r}all fines of 10^{li}, or else to " bee excluded."

Hawkridge's Voyage, ? 1619 (pp. 248-259). As stated in the long foot-note on pp. 248-249, Foxe's account of this voyage (which is the only known reference to it) gives no indication of the year in which it was undertaken or as to the names of those who promoted it. Owing, however, to certain statements made by Rundall in 1849 (*Voyages towards the North-West*, pp. 150-151) all later writers have assigned the voyage to the year 1619, and have credited Sir John Wolstenholme with having been the chief promoter of it. It appears, however, on a careful examination of the original records of the East India Company, upon which Rundall based his statements, that these statements are totally unreliable, and that there is no evidence whatever contained in the records to support either of the assumptions indicated above. In the first place, it is noticeable that in the record of Sir John Wolstenholme's application to the Company for a grant in aid of an intended voyage north-westward, upon which Rundall relies, no reference to Hawkridge appears at all, the intended expedition being that of a certain Captain Bullock, of which nothing is known—a fact which Rundall suppresses. Furthermore, even if it could be shown that the record in question does relate to Hawkridge's voyage, the date given for it by Rundall (1619) is certainly wrong. Sir John Wolstenholme's application was made (*Court Minute Books*, vol. iv, p. 114) on "January 20th, 1617", which would, of course, be 1617-18, not "1618-19", as given by Rundall. Moreover, the Court Minute Books and the volumes of "Original Correspondence" afford ample evidence that Hawkridge's expedition did *not* sail either in 1618 or 1619; for we find clear proof that Hawkridge was in command of vessels belonging to the Company in the East Indies and elsewhere, from the early part of the former year till the close of the latter. In the face of these facts, one can only surmise that Hawkridge's voyage took place (if it took place at all) in the preceding year (1617), which would be the year following the return of Bylot and Baffin from their second north-west voyage. It may be, however, that the intended expedition of Captain Bullock in 1618 actually took place, and that Foxe has wrongly assigned it to Hawkridge. Foxe's information about this voyage is so unsatisfactory that one is prepared to accept any such surmise. The point will be considered more fully in my intended work on the voyage of Jens Munk (see p. liii, *note*), which took place in 1619.

With reference to the Certificate printed on p. l (*note*), Mr. G. W. Waddington, of Grosmont, points out that it "was probably intended to obtain the Bishop of London's Brief for Collections to be made in churches in his diocese, or for private collection by authority of Quarter Sessions. A Privy Council Brief [he says] ran throughout the Kingdom."

Luke Foxe's Voyage, 1631 (pp. 261-445). On further investigation, I find that Foxe's reference in his Preface (p. 11) to "little Mr. Jeffery [Hudson], that pretty courtier", and "Mr. Evans, his Maiesties great porter", are perfectly intelligible. William Evans, the King's porter, was a Monmouth man, and 7 ft. 6 in. in height. There is a notice of him in Fuller's *Worthies*. Upon one occasion, at a court masque, he drew out of his pocket Jeffery Hudson, the Queen's dwarf. This individual was born in 1619 at Oakham, in Rutlandshire, where his father was a butcher. He led a very chequered life, dying in 1682. Scott has introduced him into *Peveril of the Peak*. Until recently, there might be seen over the entrance to Bull-head Court, Newgate Street, a curious old sculptured stone sign of "the King's Porter and Dwarf", which commemorated the two individuals above-mentioned. Fuller notices of them and of the sign may be found in Larwood and Hotten's *History of Signboards*, in Norman's *London Signs and Inscriptions*, and in the *Dictionary of National Biography*.

Since the statement on p. lvii, to the effect that the name of Foxe's wife was unknown, was printed off, the matter has been set at rest by the discovery, by Mr. G. W. Waddington, of Grosmont, near Whitby, of the records both of the granting of the marriage licence and of the marriage itself. In Mr. W. Paver's "Abstract of Marriage Licences granted by the Ecclesiastical Court of York from

1567 to 1714", now in the British Museum (Add. MS. 29,667), appears the following entry[1]:—

"1613. Luke Fox and Ann Barnett of Whitby, at Whitby."

Mr. Waddington has also been good enough to search for me the Whitby Parish Registers, wherein, among the marriage entries, he has discovered the following:—

"1613. Luke Foxe and Anne Barnard, May 13."

Mr. Waddington states that the spelling Barnard, not Barnett, is undoubtedly correct. He adds that the Barnard family were gentry, and owned the Abbot House property in Goathland (see Young's *History of Whitby*, p. 361), which they held till about 1720. It is traditional that Sir John Barnard, Lord Mayor of London in 1737, was of this family. Mr. Waddington adds that no mention of children of Luke and Anne Foxe occur in the Whitby Registers up to 1637; nor does an entry of the latter's burial appear up to 1649, when a lapse of ten years occurs.

Mr. Waddington suggests that the fact of Foxe's Journals passing into the hands of Lord Valentia may possibly be accounted for as follows:—Supposing the Journals to have remained in the possession of Foxe's widow and others for a number of years, and then to have come into the hands of the

---

[1] Printed in the *Yorkshire Archæological and Topographical Journal*, vol. xii, p. 279.

Linskill family of Whitby, they may have passed from them to the family of Lord Valentia, with which some acquaintance or friendship very likely existed; for, on October 17th, 1853, William Linskill married the Hon. Frances Annesley, second daughter of Viscount Valentia.

For a copy of Foxe's will, I have searched in vain at Somerset House.

With reference to Foxe's discovery in Port Nelson, in August 1631 (see pp. ci and 344), of the relics of the vessel abandoned there by Button in July 1613 (see p. 167, *note*), it is interesting to note that these relics were still there nearly fifty years later. Oldmixon says (*The British Empire in America*, vol. i, p. 391) that, when the Sieur des Groseilliers, in August 1673, arrived at Port Nelson to establish there a post for the Hudson's Bay Company, he saw "the relicts of Sir Thomas Button's ship, and one of his company, Mr. Cole, brought home a piece of shot, a piece of her bulk-head, and a small piece of cable, which had lain there about sixty years."

Thomas James's Voyage, 1631-32 (pp. 447-627). With reference to the imaginary Cape Monmouth alluded to on p. 493 (*note*), Mr. Ivor James points out that Captain James probably named it, not in honour of the Duke of Monmouth, but after the County of Monmouth, of which he was apparently a native (see p. ccvii). James named other rivers, capes, and hills after other local geographical features.

Mr. Ivor James has also been good enough to inform me that Sir John Winter (p. 529) was probably connected, and certainly had dealings, with the James family, which was probably the reason why Winter's Forest was named after him. Mr. Ivor James informs me that Mrs. James (daughter of Thomas Powell of Preston Court, and sister of the first Lady Coke), widow of Edward James (a son of Alderman Thomas James) had leasehold interests in the Forest of Dean, which were assigned to Sir John Winter just before the date of James's voyage. "Winter's Forest" was doubtless named in allusion to this.

CAPTAIN LUKE FOXE.

As this work was on the point of being issued, the following additional information concerning Foxe was kindly forwarded by Mr. Edward S. Wilson, the Secretary to the Trinity House at Hull :—

"Richard Foxe, the father of Luke Foxe, was a Master Mariner, and was elected a Younger Brother of this Corporation in 1581 ; was made a Steward of the Corporation in 1585 ; and afterwards became an Assistant, which entitled him to a seat at the Board along with the Elder Brethren."

"Luke Foxe was trained to the service his father had followed, and he also became a Younger Brother of this Corporation."

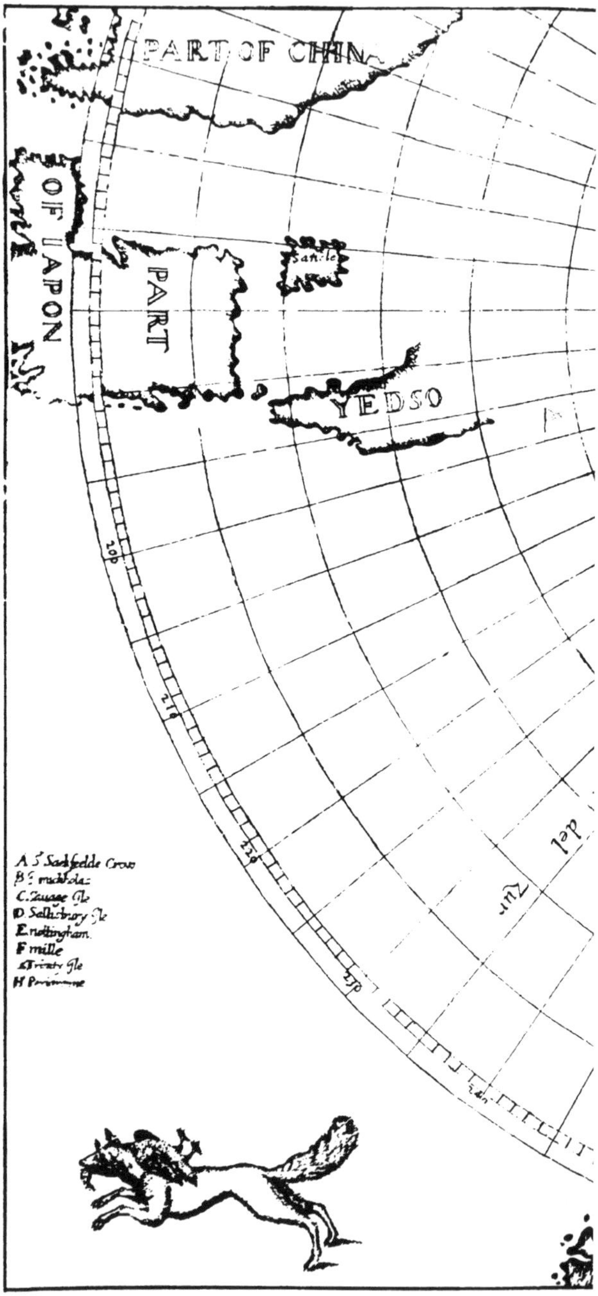
PART OF CHINA
PART OF IAPON
PART
YEDSO
A. S. Sackfeelde Crow
E nottingham.
F mille

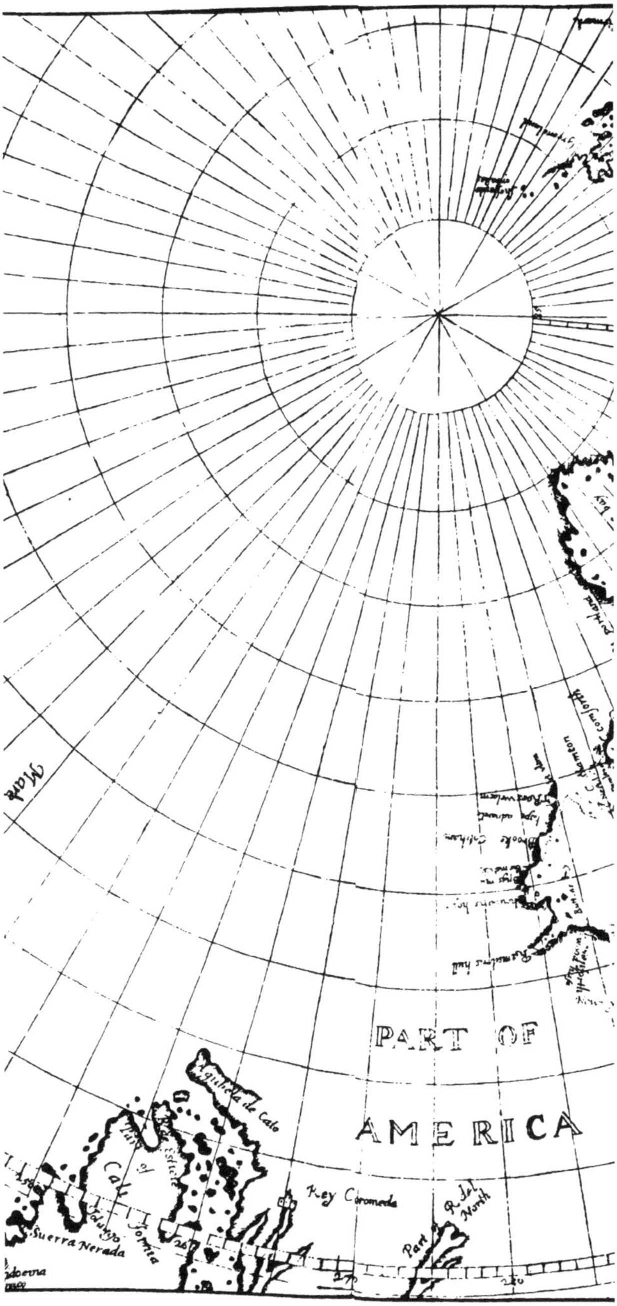

PART OF
AMERICA
Key Coromeda
Suerra Nerada
Mare

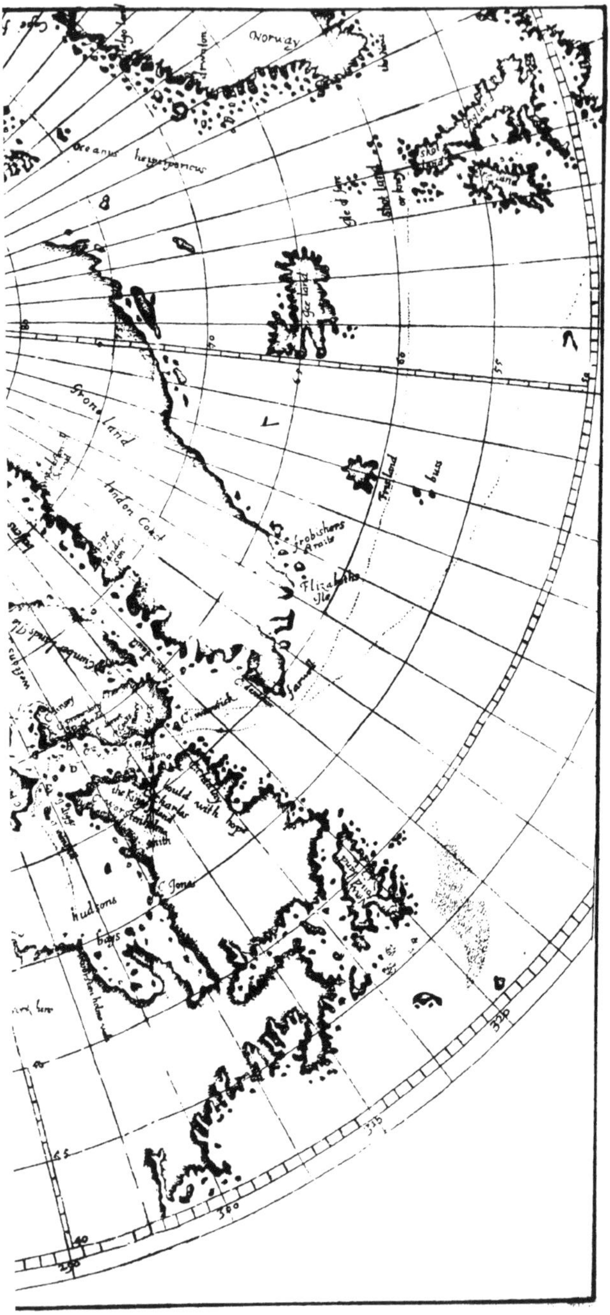
Norway
Oceanus hyperboreus
Groneland
London Coast
frobishers straits
Frisland
buss
hudsons bays

NORTH-WEST FOX.

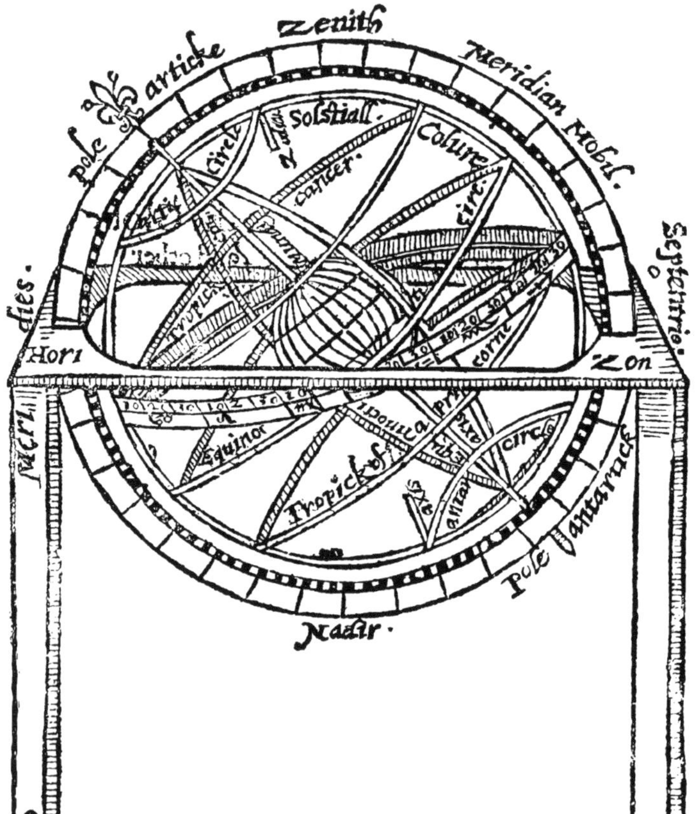
Zenith
Pole Articke
Meridian Mobil.
Solstiall
Colure
cancer.
Hori
zon
Equinoc
Tropick of
Pole antartick
Nadir.
The Meridian Line.

NORTH-VVEST FOX;

OR,

Fox *from the North-west passage.*

BEGINNING

VVith King ARTHVR, MALGA, OCTHVR, the two ZENIS of *Iseland*, *Estotiland*, and *Dorgia;* Following with briefe Abstracts of the Voyages of *Cabot*, *Frobisher*, *Davis*, *Waymouth*, *Knight*, *Hudson*, *Button*, *Gibbons*, *Bylot*, *Baffin*, *Hawkridge* : Together with the Courses, Distance, Latitudes, Longitudes, Variations, Depths of Seas, Sets of Tydes, Currents, Races, and over-Falls; with other Obſervations, Accidents and remarkable things, as our Miſeries and ſufferings.

Mr. IAMES HALL's three Voyages to *Groynland*, with a *Topographicall deſcription of the Countries, the Salvage's* lives and Treacheries, how our Men have beene ſlayne by them there; with the Commodities of all thoſe parts, whereby the Marchant may have Trade, and the Mariner Imployment.

Demonſtrated in a Polar Card, wherein are all the Maines, Seas, and Ilands, herein mentioned.

With the Author his owne Voyage, being the XVIth; with the opinions and Collections of the moſt famous Mathematicians, and Coſmographers; with a Probabilitie to prove the ſame by Marine Remonſtrations, compared by the Ebbing and Flowing of the Sea, experimented with places of our owne Coaſt.

By Captaine LVKE FOXE *of* Kingſtone *vpon* Hull, *Capt.* and Pylot for the Voyage, in his Majeſties Pinnace the CHARLES.

Printed by his Majeſties Command.

LONDON,

Printed by B. ALSOP and THO. FAVVCET, dwelling in *Grubſtreet.*

1635.

TO THE KING'S MOST
SACRED AND EXCELLENT
MAIESTIE.

MOST MIGHTY MONARCH,
and
MY DREAD SOVERAIGNE.

SEeing that it hath pleased You to examine the Endeavours of those formerly imployed towards the search of the *North-west* Passage, and that it hath bin Your Highnesse pleasure to examine me, Your meanest subject, concerning the hope of a further Discoverie ; (how it formerly had fayled, and what better helpes could now be produced), appointing me [to the command of] Your Highnesse Pinnace, the CHARLES, with Your Royall Instructions for my better proceeding : which unto me was the greatest Honour that ever Subject received (and at my Returne continuing Your Royall and Gracious favour, commanding me unto Your Royall Presence to deliver my Accompts) hath since imboldened me to examine with my selfe thus farre, that hereby (as duty obligeth) I might bring some better satisfaction to your Royall demand ; which was, *Whether there were a Passage or no, and were I capable thereof at my being neere the same?* But, since that, meditation and contemplation hath given me more insight thereinto than Speculation at that instant could; For it was not possible for me to frame any Conclusion untill I had circuted the Bay[s] of *Hudson* and *Button*, whereby I might find from whence a new Tyde (which I found there) should come; and heere SIR, I prostrate my Accompt in particular, which then I presented in generall.

I begin with King ARTHVR his Conquests, and so proceed to all those Discoveries (that I can find Antiquity hath preserved) towards the *North-west*, untill this Your owne time; to show how those Maynes, Ilands, and Continents they have discovered, doth, like dew from Heaven, descend upon Your Royall Throne; so as, most Dread Soveraigne, the true Right thereto is Yours, which I pray may be augmented unto the furthest bounds of the East and W. Ocean.

I doe not onely bring to show, but put your Majesty in mind of what you know: knowing that your Majesty, being called by more weighty affaires of State, may passe by (unremembred) such small Trifles as this; yet, observing Your Highnesse willingnesse to promote Navigation and those discovering Enterprises, hath made me thus presume.

I have endeavoured [to show] the probability of a Passage, praying unto Almighty GOD, to set it as a faire Iewell in your Royall Crowne, and bring [it] to passe in those happie dayes which wee doe enjoy under your Gracious and godly Gouernment; Rather than some Forraigne Prince or State should advance and finde the same. These shall bee the prayers, with [others for] Your long life and prosperous Raigne,

Of Your humblest Subiect
and Servant,

LVKE FOXE.

The Preface to the Reader.

GENTLE READER, expect not heere any florishing Phrases or Eloquent tearmes; for this Child of mine, begot in the North-west's cold Clime (where they breed no Schollers), is not able to digest the sweet milke of Rethorick, that's food for them.

The Parent, not being able to maintaine it any longer, it being now past the Cradle, hopes the Publike will willingly bestow Education of the Infant for his better breeding; for which the Father prayeth, and the rather because the Midwife perswades him that the Child lookes like himself. There were some desired to have had him come abroad like the Lapwing, before the shell was from over his head[1]; but the Nurse, more carefull of the tenderly-borne Infant, procrastinated the time, supposing that, after birth, it would strengthen apace in this temporate Clime, and in time be the better able to Iourney; but in Conclusion its turned to Discoverie, not of a Passage, but of the way to find one, to come to the Ile *Iapon*, *China*, and the Orientall *India;* of which, hearing that divers inquired after (for who would not share in such Treasures as there's to be had) as what hath FOX done? others, how farre hath hee beene? hath he beene as farre as any man? if there be a Passage, how chance hee hath not found it? if not, why doe they search after it? and, in fine, hath hee made a Voyage or no? The meaning of this last [query] I know not, nor I thinke they neither; but for that the

[1] See the Introduction for an explanation of this phrase.—C.

most desire to know what I have done, and how farre I have bin, I answer, as the Old women tells tales, *Further and further than I can tell;* and for the rest, I referre them to mine owne Iournall. For I hope it may satisfie those that are indifferent, and that stand well affected to those Discovering enterprises, for whose cause, and the good or content of others, it was chiefly compiled. I doe confesse that they have not done me much amisse in causing me to show my selfe; for hereby they shall know that before this vndertaking I was not ignorant of what my predecessors had done before I came, which was no small helpe, both to my safety and proceeding. For some will Calumniate out of malice, some for that they dislike all things but what themselves or friends doth; but I hope the judiciall Sea-man will stand in my defence. For amongst those inquisitors (especially the never-satisfied-curious), I finde few that doth or will know what they inquire after; others, having other imployments, rest content with the verball Report of their owne Times, rather than looke after the Labour and Experiences of so vnpleasant and, as it is held, vnprofitable a Voyage. [As] for these and others, I leave [them] to peruse what shall follow, wherein the desirous may bee satisfied how every Successour (as his fancy, direction, or opinion guided) thought to rectifie himselfe by others mistaking, or to finde the Passage in those places left vnsearched or [im]perfectly discovered. For [there is] no question but this Voyage might long since have attained his full perfection if the first beginners had beene yearely continued, which, since Captaine DAVIS, was not; but supplyed alwayes (after some yeares of breathing) by Greene men or those who (in that time) had forgot their experience; nay, I may avouch that, if this course had beene taken, and private ends had [not] beene wanting, that since Mr. HVDSON his first Voyage, 1610, the Passage had been sayled through before 1618, and done at one quarter of the Charge. And howsoever it was not my fortune to finde [the Passage], yet I praise GOD for that I have brought home the newes thereof, though I have left it for him whose time God shall be pleased to ripen for the same. For it is a wonder that a Voyage of such Consequence as this, wherein all the parts of Navigation may be practised, should so long lye raked in the embers, and no further search made thereinto (considering that, within these few yeares, there hath bin more excellent Secrets in Navigation found out, then hath beene in all the former Ages), it being (no question) more then the difficultest halfe Discovered; to incourage which, thou shalt reade heere how great Princes have taken pleasure to increase their Dominions (in Person) by Navigation, as also that our Ancestors were not so simple as we make them; and that this Hunting of the Whale (which as yet wee make rare) was in practice in King ALFRED'S time, related by OCTHER, who cald him Lord, as King ARTHVR'S Successour; which OCTHER sayled about the North Cape of *Finmarke*, as appeareth by his owne Narration; and this was long before RICHARD CHANCELOR, whom wee make *Prima*. I begin with those Princes, to show the Honour and Renowne of our Nation, and so proceed to all that went before me North-Westwards, out of a coniecturall necessity [by] which I conceive them to bee as Appendices to the Iournals of these moderne times.

It will be obiected that many of these Abstracts are taken out of other Bookes, and that those are the Voyages of other men. I answer, It is true; the most of them are. For what are all those of Mr. *Hackluits*, and Mr. *Purchas*, but the Collections and preservations of other men's labours? For who can speake or write that which was never done before; and I doe confesse my selfe to be infinitly bound vnto them and others for their paines; for, if they had not bin thus carefully stored vp, these had beene wanting, as are divers others. For in searching thou shalt finde that the workes of many brave men are buried in oblivion, whose Names are devolved and but something brought vnto vs through the succession of Ages; and, though part of those workes have escaped, the wracke of the rest have perished in the Gulfe of time, which hath swallowed the best of many men's endeavors; and, [as] for those extant, thou canst not come vnto [them] without labour and charge; all which I have brought to thy hands almost for nothing, considering the charge of Bookes of great prise, the labour in search and Quotations; besides, heere is much that never came in Print as yet, being very difficult to be had. And, though I have abstracted those Workes of my Predecessors, yet I have interlaced my owne experience, and beene curious not to leave out anything of theirs or my owne, wherein I could thinke my remembrance might avayle thee. And, although wee all have sometimes stumbled, yet wee may bee excused, considering the long and strange Iourney wee had to goe; and, who followeth, doth but persevere by our example; nor have I, like the Poet, feigned a Storme where I had none. Nor doe I hold that man fit to take charge of Voyages remote, especially Northeast or [North] Westward, and be Ignorant of those Abstracts and Iournals following. For I know knowledge will conceive them to be good Preparatives. Howsoever, in them I have done my best, and whosoever will amend them, I shall take him for my friend.

I have restored all the names of Capes, Headlands, and Ilands, formerly given by Captaine DAVIS, Mr. HVDSON, and Sir THOMAS BVTTON (which since have beene infringed vpon) vnto their first appellations, both in my Booke and Map. I doe confesse that there be many who might more fitly have vndertaken this labour then my selfe, if they had been so well acquainted with the action; For we have many Practitioners whom we call Mathematicall Sea-men, who that, after a little practice in Art, purposeth then to goe to Sea and Travell; therefore carefully busying themselves about perusing the Cosmographicall Maps, and with great industry studied the bookes of Navigation, and, to better their knowledge, have enquired of many excellent Mariners for their better satisfaction in the difficultest parts thereof, thinking that they are now fit to set sayle and launch foorth to Sea, conceiting of themselves to bee farre more able then the Sea-man; yet, when they have taken the charge of the Ship vpon them, and is hoyst forth of the quiet port into the ruffe and boisterous Ocean, where they shall behold many hideous mountaines of high threatning billowes, and raging waves, tempestuous gusts, with hayle, raine, and thunder, Shifts of windes, and counter Seas, Currents, Races, Sets, and Over-fals, being deprived of Sun, Moone and Starres for long season, they will then thinke that they onely dreamed before, when they imagined of the Course of the Seas, and that their Bookes were but weake Schoolemasters; that the talke of Art were farre short of the Practice, when, at beholding, the Starres which they thought to have vsed as guides and directions, seeme now as they threatened their ruine and destruction; nay, when you shall looke forth and tremble at the rising of every wave, and shall be agast with feare to refraine those Rockes and dangers which lye hid within the Sea's fairest bosome, together with the greatnesse of the Ocean, and smalnesse of their Ship, for want of experience to handle, not knowing how to shun, they will then thinke that the least gale is of force to overthrow them, and know that Art must be taught to practice by long and industrious vse.[1]

For it is not enough to be a Sea-man, but [it is necessary] to be a painefull[2] Sea-man ; for a Sea-bred man of reasonable Capacity may attaine to so much Art as may serve to Circle the Earth's Globe about ; but the other, wanting the experimentall part, cannot ; for I doe not allow any to be a good Sea-man that hath not vndergone the most Offices about a Ship, and that [hath not in] his youth bin both taught and inured to all labours ; for to keepe a warme Cabbin and lye in sheets is the most ignoble part of a Sea-man ; but to endure and suffer, as a hard Cabbin, cold and salt Meate, broken sleepes, mould[y] bread, dead beere, wet Cloathes, want of fire, all these are within board ; besides Boate, Lead, Top-yardes, Anchor-morings and the like.

For, as Oyle Paper layd vpon the obiect, makes it more transparant, so doth Experience show Art [how] to see without spectacles ; and, concluding this point in starting (for I have no other Coyne, and but small store of that), to those who saith that I am but a North-Countrey Coaster and hath but been brought vp in small Vessells, I wish their pride to know that God's mercy is over all his Creatures, and [He] hath created as good men in the North as in the South. Let no man bee disdained, misprised, or vndervalued for his wants (except in Vertues.) For I doe not thinke but litle Mr. IEFFERY, that pretty Courtier, can pitch his eye-sight as high into the Firmament as Mr. EVANS,[3] his Maiesties great Porter. And [as] for Sea-men, where have you better than the *Scotshmen* are, and yet [they are] North-Countrey men ?

It was also cast into my dish, that I was an Officer of the Admiralty ; a poore conceit to obiect against me for this vndertaking, and as much as to say as the greatest Civilian held to be in Christendome cannot be capable of higher Promotion, if hee belong to the Admiraltie.

---

[1] There can be no question whatever that this long dissertation on the need of practical experience in a seaman was intended by Foxe as a sneer at Captain James, who (though an educated man and a skilled mathematician) was a theoretical, rather than a practical, seaman, and for whom, accordingly, Foxe (a sailor from boyhood) felt the utmost contempt.—C.

[2] Painstaking.—C.

[3] I have been unable to learn anything about these gentlemen. Probably they were officials about the court, and not of great eminence. —C.

I have also placed a Polar Map or Card,[1] that this Discoverie may be the better vnderstood, and for that I did desire to give satisfaction by Demonstration of all [things] treated of in the Booke ; for, otherwise, another proiection could not have contained it but at vnreasonable diversity ; and, because I cannot describe all the Names in *Fretum Hudson*, of Capes, Ilands, and Bayes at length in Letters, in respect of the smalnesse of the Degrees of Longitude, I have inserted them in a table by the letters of the Alphabet, as thou shalt find, beginning with A, b, c, d ; and [I have] tracted my owne way and discovery foorth and home in small prickes.

I acknowledge it[2] to be but rough-hewen, like Shipwright's timber, but what it wants in smoothnesse or forme, let thy good Report licke it into. I trust I shall not need to vse the fashion of Authors, to make any doubt of carping *Momus* or rayling *Zoilus*,[3] seeing I have intended thy Good, without any Reservation to my selfe. But I feare mee I have held thee too long in this place, like him who purposeth to take a long Iourney [and] stumbles vpon his owne threshold ; but have Patience, for I had rather be in fault then want. I have here prostrated my duty to my King, and my service to my Countrey. Craving thy favourable acceptance, I rest,

Thine, in all welwilling,
LUKE FOXE.

From Kingston *upon* Hull,
this first of Ianuary,
1635.

---

[1] An old form of "chart"; originally the card on which the points of the mariner's compass were marked. (Cf. Fr. *carte*.)—C.

[2] It is not very clear whether Foxe here refers to his book or to his map, but the apology applies equally to either.—C.

[3] The idea which Foxe intended to express in this passage is by no means clear. Momus, in mythology, was the god of buffoonery and jests. He is credited with a carping and satirical disposition. Zoilus was a grammarian who attacked Homer and other writers so violently that his name became proverbial for a railing and malignant critic.—C.

CERTAINE TESTIMO-nies concerning King *Arthur* and his Conquests of the North regions[1]; taken by Mr. *Hacklute* out of the Histories of the Kings of *Brittaine*; written by *Galfridus Monamutensis*; newly Printed at *Hedleberg*, 1587.

IN the yeare of CHRIST, 517, King *Arthur*, in the second yeare of his raigne, having subdued all partes of *Ireland*, sayled with his Fleet into *Iseland*, and brought it and the people thereof into subjection. The rumour being spred throughout all the other Lands, that no Countrey was able to withstand him, *Doldanius*, King of *Gotland*, and *Gunfacius*, the King of *Orkney*, came voluntary unto him, and yeelded their obedience, promising to pay him Tribute. The winter being spent, hee returned

---

[1] It cannot be maintained that the opening pages of Foxe's book, as far as the first voyage of Sir Martin Frobisher, are of much, if of any, value. They would not have been reprinted, had it not been felt that, in reproducing the work, it would be a pity not to reproduce it *entire*, especially as the unimportant passages are not of great length. They consist merely of extracts from some of the early chroniclers,

into *Britaine ;* [and] established his Kingdome in perfect peace. He continued there for the space of twelue yeares. (*Lib.* 9. *Cap.* 10.)

After that, King *Arthur*, sending his Messengers into divers Kingdomes, hee summoned such as were to come to his Court, aswell out of *France*, as out of the adjacent Ilands of the Sea, and a little after, from these adjacent Ilands came *Guillaumarius* King of *Ireland*, *Malvatius* King of *Iseland*, *Doldanus* King of *Gotland*, *Gunnotius* King of *Orkney*, *Lot* the King of *Norway*, and *Aravilius* King of *Denmarke.* (*Lib.* 9. *Cap.* 12.)

A testimony of the Right and Appendances of the Crowne of the Kingdome of Brittaine, *taken out of Mr.* Lambert.[1] (*Fol.* 137. *pag.* 2.)

ARTHVR, which was sometime the most renowned King of the *Britaines*, was a mighty, valiant and famous Warriour. This Kingdome was too little for him, and his minde was not contented with it ; hee therefore valiantly

and are exceedingly confusing and unsatisfactory. In many cases, it is impossible to tell whether the words used are those of Foxe himself, of someone who compiled the extracts for him, of the original chronicler, or of some navigator whose statements the narrator is quoting. It is impossible, too, to do more than guess at the identification of the lands mentioned ; and the confusion is increased by the fact that the same lands appear more than once under different names. For this reason, but little attempt has been made to identify them. The voyages in question have, for the most part, been critically discussed in the series of the Hakluyt Society by competent writers, who have made use of the original narratives from which Foxe's abstracts are taken.

[1] Lambert of Aschaffenburg, one of the best of the early German chroniclers, was born about 1020. His most important work is the *Chronicon, sive Historia Rerum in Germania Gestarum*, an accurate and impartial record of events down to the year 1077, prefaced by a universal history, compiled from Bede and others. The manuscript of this was discovered by Melancthon in the Augustine Monastery at Wittenberg, and was first printed in 1525. He died about the year 1080.—C.

subdued all *Scantia*, now called *Norway*, and all the Iles beyond *Norway*; to wit, *Iseland*, and *Groenland*, which are appertaining unto *Norway*, *Sweveland*, *Ireland*, *Gotland*, *Denmarke*, *Semeland*, *Windeland*, *Curland*, *Roe*, *Femeland*, *Wireland*, *Flaunders*, *Cherilland*, *Lapland*, and all other Lands of the East Sea, even unto *Russia* (in which *Lapland* he placed the Eastmost bounds of his *Brittish* Empire) and many other Ilands beyond *Norway*, even under the Pole, which are appendances of *Scantia*, now called *Norway*. He planted the Christian Faith throughout all *Norway*, matched their Nobility with *Brittish* bloud; called *Norway* the Chamber of *Brittaine*, and incorporated them unto us.[1]

Another testimony out of Galfridus Monumetensis[2] *concerning the Conquest of* Malga, *King of* England.
(*Lib.* 11. *cap.* 7.)

MALGO succeeded *Vorciporius*, which was the goodliest man in person of all *Brittaine*, a Prince that expulsed many Tyrants. Hee was strong and valiant in warre, taller then most men that then lived, and exceeding famous for his vertues. This King also obtained the Government of the whole Iland of *Brittaine*, and by most sharpe battailes he recovered to his Empire the sixe Ilands of the Ocean Sea, which were before Tributaries to King *Arthur*; namely, *Ireland*, *Iseland*, *Gotland*, *Orkney*, *Norway*, and *Denmarke*.

---

[1] It seemeth K. *Arthur* passed ye North Cape of *Finmarke*.—F.

[2] Galfridus Monumetensis, otherwise known as Geoffrey of Monmouth, was born at Monmouth early in the twelfth century. He was a diligent chronicler, and at one time Bishop of St. Asaph. His most important work was his *Historia Britonum*, or *British History*, the first printed editions of which appeared at Paris in 1508 and 1517, and another at Heidelberg in 1578, as mentioned above. An English translation, by Aaron Thompson of Oxford, appeared in 1718 (London, 8vo.), and there have been various later ones.—C.

The Voyage of Octher[1] *to the Northeast parts beyond* Norway, *reported by himselfe vnto* Alfred, *the famous King of* England, *about the yeare* 890.

OCTHER sayd that the Countrey wherein he dwelt was called *Helgoland;* hee told his Lord, King *Elfreed*, that he dwelt furthest North of any *Norman;* he sayd that he dwelt towards the North part of the Land towards the West coast. And in another place hee affirmeth that there was no man dwelling towards the North from him, and affirmed the land to stretch farre to the North, yet all Desert, and not inhabited. Vpon a time, he fell into a fancy to know how farre the Land stretched Northward; whereupon hee tooke his voyage directly North, along the Coast, having alwayes the Desert land upon his Starboard, and upon the Larboard the maine Oceane, and continued his course for the space of 3 days; in which space, hee was come as farre towards the North, as the Whale-hunters used to travell.[2] Hee proceeded to the North, as farre as hee was able to sayle in other 3 days, at the end whereof he perceived the Coast turned toward the East, or else the Sea opened with a maine Gulfe into the land, hee knew not how farre. Well he wist and remembred that he was faine to stay till he had a Westerne wind and somewhat Northerly, and thence hee sayled plain East along the Coast still, so farre as hee was able in 4 dayes, at the end of which time, hee was compelled to stay, untill he had a full Northerly

[1] Octher, a Norwegian who travelled far and wide, came in one of his excursions to the court of King Alfred the Great, and related his voyage round the most northerly point of Europe. His relation was incorporated with the travels of Wulfstan, in the introduction by King Arthur to the Anglo-Saxon reproduction of the history of Paulus Orosius: *De Miseria Mundi.* It was edited by Joseph Bosworth, under the title of *King Alfred's Anglo-Saxon version of the Compendious History of the World by Orosius* (London, 1859).—C.

[2] Whale-fishing hath beene of long continuance.—F.

wind, for as much as the Coast bowed downe thence directly towards the Southward ; at leastwise, the Sea opened into the Land, that he could not tell how farre ; so that he sailed along the coast directly South, so farre as he could travel in five dayes ; and at the 5 dayes end, hee descryed a mighty River which opened very farre into the land, at the entry of which river he stayed his course, and in conclusion, turned backe againe. For he durst not enter, for feare of the Inhabitants of the Land, perceiving that on the other side of the River, it was thoroughly inhabited ; which was the first peopled Land he had found from the place of his dwelling ; whereas continually hee had a desert Wildernesse upon his Starboard-side, except in some places he saw a few Fishers, Fowlers, and Hunters, which were all *Finnes ;* and on the Larboard [he had] the maine ocean. Thus much for *Octher*, whome it appeares was the second that doubled the North Cape that we have [any account of] extant, King *Arthur* being the first.

Although I know here is something that will not bee beleeved, yet I will proceed, seeing that the more and further I doe goe on, I finde some reason to have better confidence of the former, and thus much, &c. Therefore as followeth, out of the Comments of Arngrinus Fitz Ionus.[1] 1598.

A Certaine Writer sayth hee hath put downe the distance betweene the mouth of *Elbe* and *Baczend* in the South part of *Iseland*, to be 400 leagues ; from whence, if you will accompt the difference of Longitude to the Meridian of *Hambrough*, *Iseland* must have none of the

[1] He was an "honest and learned young man", appointed by Gudbrandus Thorlacius, Bishop of Holen, in Iceland, to prepare a treatise or commentary on the country, "taking his proofs, not out of the vaine fables of the people, but from his owne experience and many other mens also of sufficient credit." He says, among other things, that "the distance is not so infinite as has been imagined ; that they are

Longitudes accompted of, as 5.7.6.20.30.28.325. according to several Authors; for I am able to prove by 3 several Voyages of *Hamburgers*, that it is but 7 dayes sayle from *Iseland* to *Hambrough*; besides, all the Ilands which for the abundance of Sheepe are called *Farrier*,[1] as likewise the Desert Shoares of *Norway*, are distant from us but two dayes sayling, wee have 4 dayes sayling into habitable *Greenland*, and almost in the same quantity of time, wee past over to the Province of *Norway* called *Stad*, lying betweene the Townes of *Oridrosia* or *Trondon*, and *Bargon*,[2] as we find by antient Records of those Nations.

Iseland hath bin called by three names one after another; for one *Nuddocus*, a *Norvegian* borne, who is thought to bee the first Discoverer thereof, as he was sayling towards *Farra* Ilands, through a violent Tempest, did by chance arrive at the East shore of *Iseland*, where beholding the Mountaine-tops covered with Snow, [he] called this Iland *Snowland*.

After this, one *Gardarus*, being moved with *Nuddocus* reports, went to seeke, and when hee had found, hee called it *Garders* Ile.

After these two, one *Flok* went into this Land, and named it after the Ice hee found there, *Iceland*.

This land hath some years no Ice at all, as in 1592, but the Sea is open for most yeares from *Aprill* to *Ianuary*, and in *May* all is driven to the West, and this Land hee inhabited about the yeare of our Lord, 874.

---

but seven days sailing from Hamburg, two from the Faroe Isles, four into Greenland, and four to Norway." He describes graphically the superstitious horror with which the volcano was regarded as the place of torture of the lost; and gives information respecting the inhabitants, who were, he informs us, converted to Christianity by Adalbert, Metropolitaine of Hamburg, in the yeere of Christ, 1070. (See *Hakluyt*, vol. i, p. 550.)—C.

[1] The Faroe Islands.—C.

[2] Drontheim and Bergen.—C.

In the yeare 1591, there was a ship of *Germany* lay laden with Copper 14 dayes in the harbour of *Vopnaford;* in *November*, she set sayle and departed; this was some ship of the *Balticke* Sea, or *Hambrough*, which durst not goe through the Narrow seas for [? fear of] *Spaine*.

The Commentor *Arngrinus* sayth, in the yeare of Christ 874, *Iseland* (being indeed discovered before that time, as is aboue mentioned) was, the first of all, inhabited by certaine *Norvegians*. Their Chieftaine, was one *Ingulphus*, from whose name the East Cape of *Iseland* is called *Ingulfe Hoffdie*. These planters are reckoned vp by name in our Records, saith he, more then to the number of 400 [al-] together, with those of their blood and kindred, and great families besides; neither onely is their number described, but it is plainely set downe what Coasts, what Shoares, and what inland places each of them did occupie and inhabite, and what names the inhabitants did giue vnto Straights, Bayes, Harbours, Necklands, Creekes, Capes, Rockes, Crags, Mountaines, Hills, Valleyes, Hammockes, Springs, Flouds, Rivers; and, to be short, what names they gaue vnto their Granges and houses: whereof many at this day are [still] received and vsed. Therefore the *Norvages*, with their company, peopled all the habitable places of *Iseland*.

The voyages of the Brethren Mr. Nicolo *and* Anthonie Zeni,[1] *Venetians, collected out of their owne Letters by Mr.* Francisco Maritino.[2]

N*Icolas* caused a ship to be made ready at his owne charge in *Italy* [in] 1380, with intent to see *England* and *Flanders*, but was, by tempest, cast vpon the Ile of *Freesland* where he had beene cruelly entreated had it not beene

[1] The Zeno family belonged to the highest nobility in Venice, and was celebrated for the performance of great actions, members of it having been chosen from time immemorial to fill the highest offices of

for *Zichmni*, Lord of certaine Ilands called *Porland*, lying on the South of *Freesland*, being rich and populous. He was also Duke of *Sorany*, lying ouer against *Scotland*.

This Prince, being arrived vpon *Freesland* to make warre

State. About the year 1200, Marino Zeno assisted in the conquest of Constantinople, and was *Podesta* of that place from 1205. His son, Pietro, was the father of Rimieri, who was Doge of Venice for seventeen years from 1282. He adopted his brother Marco's son, Andrea, who became afterwards captain-general of the Venetian fleet, and the father of Pietro, surnamed *Dracone*, from the dragon which he bore on his shield. Pietro had three sons: Carlo, High Admiral of Venice, born about 1334; Nicolo; and Antonio. The narrative of the explorations of the brothers is taken from letters written by Nicolo to Antonio, and (after the latter joined his brother) from the letters of them both to the elder brother Carlo. It is asserted that a complete chronicle had been prepared by Antonio, which a descendant, born in 1515 (and also named Nicolo Zeno), partially destroyed, not knowing the value of the papers; but subsequently he was able, from the remainder, to compile the narrative and to publish it, as we now have it, in 1558. He found, also, in the Zeni palace, a map, rotten with age, illustrative of the voyages. The story of the Zeni has puzzled many, and is rejected by some as a tissue of improbabilities. John Reinhold Forster, the distinguished associate of Captain Cook, was the first to assert the truth of the narrative (*History of the Voyages and Discoveries made in the North*, pp. 178-209; London, 1786). This he attempted to prove by locating every place mentioned, though some of his conjectures were doubtless misleading. The travels have been edited, with notes and introduction, by Mr. R. H. Major, F.S.A. (Hakluyt Society, London, 1873). Both Forster and Major concur in believing the Prince Zichmi to be Henry Sinclair, who in the year 1379 was invested by Hakon, King of Norway, with the Isles of Orkney. Zichmi they consider the Italian form of his name. For remarks on the probable identity of the numerous localities mentioned in the following narrative of the voyages of the Zeni, the reader must refer to the two admirable works above-mentioned.—C.

2 (p. 19). Francisco Marcolini (or, as Foxe erroneously calls him, Martini) was an Italian printer, and was born at Forli early in the 16th century. A translation of his account of the discoveries of the brothers Zeno appears in *Hakluyt* (vol. iii, p. 121), whence, no doubt, Foxe obtained the following information.—C.

vpon the same, vnderstanding[1] the shipwracke, came presently, and hearing by the Latine tongue that he was of *Italy*, he received him into protection with great joy.

This Prince, having the last yeare given the overthrow to the King of *Norway* in some of his *Signiories*, was come to *Friesland* also to take the same from the said King, whereof he was Lord, and to that purpose, vnderstanding that Mr. *Nicolo* had great judgment in Sea and Martiall affaires, he gaue him commission to goe aboord his ships, commanding the Captaine to honour him in all things, and to vse his counsell. His Nauie consisted of 30 saile, whereof two rowed with Oares.

With these small Barkes, and one ship, they sailed to the Westward and won *Ledovo* and *Ilose*, and divers other small Islands, and turned into a Bay called *Sudero;* in the Hauen of the Towne called *Samstoll*, they tooke certaine small Barkes laden with fish.

Here they found *Zichmni*, who came thither by land, conquering all the Country. They sailed to the W., by another Cape or Gulfe, and conquered all the Ilands they found, to the Signorie of *Zichmni*. These Seas, for as much as they sailed, were in manner nothing but shoales and Rockes, so as it was thought amongst them that the fleete had perished, if it had not been for the skill and knowledge of Mr. *Nicolo* and his men, who had beene brought vp in the practise of Nauigation all their liues.

At the counsell of Mr. *Anthonie*, they goe on Land, when they heard of the good successe of *Zichmni* in his warres, and that all the Iland by Embassadors was yielded vnto him.

At their meeting, the Prince gaue Mr. *Nicolo* the honour of Knighthood, & graced him as the preseruer of his fleete, and rewarded his men. In Triumphant manner, they goe

---

[1] Presumably *notwithstanding* is intended.—C.

towards *Friesland*. The chiefe Citty thereof is scituate on the S. E. side within a Gulfe or Bay, as there are many in that Iland. In this Gulfe or Bay is such abundance of fish taken, that many ships are laden to serue *Flanders*, *Brittaine*, *England*, *Scotland*, *Norway*, and *Denmarke*.

Thus much is taken out of a Letter that Mr. Nicolo *sent to his Brother* Anthonie, *requesting that he would seeke some meanes to come to him.*

MAster *Anthonie* furnisht a ship, and after great danger, with great joy, arrived with his Brother, where he remained 14 yeares, 4 yeares with his Brother and 10 yeares alone; where they so behaued themselues, that *Nicolo* was made Captaine of *Zichmnies* Nauie. They set forth for the enterprise of *Estland*, being betweene *Friesland* and *Norway*; but by storme of wind, they were driven vpon certaine shoales, where a great part of their Fleete was cast away. The King of *Denmarke*, comming to rescue *Estland*, his Fleete vtterly perished, which *Zichmni* having notice of by a ship of the Enemies, he determines to set upon *Iseland*; which, together with the rest, was subject to the King of *Norway*; but he, finding the Country well fortified, his small Fleete were glad to retire. In the same channels, he assaulted the other Iles, called *Iseland*; which are 7: *Talas*, *Broas*, *Heance*, *Trans*, *Mimant*, *Dambere*, and *Brest*; and having spoiled them all, hee built a Fort in *Brest*, where he left Mr. *Nicolo*, with men, small Barkes and Munition, and with the few ships he had left, he returned to *Friesland*.

Mr. *Nicolo*, in *July*, with 3 small Barkes, sailes Northward, and arrives in *Engroneland*, where he found a Monasterie of Friers, neare a hill that casteth forth fire, like *Vesurius* & *Etna*.

They subsist specially, by a Fountaine of water, issuing

from that Hill, whereby the Monastery and all other buildings are refreshed and warmed at their pleasure, as also their Gardens thowed and watered; their stones for building are cast out of this combustable forge, which being once cooled will not dissolue, but taken hot and cast into Water, maketh a marvailous strong lime.[1] The people haue these devoted men in great veneration.

They feed upon wild Fowle and Fish, for the Bay or Haven, by reason of the heate of this water, doth neuer freeze; by meanes whereof, there is such concourse of Sea-fowle and abundance of fish, whereby all the people are fed and maintained, and 1000 imployed about taking the same. Hither, in Summer time, which lasts but three Moneths, come many Barkes from the Ilands thereabout, and from the Cape about *Norway*, and from *Trondon*,[2] and exchange all needfull commodities for fish, which they dry with the Sunne or cold. To this Monastery, resort many Friers of *Norway*, [and] *Suetia*, but the most part are of *Iseland*. There are continually in that part, many Barkes, which are kept there by reason of the Sea being froze, waiting for the Spring to dissolue the Ice. Their Boates are made like vnto Weavers shuttles, [by] taking the skins of fishes, and sowing them together in many doubles, they fashion them with the bones of the same fish, and make them so sure and substantiall, that it is marvailous to see how in tempests they will shut themselues close with in, and let the sea and wind carry them they care not whither, without either breaking or drowning; and, if they chance to be driven vpon the Rocks, they remaine sound without any bruise at all, and they haue, as it were, a sleeue in the bottome, which is tyed fast in the middle; and, when there commeth any

---

[1] The next paragraph but one seems as if it should have been inserted here, with the exception of its last sentence. Probably it was misplaced by a printer's error.—C.

[2] Drontheim.—C.

water into the Boat, they put it into the one halfe of the sleeue; then, fastening the end with two peeces of wood, and loosening the band beneath, they convay the water forth of the Boat, and this they doe as often as they have occasion.

Moreover, this Water is of a Sulphurous brimstone nature, and is conveyed into vessels of Brasse, Tinne, and Stone, and so imployed to all vses, to warm their houses, or boile their meate, without any stinke, or noisome smell. The Friers of this Monastery speake all the Latine tongue; and this is all that then was knowne of *Engroneland.*

Mr. *Nicolo* dyed in *Friesland,* and, being dead, *Anthonie* succeeded him, both in Goods and Dignitie; and, albeit he attempted and made great supplication, yet could he not obtain licence to returne to his Countrey. For *Zichmni* intended to make himselfe Lord of the Sea, vfed his counsell, and determined to send him with divers Barkes to the West-wards; for that certaine Fishermen of his, had discovered certain Ilands, very rich and populous which discovery Mr. *Anthony,* in a Letter to his Brother *Carolo,* recounteth thus:—

SIx and twenty yeares since, there departed 4 fisher Boates, the which a mighty tempest tossed too & fro divers dayes when at length they discovered an Iland called *Estotoland,* lying West-ward above 1000 miles from *Freesland,* upon which one of the Boates was cast away, and the 6 men therein were taken and carried to a faire and populous City, where the King sent for many Interpreters, but none could vnderstand their language, but one who spoake Latine, who was also cast there by chance, who vnderstanding their case, rehearsed it to the King, who caused them to stay in his Country 3 years, in which time they learned the Language. One of them was in divers parts of the Iland, and reports it to be Rich, and aboundeth with all Commodities of the world. It is [a] little lesse then *Iseland*, but farre more fruitful ; in the middle, is a Mountaine, from whence there Springs 4 Rivers that passe through the whole Countrey.

The Inhabitants are very witty people, and have all Arts and faculties, as we have ; and it is credible that, in times past, they have had traffique with our men ; for he said he saw Latine books in the King's Library, which at that present they did not understand. They have a peculiar language and letters or characters to themselves. They have Mines of all manner of mettals, but they abound with Gold. They have trade in *Engroneland*, from whence they bring Furres, Brimstone, and Pitch. He saith that, to the Southward, there is a great populous country, very rich in gold ; they sow Corne, and make Beere and Ale, and use it as we do wine ; they have mighty great woods. There are many Cities and Castles. They build small Barkes, and have sayling, but they have not the Loadstone, nor know not the use of the Compasse ; wherefore the Fishers were had in great estimation, insomuch that the King sent them Southward to a country they call *Dorgio;* but in that voyage they had such weather that they thought to have perished in the sea; but, escaping that [peril], they fell into another more cruell ; for they were taken in the Country and the most part of them eaten of the Salvage people, which fed upon Man's flesh as the sweetest meat.

But this Fisherman [before mentioned] escaped, by teaching them to take fish with nets in the Sea and in fresh rivers, whereby he was honoured and beloved of all.

The Lords of the Country fell out about him ; and, in 13 yeares he was there, he was sent to at least 25 of them, from one to another, to teach his Art of Fishing, so that he was in all parts of the Country ; which, he saith, is very great ; the people very rude, and voide of goodnesse. They all goe naked, neither have they wit to cover their bodies with the skins of beasts they take by hunting, whereby they are miserably vexed with cold. They have Lances, sharpe at the point; they have Bowes, and strings made of beast strings; they are a cruell people, and have certaine lawes amongst themselves. To Southwards, the people have more civility, the aire being more temperate, and [they] have Cities and Temples with Idols. There they sacrifice men to Idols, and afterwards eate them; they have some knowledge of gold and silver.

Now this Fisherman had an intent to get home; but his companions, being in despaire thereof, never offered to attempt their freedome, and therefore stayed there; whereupon he travailed through the woods towards *Dorgio.* The next Lord knew him, and he was conveyed from one to another, until at length he came to *Dorgio*, where he stayed three yeares; and, hearing of divers Boates that were arrived there, he went to the sea-side, and, asking of whence they were, they said of *Estotoland.* He, being glad, requested to be entertained by them and carried to their country, which they gladly granted and used him for their Interpretor.

After that, he frequented that trading with them, and became very rich, wherewith he furnished a barke, returned into *Friezeland*, and to his Lord made this report of that wealthy Country.

He is credited here, for that the Mariners doe affirme his reports to be true, wherefore this Lord is resolved to send me forth with a fleete toward these parts, but it was not so; for *Zichmni* went himselfe; and, concerning their proceedings, I have a Letter in forme following:—

OVr preparation for *Estotoland* was begun in an unluckie houre; for, three dayes before our departure, the Fisherman dyed that should have beene our guide. Notwithstanding this, [the] Lord would not give over the enterprise, taking the Marriners that returned with him. Making our navigation to the Westward, we discovered Ilands subject to *Friesland*, and having passed certaine shelves, we stayed at *Ledovo* 7 dayes.

The first of Iuly, we arrived at *Ilose*. We stayed not there, but had a vexing storme at Sea for 8 days, not knowing where we were. A great part of their barkes were cast away. The weather being faire, they gathered up the pieces of broken barkes ; [and,], sayling on with prosperous wind, they discovered land at W.

They tooke harbour, and an infinite company of armed men came to defend their Iland. *Zichmni* makes signes of peace, sends 10 men, of ten severall languages, but they could understand none but one that was of *Iseland*, who informed them that the Iland's name of *Icaria*, named after the name of the first King of that place, whom they say was sonne to *Dedalus*, King of *Scotland*, who conquered that Iland and left his sonne there for King, and [that] they called this Sea the *Icarion* Sea, in remembrance of that King [who], in further search, was drowned there. They would not suffer us to land, but onely they would receive one man to learne the *Italian* tongue, as they had received the other tenne.

He, espying a harbour on the East side, put therein for wood and water, but the inhabitants assaulted, slew, and maimed many of them, enforcing them away ; and, sayling about the Cape of the Iland towards the North, we found many shoalds, amongst the which we were ten dayes, continually in danger of losing our whole fleete.

Zichmni, seeing he could not prevaile against the force [of the inhabitants] of the Ilands, sailed 6 dayes towards the West ; but the wind changing into S.W., he sayled 4 dayes with wind a-poope, with a growne sea. He discovers land, not knowing what land ; and, rowing to it with Oares, they finde a good harbour, and see farre off a Mountaine, that cast forth smoake. *Zichmni* sent forth an hundred Souldiers to search what people inhabited it, the whiles he takes in wood and water, catches great store of fish and Sea fowle, with such aboundance of Egges, that his men, halfe famished, were filled therewith. This was in *Iune*, ye ayre so temperate and pleasant as impossible to expresse. They finde no people; they name the haven *Trine*. The Souldiers, returning backe, had bin through the Iland, finds the fire naturall that the hill's bottome produced, and that there was a Spring which issued water like pitch, and run into the Sea; and that there dwelt a multitude of people of small stature, hiding themselves in caves in the ground. *Zichmni*, taking liking to the soyle, stayes there with such as were willing; the rest, he sent away in the ships, wherein, unwilling, I was Captaine.

Anthonie sayles 20 dayes East without sight of any land; altering his course S.E., in 5 dayes he discovered *Neome*; he knowing the Country, had sailed past an Iland as he found. He victualled here, the Iland being subject to *Zichmni*. And in 3 dayes he sailes to *Friesland*.

What followed after this Letter he knew not, but by coniecture, that *Zich*: built a City, and discovered on both sides of *Engroneland*, and that he see it particularly described in the Sea carde, but the narration is lost. The beginning of his letter is thus:

COncerning these things that you desire to know of me, as of the Manners and Customes, of the Beastes, and Countryes adioyning. I have made thereof a particular Booke, which, by God's helpe, I will bring with me; wherein I have described the Country, the monstrous Fishes, the Lawes and Customes of *Friesland*, *Iseland*, *Eastland*, the Kingdome of *Norway*, *Estotiland*, *Dorgio*, and the end of the life of Mr. *Nicolo*, with his discoverie, and the state of *Engroneland*. I have also written the Life and Acts of *Zichmni*, a Prince as worthy of Immortall memory as any that ever lived, for his great vigilancy and singular humanity: Wherein [also] I have described the discovery of *Engroneland* on both sides, and the Citty he builded; and, further, I hope to be with you very shortly, to satisfie you by word of mouth.

All these Letters were written by Mr. Anthonie, *to Mr.* Carlo *his brother.*

FOr the more credit and confirmation of the former History of Mr. *Nicholo* and Mr. *Anthonie* (which for some few respects may be called into question), I have hereunto annexed the judgment of that Famous Cosmographer *Abraham Ortelius*,[1] or rather the yielding or submitting of his judgment thereunto, who in his *Theatrum Orbis*, *Fol.* 6, next after the Mappe of *Mar del Zur*, borroweth proofe and authority out of this Relation, to shew that the N.E. part of *America*, called *Estotiland*, is, in the Originall, alwaies affirmed to be an Iland: [and] was, about the yeere 1390, discovered by the foresaid *Venetians*, above 100 yeares before *Columbus* set sayle

[1] Abraham Ortell (more commonly known as Ortelius), a very eminent early Dutch geographer, was born the 4th of April 1527, at Antwerp. He received an excellent education in Greek, Latin, and mathematics, from his father, a well-to-do merchant, and he made extensive travels in Germany, Italy, and Great Britain, forming a valuable collection of books, coins, and other objects likely to assist him in his work. He was the first to conceive the idea of an atlas, such as we now have; and in 1570 appeared the first edition of his *Theatrum Orbis Terrarum*, which was at once translated into French, Italian, and English, and ran through numerous editions. The labours of Ortelius were indefatigable, as is shown by the posthumous edition of 1603, in which, out of the thirty-eight charts, twenty-nine bear his own signature. Ortelius published several other valuable works of cartography. He was appointed geographer to the court of Spain by Philip II, to whom the *Theatrum* was dedicated; he was a stanch Catholic unto the end of his life. He died June 28th 1598.—C.

for these Westerne Regiones, and that the Northerne Seas were even then sailed by our *Europian* Pilates.

This writer,[1] acknowledging the originall copies of the *Zenys* Letters were by him carefully torne in peeces in his youth, which losse he now grieued at, I doubt in this, he was enforced in many things to patch vp, as his memorie would serve, so as there may be some likelyhood of vntruths. Howsoever, I doe beleeue the first Copies were true, though this is subiect to mistakings.

To continue antiquity may be to some purpose, seeing there may be some vse thereof by the painfull Seamen; otherwise it may something satisfie the never-satisfied *Curious*, who enquire after the beginning of things; and herein Mr. *Hackluit's* paines doth deserve great commendations: who recordeth, in his *English Voyages*, that *Madoc*, sonne of *Oween Guyneth*, Prince of North *Wales*, left the Land in contention between his Brethren, and prepared certaine ships with Men and Munition, and sought adventures by Sea, sailing West, leaving the coast of *Ireland* so farre North, that he came to a Land vnknown, where he saw many strange things.

This must needs be some part of this Countrey, of which the *Spaniards* affirme themselves to be the first founders since *Dannos* time, whereupon it is manifest, that this Country was by *Brittanies* discovered long before *Columbus*.

Of *Madocs* returne, there be many fables, but he did return, and declared of the fruitfull Countryes he had seen without Inhabitants, and, on the contrary, what barren and wild ground his Brethren and Nephewes did kill and murder one another for: He prepared a Navie of ships, got with him such men and women, as were desirous to live in quiet, and, taking leave of his friends, tooke his journey thither againe.

---

[1] This refers, not to Ortelius, but to the younger Nicolo Zeno. —C.

Therefore it is to be supposed that he and his people inhabited part of this country; for it appeareth by *Francis Lopez de Gomara*[1] that, in *Acuzamill* and other places, the people honored the Crosse, whereby it may be gathered that Christians had been there before the coming of the *Spaniards*.

This *Madoc*, arriving againe in that West Countrey, vnto the which he came 1170, left most of his people there, and returned backe for more of his owne Nation, acquaintance[s], and friends to inhabit that faire and large countrey, went thither againe, with 10 saile, as I find noted by *Guyten Owen*. I am of opinion, that the Land whereunto he came was some part of the *West Indies*.

As concerning *Sebastian Cabot*,[2] I cannot find, that he was

---

[1] Francisco Lopez de Gomara was a Spanish historian. He was born at Seville in 1510, and was, for some time, Professor of Rhetoric at Alcala. Ticknor (*History of Spanish Literature*, vol. ii, p. 36; ed. 1888) describes him as "the oldest of the regular historians of the Old World". He wrote a *History of the Indies* (Saragossa, 1552), and a *Chronicle of New Spain* (1554), with other works of minor consequence; but his information is by no means trustworthy. He was never in the countries the history of which he wrote.

[2] Sebastian Cabot was the second son of John Cabot, a Venetian pilot who settled in Bristol about 1472. He was probably born, or removed, there at a very early age. Hakluyt, Forster, and many other writers, have supposed him the first discoverer of Newfoundland; but it is to be feared that nearly, if not quite, all the discoveries usually assigned to Sebastian Cabot are really those of his father, and a contemporary manuscript says he was never in that land at all. His fame as a cartographer attracted the attention of Henry VIII, and he afterwards entered the service of King Ferdinand as map-drawer. His famous *Mappemonde* was probably printed at Antwerp. It embodies his own and his father's discoveries in North America, as well as those of the Portuguese and Spaniards down to his day. It served as the model for the *Theatrum Orbis Terrarum* of Ortelius (see note to p. 29). After the death of Henry VIII, Cabot returned to England, and received a pension. He was one of the first to notice the variation of the compass, and he was nominated governor for life of the "Company of Marchants Adventurers of England for the Discovery of Lands, Territories,

any further Northward then the 58 Degree, and so returned along the Land of *America* to the S.; but, for more certainty, heare his owne Relation to *Galeacius Butrigarius*, the *Popes* Legate in *Spaine*. Vnderstanding, saith he, by reason of the *Spheare* that, if I should saile by the way of North-west, I should by a shorter tract come to *India*, I caused the King to be advertised of my device, who immediately caused two *Caravels*[1] to be furnished, with all things needful, which was, as neere as I can remember, in the yeare 1496. In the beginning of Summer, I begun to saile North-West, not thinking to find any other Land then that of *Cataia*, and from thence to come in to *India*.

But, after certain dayes, I found the Land run to the North, which was to me a great displeasure; neverthelesse, sailing along the coast, to see if I could finde any Gulfe that turned, I found the Land still to continue continent to the 56 Degree vnder the Pole; and, seeing that there the Coast turned toward the East, dispairing to find the passage, I turned backe againe, and sailed downe by the Coast of that Land toward the Equinoctiall. *This much from himselfe.*

But heare how *Baptistie Ramusius*, his country man, how he flattereth him thus. It's many yeares since it was written vnto me by *Sebastian Cabota*, our Countryman, a Venetian, a man of great experience and very rare in the art of navigation, and the knowledge of Cosmographie, who sailed along and beyond this coast of new France, at the charges of Henry

---

Isles, Dominions, and Seigniories unknown", commonly called the Muscovy Company, which was incorporated on the 26th of February 1555. The date of Cabot's death is uncertain, but it was probably in or soon after 1557. With the exception of the engraved map of 1544, no literary relics of Cabot remain. The rib of a whale in the entrance of St. Mary Redclyffe Church, supposed to have been placed there as a trophy of the discovery of Newfoundland, and a street near the church named Cathay, are all that Bristol now has to show of this distinguished inhabitant.—C.

[1] Old-fashioned, lateen rigged, Portuguese dispatch-boats.—C.

the 7, King of *England*, and he advertised me that, having sailed a long time West by North, beyond those Islands into the Latit. of 67 Deg. and ½ vnder the North Pole; and, on the 11 day of *June*, finding still the open Sea, without any impediment, he thought verily by that way to have passed on still to *Cataia*, which is in the East, and would have done it, if the Mutinie of the shipmasters and Marriners had not hindered, and made him returne home from that place.

Out of Robert Fabian's *Chronicle*.[1]

THere is also mention made of *John Cabota*, *Sebastian's* Father; by his meanes, in the 13 yeare of *Henry* the Seaventh, at the same Kings charge, one ship was victuailed from *Bristow*, wherein divers Marchants of *London* adventured smal stockes, to search for an Iland he said was very Rich and replenished with great Commodoties: In his company went from *Bristow* 3 or 4 small Barkes, laden with coarse Cloth, Caps, Laces, Points, and other trifles; they departed the beginning of *May*, but were not heard of in that Maiors time.[2]

---

[1] Robert Fabyan, an alderman of London, presents us with the rare instance of a citizen and merchant, in the fifteenth century, devoting himself to the pleasures of learning. The place and time of his birth are uncertain, but his family were people of substantial respectability in Essex. Having busied himself with the affairs of the city of London for some time, he, in 1502, resigned his alderman's gown, not wishing to take the mayoralty, and retired to his mansion at Theydon Garnon, Essex. He was conversant with French and Latin, and wrote poetry. The first printed edition of his *Chronicle*, which he calls *The Concordance of Histories*, appeared in 1516, edited by Pynson; three other editions shortly appeared; and in 1811 a revised edition, with biographical preface by Henry Ellis, was published.—C.

[2] The Lord Mayor's year of office.—C.

Out of the same Chronicle.

WIlliam Purchase, being Maior,[1] three Saluages were taken in New-found-land and brought to the King. They were cloathed in Beast's skinnes, and eate raw flesh, and spake such speech, that no man could vnderstand them, and in their demeanor [were] much like to bruite Beasts; of which, two yeares after, I saw two apparrelled after the manner of *Englishmen* in *Westminster* Parish. I could not discerne them from *English*, vntill I had learned what they were.

An extract taken out of the Mappe of Sebastian Cabota, *cut by* Clement Adams.[2]

IN the yeare of Grace 1497, *John Cabot*, a *Venetian*, and *Sebastian*, his Sonne, with an *English* Fleete, set [sail] from *Bristow* [and] discovered that Iland, which before that time no man had attempted, on the 24 day of *June*. This Land hecalled *Prima Vista*: that is to say, first scene. That Iland lying out before the Land, he called the Ile of St. *John*, because he discovered it vpon that day of St. *John Baptist*. The inhabitants of this Ile vse to weare Beasts skinnes. In their Warres, they vse Bowes, Arrowes, Pikes, Darts, Wooden-clubs, and slings. The soile is Barren in some places and yeeldeth little fruit, but [the country] is

---

[1] William Purchas, mercer, was Lord Mayor of London in 1497-8 (see Fabyan's *Chronicle*, ed. 1811, p. 686). But Foxe has here followed Stow, who gives, in his *Annals*, an account of the exhibition of savages. Fabyan contains no mention of Cabot at all; and, in the *Memoir* of Sebastian Cabot, p. 229 (London, 1831), the idea of his bringing aborigines to England is dismissed.—C.

[2] The map was not in Foxe's own possession, as a hasty reader might suppose. With a view to economy of space, Foxe omitted the end of the heading: "An extract taken out of the map of Sebastian Cabot, cut by Clement Adams, *concerning his discouery of the West Indies, which is to be scene in Her Majesties Priuie Gallery at Westminster and in many other ancient merchants houses*" (Hakluyt).—C.

full of white Beares and Stagges, farre greater then ours. It yieldeth plenty of Fish, and these very great, as *Seales* and *Sammons.* There are Soales of a yard in length, but especially there is great plenty of that kind of Fish which the Salvages call *Bacculaos.*[1] There are also bred Howkes and Eagles.

Another Testimony of the Voyage of Sebastian Cabot *taken out of the third Decade of* Peter Martyr *of* Angleria.[2]

THe North seas have bin searched by one *Sebastian Cabot*, a *Venetian* borne. He furnished 2 Ships at his owne charge, (and first) with 300 men directed his course towards the N. Pole, that even in the Moneth of *July*, he found monstrous heapes of Ice swimming on the Sea, and in a manner continuall day-light. Thus, seeing these heapes of Ice before him, he was enforced to turn his sailes and follow the W., coasting the shore. He was thereby brought so farre into the South, by reason of the land stretching Southward, that it was there almost equall in latitude with *Fretum Herculeum.*[3]

---

[1] A name still sometimes given to the codfish found on the coasts of Newfoundland.—C.

[2] Pietro Martire d'Anghiera (commonly called Peter Martyr Anglerius in English) was born in February 1457, at Arona in Italy. He came of noble family, and was educated in Rome, where he made the acquaintance of many literary men. In 1487, he accompanied the Count of Tendilla, then Spanish Ambassador, to the Court of Ferdinand and Isabella, where he was well received. There he remained until his death in 1526, receiving many high honours and preferments. His chief literary works are his *Opus Epistolarum* (Alcala, 1530), a series of letters; *De Rebus Oceanicis et Novo Orbe* (1511), an account of the discovery of the New World; and *De Legatione Babylonica*, a narrative of his visit to the Sultan of Egypt in 1501. These are all works of high historic value. Notices of him are to be found in Prescott's *History of the Reign of Ferdinand ana Isabella*, Prescott's *History of the Conquest of Mexico*, and Bernay's *Petrus Martyr Anglerius* (Strassburg, 1891).—C.

[3] An old name for the Strait of Gibraltar.—C.

As he travelled by the coasts of this great Land which he named *Bacculaos*,[1] he saith hee found the like course of waters toward the West, but the same running more soft and gently than the swift waters which the *Spaniards* found in their Navigation South-ward. Wherefore it is not onely more like to be true, but ought also of necessity to be concluded, that betweene both the Lands hitherto vnknowne there should be certaine great open places, whereby the waters continually passe from the East to the West. Which waters I suppose to be driven about the Globe of the Earth by the incessant Motion and impulsion of the Heavens, and not to be swallowed and cast vp againe by the breathing of *Demogorgon*,[2] as some have imagined; because they see the Sea by increase and decrease to ebbe and flow. *Sebastian Cabota* himselfe calls those Lands *Bacculaos*, because that in those Seas thereabout he found so great multitude of certaine bigge Fish, much like vnto *Tanis*, which the Inhabitants call *Bacculaos*, and that they sometime stayed his ship. He found the people also of these Regions covered with Beasts' skins. He saith also he saw great plenty of Copper. And this is all of Note out of this Author.

Francis Lopez de Gomara, *out of his Generall Historie of the* West Indies.

HE which brought most certaine newes of the Countrey of *Bacculaos*, saith *Gomara*, was *Sebastian Cabot*, a *Venetian*, which rigged 2 ships at the Cost of Henry the 7, King of *England*, having great desire to Traffique for the Spices, as the *Portugales* did. He carried with him 300 men, and tooke the way towards Iseland, from beyond the Cape of *Labrador*, vntill he found himselfe in 58 Deg. and

[1] A name given to Newfoundland and the adjacent islands from the codfish taken there (see p. 35).—C.

[2] Demogorgon, in mythology, was a malignant demon, who was supposed to inhabit the bowels of the Earth.—C.

better. He made relation that, in the Moneth of *July*, it was so cold and the Ice so great, that he durst not passe any further. The dayes were very long, in manner without night. He returned and refreshed at *Bacculaos*, and after [-wards he] sailed along the Coast Southward vnto 38 Degrees, and from thence he shaped his Course to returne into *England*.

An abſtract of the Courſes, Diſtances, Latitudes, Longitudes, Variations, Depths, and other Obſervations; as alſo the ſeverall proceedings, Diſcoveries, Accidents, and remarkable things of the Captaines, Maſters, Pilots, and others formerly imployed for the ſearch of the *Northweſt* Paſſage to the *Eaſt India*.

The History of Sir Martyn Frobrisher's *Voyage*, 1567.[1]

WAs 15 yeares in noting and bringing up the Adventure, before hee did attempt the same, which was brought to passe by the helpe of the right honourable *Ambrose Dudley*, Earle of *Warwick*. The

[1] The date of Sir Martin Frobisher's first voyage is wrongly given by Foxe, who prints it 1567, instead of 1576. The following narrative by Foxe is abstracted from that given by Hakluyt. For criticisms upon, and further information about it, *vide* Admiral Collinson's *Three Voyages of Martin Frobisher* (London, Hakluyt Society, 1867), and Rundall's *Voyages* (pp. 7-15).—C.

Expedition was prosecuted in two small Barques of 25 tonnes the piece, *viz:* The *Gabriel*, and the *Michael*, and one Pinnace of 10 Tonnes.

He departed from *Blackwall*, Iune 15; went by the North. The 15 of Iuly he had sight of a ragged land, he tooke to be *Freezeland*[1]; durst not approach the same for Ice and Fogge, and thereabouts he lost the sight of his small Pinnace by storme, which he thought the Sea had devoured, wherein were onely foure Men.

The *Michael*, misliking the matter, slipt back for *England*, with report that the Captaine was lost at Sea.

The worthy Captaine, notwithstanding he had sprung his Mainmast in the storme, proceeded on W. Northwards, and [on] the 20th, he had sight of a high land he named Queene *Elizabeth's Foreland*, and, sailing more Northerly along the Coast, descried another Forland, with a Gut, Bay or passage Westward. He met great store of Ice along the Coast; was crost with winds, that he could not thwart those streights. In few dayes, he discernes the Ice to be well consumed, either ingulfed therein by Indrafts, or else set so[2] Southwards by Currents. He enters the 21 [of July]; sayles Westwards 50 leagues, with land on both sides; imagines the one the maine Continent of *Asia*, the other, on larboard, the firme land of *America*. He named the Straight *Frobrisher's straight*.[3] After he had sailed 60 leagues up,

[1] This was obviously not the Freezland of the Zeni, but the southern part of Greenland.—C.

[2] This is, of course, a misprint for *to*.—C.

[3] It is necessary to bear in mind that, until long afterwards, Frobisher's Strait (which is really identical with what is now known as Lumley's Inlet, on the west side of Davis' Strait) was supposed to run *through Greenland*. It is so shown on the maps of both Foxe and James, and on many other maps, even of a much later date. Frobisher doubtless had with him a copy of the Zeno map; and, having encountered what he erroneously thought to be Freezland of that map (which must really have been Cape Farewell), would naturally think, when he

he went on land; found signes where fire had been made, and Deere, [al]so mankinde, as he was faine to defend himselfe against them. The people [who] resort to him in their Canoes of Leather, had like to have stolne his Boate from him before he was aware. They came on Shipboard; brought him Salmon, Flesh, and Fish. They appeared to be nimble of their joynts and strong. They fall to trade for Fish, Seale, coates of Seale skinnes, and Beares skinnes, for bels, looking-glasses, and other toyes. They intercept his boate with 5 men; ne'er till this day heard of. The Captaine by the ringing of a bell intices one of the *Salvages* to him, as though he would give it him; he lets the bell fall into the Sea as the *Salvage* should have tooke it; he takes him by the hand and pulls him with his boate into the Ship. Whereupon, in despight, the *Salvage* bit his tongue in twaine; yet he lived till he came into *England*, and then dyed of cold he had taken at Sea. With this prize, he returnes for *England;* arriveth in *Harwich* the 2 of October.

He commandeth his Company, at their landing in the foresaid Countrey, to bring away something with them of the first they could lay their hand upon, in token of possession in the right of Queene *Elizabeth*. Whereupon, some brought Stone, some flowers, some grasse; one brought a peece of stone much like to Sea-coale[1] in colour, but by weight seemed to be Mettal or Mynerall; but as yet not esteemed of, but from the place from whence it came.

Of this bright Stone, he gives a piece (as of something brought from that farre Country) to one of his Adven-

---

sailed on and encountered more land to the westward, that he had come to the east side of Greenland. Hence this unfortunate and long-maintained geographical blunder.—C.

[1] An old name for coal (formerly spelled *cole*), which was so called because it was usually brought by sea.—C.

turer's wives, who, throwing the same into the fyre to burne, and after it was quenched with vineger, it glistered with a bright Marquesset[1] of Gold ; whereupon, the matter being called into some question, it was tryed by certaine Goldsmithes in *London;* who, upon essay made, gave out that it held Gold, and that very richly for the quantity ; and the said Goldsmithes promised great matters if any quantity thereof could be had. Which hopes produced a second Voyage.[2]

This [first] Voyage they gave names to Sounds, Ilands, Bayes, Capes, Streights, &c, as *Elizabeth Forland*, Cape *Labradore*, *Gabriel's* Iland, *Prior's* Sound, *Thomas Williams'* Iland, *Bourcher's* Iland : *Frobrisher's Streights*. He describes the Country people to be like the *Tartars*, with long blacke hayre, broad Faces, flat Noses, tawnie coloured, [and] wearing Garments of Seales skinnes. He arrived in *Harwich* the 2 of *October*.

Navigation of Christopher Hall, *Master, with* Frobrisher, *the first Voyage.*[3]

SEcretary *Woollye* was sent from her Maiestie to give charge to the Company of the ship to obey their Captaine, and be diligent in all things. He[4] observes at *Gravesend*, and found the Latitude 51 deg. 23 min. *Variation* 11 deg. ½.

June 24 he had sight of *Faires Ile*. The 25, he had

[1] Ash's *Dictionary* gives Marquesite or Marcasite as "a compound inflammable metallic body, frequently found in mines."—C.

[2] In Rundall's *Voyages* (p. 14) will be found a very different account of the experiments tried upon this stone, taken from a MS. by Michael Lok (*Cottonian MSS.*, vol. viii, No. 41).—C.

[3] Hall sailed with Frobisher as master of the *Gabriel* on the first voyage, and of the *Ayde* on the second voyage. Hall's narrative here following is abstracted by Foxe from the account given in *Hakluyt* (1589 ed., pp. 615-622).—C.

[4] Hall, not Woolley, is evidently here referred to.—C.

sight of the *Swinborne* in *Shotland;* the S. most Cape beares N.N.W., *Fairely* at the same time W.S.W.; sailing from these bearings with *Swinborne,* he had depth 60, 50, 40 Fathams. Lat. 59. 46. *Fairely* bearing W.N.W. 6 leagues off him, he had depth 59 and 46 fathom, and went into *Tromius* Sound in *Shotland* to stop a leake. 14 Leagues West South-wards from *Faire* Ile he had variation W. 11 deg. 9 min.

Iuly 11. He had sight of *Friesland,* sharpe pinacled High-Land, and covered with snow, bearing W.N.W. The Captaine attempted to get on Land, but could not for great store of ice. And had no ground at 120 Fatham.[1] Sailing from thence 20 leag. S.W., observing the Sunne in the Meridian 52 deg.; iudges the variation W. 2 points and a halfe.

The 28 was foggie, but at the breaking vp thereof, he had sight of Land, supposed *Labrador.* Found great store of Ice about the Land, but he had no ground at 100 Fath. a Cables length off shore. And a great Current sets S.W. and N.E., 1 League ½ in the houre. The tide sets to the shore; and then, one great peece of Ice breaking, made a noyse as if the Cliffe had fallen into the Sea.

August 10. The floud sets S.W. along the Land, and it flowes so.

11. Latitude here 63 ⅛. This day they enter the Streights, and set saile for *Gabriel's* Iland, then distant 10 Leagues.[2]

13. They enter within a Sound in a Sandy Bay. The Land beares E.S.E.; depth 8 Fatham. A S.E. Moone, full Sea; they name this Sound *Prior's* Sound, distant from *Gabriel's* 10 leag.

16. Calme and faire. In 2 houres the Ice was froze about the Ship a quarter of an inch thicke.

---

[1] Lat. 61.—F.

[2] A mistake.—F.

The 19, the Captaine and he went on shore vpon an Iland, with 8 men; and, from the top thereof, they had sight of 7 Boates which came rowing from the East side to the Iland. Then they returned on shipboard and sent their Boate with 5 men to see which way they tooke, and so with a white Cloth or waffe,[1] brought one of their Boates with their men in her along the shore, rowing after the Boate vntill they see the ship, and then they rowed on shore, and he followed and gave every one of them a threed Point[2] and brought one of them aboard, where he did eate, and drinke, and then carried him ashore againe. Whereupon the rest, being 19 Persons, came on boord, but he could not vnderstand their Language. They be like *Tartars*, with long blacke haire, broad faced, flat nosed, and tawny Coloured, wearing Seale skinnes, and so doe the women, nothing differing; but the women in the Face hath blue stroakes downe the Cheekes and about the eyes. Their Boates are made of Seale skinnes, with a wooden keele within them, much like vnto a *Spanish* shalop, save onely they be flat-bottomed and sharpe ended.

20. They went on Land vpon the East side of the Iland with 4 men more in the Boate, where they see their houses. The people came to them calling (and rowing). One of them came into their Boate; they carried him on board & gave him a Bell & a knife. The Captaine commanded 5 men to set him on shore in an Iland, & not amongst their Company, but they, not regarding, went to them, who surprised their Boat and themselves, never as yet heard of.

21. The next day, they shot off a Falken-gun,[3] and

---

[1] This word does not occur in Hakluyt. I cannot explain its origin.—C.

[2] Presumably a *needle* (thread-point). In Hakluyt the word is *thredden-point.*—C.

[3] A small cannon (now obsolete), having a length of about 7 feet, a diameter of bore of 3 inches, and throwing a ball of nearly 3 lbs.

sounded a trumpet to heare from their men, but were not answered. This morning the Snow was one Foote thicke vpon the hatches.

22. They went to the place where their men were lost, and had sight of 14 Boates, and some came neare them, but they could heare nothing of their men.

26. Returnes homeward. At 8 a clocke in the night was thwart of *Gabriel's* Ile, and had Cape *Labradore*, as he supposed, W. 10 Leagues off.

September 1. Had sight of *Friesland*, 8 leag. off. From this day to the 6, they run along *Iseland*.

25. Sight of *Orkney*.

October 1. Anchor at *Yarmouth*.

Sir Martin Frobrisher *his* 2 *Voyage*, 1577.[1]

DEparted from *Blackwall* the 26 of *May*, with 3 ships; to wit, the *Aide* of the Queenes [Service], burthen 180 Tonnes; the *Michaell;* and the *Gabriell*, accompanied with 140 Gentlemen, Soldiers, and Saylers, victualed for halfe a yeare.

He went by the North. The 7 of *June*, arrived at *Orkney*, in the Iles of *Scotland*.

8. . He departs from thence; sailes betwixt W. and N.W. vntill the 4 of *July*, 26 dayes saile; from thence they meete with much Drift-wood, as they suppose from *New-found-land*, and [were] driven over with the Current which, they say, sets from the W. to the East.

The 4 of *Iuly*, they had sight of *Friesland*, 10 or 12 leagues off, and great store of Ice, 30 or 40 fatham aboue water, they supposed [fast] on ground, though they could

weight, with a point-blank range of 130 paces, and a random one of 1500.—C.

[1] The narrative of this voyage, and the account of the Esquimaux are both taken from *Hakluyt* (1589 ed., pp. 622-630).—C.

scarce sound the bottome for depth. The Generall attempteth to goe on Land but cannot. They coast it 4 dayes; sees no signe of habitation; Yet, by Birds which in Fogs had lost the land & came to the Ships, they suppose the Country to be more habitable within, then outward shoare maketh shew or signification.

The 8, they depart from thence.

16. He comes to the making of the Land, named the yeare before by him, the Queenes *Forland*, being, as they judge it, an Iland lying neere the supposed continent of *America*. Another Iland lying upon the *Asian* side, called *Hall's* Iland; betwixt which two Ilands goeth in *Frobisher's straights*, or the unknowne passage into the Sea of *Sur*. He doth suppose that the Ice of this Coast is carried by some contrary E. or W. tyde or current upon the Coast of *Freezland;* causing that Country to be farre more intemperate than other countries farre more North.

At their first entrance, they found the Straight mured with Ice. The Captaine, with his Pinnace, past twice through, before he durst hazzard in the great Ships.

They goe on land. The people seemed to be joyfull thereof; they embrace, and the Captaine laid hands on them; but they escape through nimblenesse, and defends themselves with their bowes and Arrowes. He tooke one; all the rest escaped.

They put their Ships into the Straights all full of Ice; they made 14 bourds in one watch, to refraine the Ice. The lightnesse of the night did them much comfort and helpe for sight, and this hazard they made for safegard of their Captaine and Master, who were on land.

17. Being the day following, the Capt. came on board with report of great riches hid in the bowels of that Continent.

Within 3 or 4 dayes after they had been in the Streights, the W. and N.W. winds disperst the Ice. The 19, they enter without impediment. The 20 they found a good harbour, and names it *Iackman's* Sound,[1] anchoring the Ships there.

The Generall marches up into the land; takes possession in the Queenes name; and imployes his men about the businesse they came thither for.

Whilest they continued in this harbour, they kept watch continually with boates and roapes, ready to hale and towe away the Ice, which otherwise might have driven thwart the Ships, with Ebbe and flood.

The Generall findes not commodity answerable to his Expectation in the supposed *America*; leaves the Ships; coasts on to the supposed *Asia*. The stones on land and Sand in Sea, sparkle like Gold on both sides (if all be Gold that glysters). Upon the west shore, they found a dead Fish floating; it proved by the horne to be a Sea *Vnicorne*[2]; the Spiders put therein dyed.

The Generall, in further search, findes Gold oare, as he supposed, with a good harbour, and returnes to the Ships. By the way, he espies a Tent covered with Seale skins; the people was fled. He leaves glasses, bels, and knives therein; onely tooke one Dog and nothing else; leaves a letter, with pen, inke and paper, for his men (which was tooke from him the last yeare) to write if they were living.

Their men march up againe, and found their Tents were

---

1 *Iackman*, Mr's. Mate.—F.

2 The Narwhal (*Monodon monoceros*), or "Sea Unicorn", one of the strangest forms among the Cetacea, is a native of the Arctic Seas, ranging up to very high latitudes. It has, however, on several occasions, occurred on the British coasts. Its most remarkable feature is its twisted tusk, or horn, which is of dense white ivory, and is generally from six to eight feet in length, but occasionally more. In the Middle Ages, it was supposed to be the horn of the Unicorn of fable, and many most extraordinary properties were supposed to belong to it. Among other things, it was thought to be an antidote to all poison. This is, no doubt, what is indicated by the mention of spiders.—C.

remooved to the foreside of the Bay. They fled by Sea. Our men incompasse them both by land and Sea. They defend themselvs with bowes & Arrowes, and fiercly assault our men. We wounded 3 with our Arrowes, who perceiving themselves hurt desperately, lept from the Rockes into the sea and drowned themselves. The rest escaped by flying into the Mountains. Onely 2 women stayed; the one for Age; the other having had a Child, was brought on Shipboard, being shot through the hayre of her head, but her child was shot through the arme, which the Chirurgion endeavouring to cure, she pluckt off the applyed salues, and, like a Dog, with licking, healed vp the Child's arme.

The man Salvage formerly taken and she [being] brought together every man with silence desired to behold the manner of their meeting, the which was more worth the beholding then can well be expressed. At their first sight, they beheld each the other very wist[ful]ly a good space, without speech or word vttered, with great change of Colour and Countenance, as though it seemed the griefe of their Captivitie had taken away the vse of their tongues. The Woman at first, very soddainly, as though she disdained or regarded not the man, turned away and began to sing, as though she minded another matter; but, being againe brought together, the Man broke vp the silence first, and, with a sterne and stayed countenance, began to tell a long solemne tale to the woman, wherevnto she gave good hearing, and interrupted him nothing till he had finished. And afterwards being growne into more familiar acquaintance by speech, they were turned together, so that I thinke the one would hardly have lived without teh comforts of the other; and, for so much as we could perceiue, albeit they lived continually together, yet they did never vse as man and wife, though the woman spared not to doe all necessary things that appertaine to a good huswife, indefferently for them both ; as in making cleane their Cabine, and in every other thing appertaining to his ease. For, when he was sicke, she would make him cleane, and kill and flea the dogs for their eating, and dresse his meate. Onely I thinke it worth the noting, the continencie of them both, for the man would never shift himselfe, except he had first caused the woman to depart out of his Cabine; and they both were most shamefast, least any of their privie parts should be discovered, either of themselves or of any other.

This Bay was named *Yorke Sounde:* The point *Bloudy Point.*

Our men returned to their Tents (seeing there was no hope to bring them to Civility) & made spoile of thē; wherein they found an old shirt, a doublet, a Girdle, and shooes of the men lost the last yeare.

This done, they returne againe to their Ship. And the third day they depart from this supposed *America.* The fourth, they came to the Generall in a faire harbour on the East side, which they named the Countesse of *Warwick's* Sound. In this place, they fully intend to lade with the supposed gold Minerall, to counteruaile the charge of their first and this second Voyage.

The Inhabitants come to them againe, and make shew that 3 of the 5 men are alive, making signes for penne and inke, and that within three or foure dayes they would returne, and bring those that were living.

Sir *Martin Frobrisher*, his Letter to the *English* Captaines,[1] taken the last yeare in *Meta Incognita.*

IN the Name of GOD, in whom wee all beleeve, who I trust hath preserved your bodies and Soules amongst those Infidels, I commend me unto you ; I will be glad to seeke, by all meanes you can

[1] English *captives.* This is clearly a literal misprint.—C.

devise, for your deliverance, either with force or with any Commodities within my ships, which I will spare for your sakes, or any thing else I can doe for you. I have on board of theirs a man, a woman, and child, which I am contented to deliver for you ; but the man of theirs which I carried away the last yeare is dead in *England*. Moreover, you may declare unto them that, if they deliver you not, I will not leave a man alive in their Countrey. And thus, if one of you can come to speake with me, they shall have either the man, woman, or child, in pawne for you ; and thus unto God, whom I trust you doe serve, In hast I leave you to him. We will daily pray for you. This Tuesday morning the 7 of August.

Yours to the utmost of my poore [endeavours],

MARTIN FROBRISHER.

[P.S.]—I have sent you, by these bearers, Pen, Inke, and Paper to write back againe, if personally you can [not] come to satisfie me of their estate.

Here their Captive [Esquimaux], being on shore, set up 5 small stickes in a Circle, one by another, with a small bone placed in the middest. They conceited thereby he would give his Countrimen to understand that, for 5 men betrayed the last yeare, he was taken prisoner, which hee signified by the bone in the middest. For, afterwards, wee showed him the Picture of his Countryman which the last yeare was brought into *England*, whose counterfeit was drawne with his Boate and furniture, both as he was in his owne, and also [in] *English*, apparell. Hee was upon the sodaine much amazed thereat, and, beholding advisedly the same with silence a good while, as though he would straine curtesie, whether should begin the speech (for hee thought him, no doubt, a living creature), at length began to question with him, as with his Companion, and finding him dumbe and mute, seemed to suspect him as one disdainfull, and would with a little helpe have growne into choller at the matter, untill at last by feeling and handling, he found him but a deceived Picture, and than with great noyse and cryes ceased not to wonder, thinking that we could make men live or dye at our pleasure.

They also make signes they have a King carried on men's shoulders, a man farre surpassing any of ours in bignes and stature. It may be thought they are vsed to traffique with some other Nation; yet, for all this faire dealing, they lay lurking to betray our men, with divers signes and raw flesh, which we got, and it served for meate for the man and woman, whose stomackes as yet could not digest the Shippes victuals. One amongst them counterfeited himselfe lame, who, being shot at with a Calliver[1] to affright him, he presently tooke to his legges and runne away. But, in all this time, they cannot heare of their men.

Their weapons are Bowes, Arrowes, slings, and darts. They have 2 sorts of Boats; the one is for one man and close deckt, shaped like a Weaver's shuttle; the other open, and will carry 14 men, more or less, and planckt with Seale skinnes.

It is thought that their habitation in Winter is farre within the land, and that they abide here in Summer onely to live upon Fish.

The 24 of *August*, after they had satisfied their minds and laden their Shippes, they depart; falls with the Land's end of *England*, and puts into *Milford* haven in the Admirall.[2]

---

[1] An old kind of hand-gun or arquebus, which was chiefly used at sea.—C.

[2] Hakluyt says they put into Milford Haven, "from whence our generall rode to the Court for order to what porte or haven to conduct the ship." In those days, the chief commander of a fleet was "called the general, and his ship was the admiral; and the second in command was the lieutenant-general, sailing in the vice-admiral. The captain conducted warlike operations, and the master was responsible for the navigation and safety of the ship, and for the merchandise; but frequently the two offices were united" (Mr. C. R. Markham in *The Voyages of William Baffin*, p. xxxv).—C.

The 2 barkes were separated at Sea by storme ; the one arrived at *Bristow*: the other came about *Scotland*, and arrived safely at *Yarmouth*. In this voyage, they lost two men. Queene *Elizabeth* named the land *Meta incognita*.

Things Remarkeable [concerning the Esquimaux *observed] in this Second Voyage.*

THey are men of large Corporature, good proportion, and of colour not much unlike the Sunne-burnt Country-man. They weare their haire something long, cut disorderly before ; their women weare their haire long, knit up with two loopes ; some of them Races[1] their cheekes, chins, and faces, whereupon they lye a colour like darke Azure.

They eate all their meate Raw, as flesh, fish, fowle, or but lightly perboyled ; with water and bloud, they will eate Ice, as we doe Suger.

For necessity, they will eate grasse like bruit beasts, without table or stoole; and, when their hands are imbrued in blood, they licke them cleane with their tongues.

They yoke their Dogs for use, as wee doe Oxen ; and, when they grow old, they fatten them to eate; and, if they dye, they doe the like.

They apparell themselves in skins of Beasts & Fowle, sowed together with the sinewes thereof, to defend them from cold.

They make their apparell with hoods and tailes, which tailes they bestow as favours to ratifie any friendship showen them. The men's garments are not so long as the women's.

They weare their hose close to their legs, from the wast to the knee, without any opening before, as well the one kind as the other. Upon their legs they weare hose of

[1] To race = to *cut* or *scratch*: an obsolete word.—C.

leather, with the fur-side inward, 3 paire at once, especially the women. In these hose, they put their necessaries they carry about them; they put also a bone into their hose from the foote to the knee, whereon their hose being drawne are held up in place of garters.

They dresse theyr skins very soft and supple with haire on. In winter, they weare the skin-side inwards; in Sommer, outward; other apparell they have none.

Their beasts, fishes, and fowles are all their meate, drinke, apparell, houses, bedding, and all their Riches.

Their houses are tents, covered with skins, pitched with quarters foure square meeting at tops, sewed together with sinewes; their entrance is alwayes o'er against the Sunne.

They have other sorts of houses, which they[1] found not to be inhabited, raised with stones and Whale-bones, and covered with skins, the entering not unlike unto an oven's mouth, which they frequent during their hunting time, and so leaves them untill they come thither againe.

They have 2 sorts of Boates; the one only for one man, the other for 16 men. They kill all their fish and flesh with darts, and it may be thought they repaire to some other place to live in winter, farther from the Sea side. They will eate theyr flesh and fish, smell it never so filthily. What knowledge they haue of God, or what Idoll they adore, is unknowne. They make signes of people that weare bright plates of Gold in their foreheads. The Country is high and mountainous, [with a] great quantity of snow lying thereon; there is little plaine ground and no grasse, but that which is like unto mosse, growing upon soft ground, such as we get turffe upon. There is no wood at all. There is great quantity of Deere, their skinnes like unto Asses; their heads or hornes doe farre exceed any, both in length and breadth, of those Countryes. Their feete are great as

[1] The explorers are evidently meant.—C.

Oxen, which, measured, were 7 or 8 inches in breadth. Their are also Hares, Wolves, fishing-Beares, and sundry sorts of Sea fowle.

The 3 *Voyage of Sir* Martin Frobrisher [1578].[1]

(*This Voyage the Queene gave him a Chaine of Gold.*)

Articles to be observed in his Fleete.

1. *Imprimis*, to banish Swearing, Dice, and Card-playing, and filthy Communication, and to serve GOD twice a day with the ordinary service vsed in the Church of *England*, and to cleare the glasse[2] according to the old order of *England*.

2. The Admirall[3] shall carry the light; and, after his light be once put out, no man to goe on head of him, but every man to fill his sailes to follow as neare as they may, without indangering one another.

3. That no man shall, by day or night, depart further from the Admirall than the distance of one English mile, and [the vessels shall keep] as neare as they may, without indangering one another.

4. If it chance to grow thicke, and the wind contrary, either by day or night, that the Admirall be forced to cast about, before her casting about, she shall give warning by shooting off a piece [of ordnance], and to her shall answer the Vice-admirall, and the Reare-admirall, each of them with a Peece, if it be by night or in a fogge, and that the Vice-admirall shall answer first, and the Reare-admirall last.

---

[1] The account given by Foxe of this voyage is condensed from the somewhat discordant narratives of it given by Thomas Ellis and Captain Best, as printed in *Hakluyt* (1589 ed., pp. 630-635). Like the two previous voyages, it was a total failure, so far as its main object was concerned. This main object was the colonisation of the land known as Meta Incognita, 120 persons being selected as colonists, and three ships being intended for their use whilst there. The expedition which consisted of no less than *fifteen vessels*, was the largest which has ever yet set forth upon an Arctic searching expedition.—C.

[2] This relates to the old custom of regulating the length of the service or the sermon by the time occupied by the sand in clearing the hour-glass. Hour-glasses were formerly to be seen in most, if not all churches, and they still remain in some country places.—C.

[3] That is, the ship of the "General" or Commander of the expedition. See note to p. 49.—C.

5. That no man in the flight,[1] discrying any saile or sailes, give, vpon any occasion, any Chase before he have spoken with the Admirall.

6. That, every night, all the Fleete come vp and speake with the Admirall at 7 of the clocke, or betwixt that and eight ; and, if the weather will not serve them all to speake with the Admirall, then some shall come to the Vice-admirall, and receiue the order of their course from Mr. *Hall*, chiefe pylot of the flight, as he shall direct them.

7. If, to any man in the flight, there happeneth any mischance, they shall presently shoote off 2 peeces by day, and by night 2 peeces and show 2 lights.

8. If any man in the night come vp and hayle his fellow, knowing him not, he shall give him his watch word, *Before the world was God;* the other shall answer him, if he be one of our flight, *After God came Christ his Sonne;* so that, if any be found amongst vs not of our company, he that first discryeth such saile or sailes, shall giue warning to the Admirall by him-selfe, or other that he can speake vnto that sayles better then he, being nearer vnto him.

9. That every ship in the fleete, in the time of Fogs, which continually happen with little winds and most calmes, shall keepe a reasonable noise with Drum and Trumpet, or otherwise, to keepe themselves cleere one of the other.

10. If it fall so thicke and misty that we lay to hull, the Admirall shall give warning with a peece and putting out 3 lights one over the other, to the end that every man may take in his sailes, and at his setting of sailes againe doe the like if it be not cleare.

11. If any man discover Land by night, that he give like warning that he doth for mischances, 2 lights and 2 peeces ; if it be by day one peece and put out his flag, and strike all the sailes he hath abroad.

12. If any ship shall happen to loose company by force of weather, then any such ship or ships shall get him into the Latitude of ——[2] and so keepe that Latitude vntill they get *Frisland*, they shall get them into the Latit. of ——[2] and to the North of ——[2]; and, being once entred into the straights, all such ship or ships shall every watch shoote off a good peece & look out well for smoake and fire, which those that get in first shall make every night, vntill all the fleete be come together.

13. That vpon the sight of an Ensigne in the Mast of the Admirall, a peece being shot off, the halfe fleete shall repaire to the

[1] Presumably an old form of *fleet*.—C.

[2] These appear to have been left blank in the original.—C.

Admirall, to vnderstand such conference as the Generall is to have with them.

14. If any chance to meete with any Enemie, that 4 ships shall attend vpon the Admirall, *viz.* The *Francis* of *Foy*,[1] the *Moone*, the Barke *Dennis*, and the *Gabriell;* and 4 upon my Lievtenant Generall in the *Judeth*, *viz:* The *Hopewell*, the *Armenall*, the *Beare*, and the *Salamander;* & the other 4 vpon the Vice-admirall, the *Anne Francis*, the *Thomas* of *Ipswich*, the *Emanuell*, and the *Michaell*.

15. If there happen any disordered person in the flight, that he be taken and kept in safe custody vntill he may conveniently be brought aboard the Admirall, there to receive such punishment as his or their offence shall deserve.

He departed from *Harwich* the 31 of *May*, 1578, with 15 Shippes, having given that Instructions for ordering of his fleete, as well for civill government as direction.[2]

He goes by the West. The 6 of *Iune*, he had sight of *Cape Cleere*, sayling towards the N.W. parts from *Ireland*.

A great Current from S.W. carried them 1 point to the N.E. of their course, which current seemed to him to continue its course towards *Norway*.

The 20 of *Iune*, the Generall discries land, and being *Frezeland*, names it West *England*. He goes on shore, and findes a good harbour for Shippes. The people fled. They Judge it part of *Meta incognita*, or *Groneland;* their boats and apparell are all alike with those of *Meta incognita*. They found in their Tents a box of small Nayles, red herrings, and divers carved things, so as they iudge them to be civill people and Artificers, or else to have trade with those that are.

23. They depart from thence and names a certain cliffe, for some resemblance, *Charing crosse*. They meete with much yce, many Whales and fogges.

The *Salamander*, a ship of his Flight, strucke upon a

---

[1] Probably *Fowey*, in Cornwall.—C.

[2] 120 Persons appointed to inhabite *Meta incognita:* 3 Ships was appointed to be left for yᵉ use of the Inhabitants.—F.

Whale with full stem, being under courses and Bonnets[1]; he makes an uglie noyse, and two dayes after they found a dead Whale, supposes it the same.

July the 2, they had sight of the Queenes *Forland;* they beare in all day; had much Ice. At night they were entred the Streight, all overcome with Ice; not froze there, but driven by the windes violence. The writer conceives the *Mary glaciale* to be a conjecture, and that no salt Sea can be froze o're with Ice, and in these places where it doth ebbe and flow above 10 Fathames.

And the Ice he met at Sea, 1000 Miles from land, all which congealed upon fresh water; he concludes they were froze in bayes and rivers, and not in the Sea.

This Ice doth shut together with winde, and open at the shifting or change thereof, as in other places, so as it is passable.

The Barke *Dennis* struck vpon a Rocke; the ship sunke, the men were saved by their boates; therein were much of the house drowned, which the men appointed to winter in *Meta incognita* should have lived in. Amidst the Ice, a storme takes them at S.E.; some takes in sailes and hulls that had roome; others make fast to the Ice; others fend off Ice with poales, oares, junkes, oken boards, and the like; so as all were put to it. Yet, for all this, their ship-sides and waeles[2] were sore torne and bruised.

The next day, the winde changed W.N.W.; the Ice dispersed. They go to Sea and meetes 4 more of their Company, who had all kept the Sea during the S.E. storme. They resolue to keepe the Sea vntill the Sunne dissolue, or the wind dispierse, the Ice out of the Straight.

---

[1] The mainsail, foresail, and mizzen were spoken of as courses. Bonnets were small supplementary sails, about one-third the size of the courses, to which they were affixed.—C.

[2] The wales, or gun-wales, are the stout planks or strakes of timber running round the upper part of a ship's sides.—C.

7. They cast about inward againe, had sight [of land]. Of what Land it should be, there was difference of opinions. Through the thicke mists, and by snow newly fallen, the habit of the Land was altered. Thinking they had bin to the N.E. of *Frobrisher's Straights;* then comming from Land by a Current comming from thence along the Coast, they were carried to S.W., off the Queen's *Forland*, more miles then they thought possible.

Here they make a peece of Land for *Mount Warwicke*, yet wonders how they should be so farre shot within the Straight without their knowledge. How be it, they confessed they found a swifter course of flood then hitherto they had observed. And here their Ships were whirled about in a moment, lying a Hull, as though they had beene in a Whirlepoole, the waters making no lesse noyse, to be heard a farre off, then the waterfall of *London* Bridge.[1]

Here they could have no observation. The Generall sends to the Shippes to know their chiefest opinions. *Christopher Hall*, chiefe Pylate, saide he had never seene that coast before, nor could not make it for any part of *Frobrisher's Streights*, although the land did lye and trent alike.

July the 10, the weather continued thicke and darke, and the fleete disperced. They were doubtfull whether to set to Sea, or to follow a doubtfull course in a Sea, Bay, or Straight they knew not, or stand along an unknowne coast where they could not discerne dangers for darke mists and thick weather.

Whereupon, some betooke themselves to Sea, as thinking that the safest Course. Others followed the Generall within that doubtfull and unknowne Streights above 60 leagues, having alwaies a faire Continent upon the Star-

---

[1] This alludes to the fact that the water, in running through the narrow arches of old London Bridge, made a great noise, and descended some distance. Davis, on the 29th of July, 1587, encountered the same current, and called it the "Furious Overfall". Many later navigators have also mentioned it.—C.

board and a continuance of an open Sea before them; so as if it had not beene for the gathering of his fleete againe, as also their lading of Ore, he both would and could have gone into the *Mare del Sur;* for the further they sailed, the lesse Ice, and 50 leagues within this Streight was none at all.[1]

This Straight hath also a great Indraft; for, by the foresaid current, the floting wracke of the barke *Dennis*, lost at the Queenes *Forland*, was brought thither along the Coast, and by the Indraft drawne in there, it being many miles distant.

They doe also affirme out of some of their best marriners' observation that, in this Straight, the flood-tyde doth runne 9 houres and ebbe 3, which may well come to passe by force of the current comming from the E. and butting upon that coast, may enforce the tyde into all indrafts and rivers, with longer flowings, untill the force of the ebbe, receiving his strength from the West Sea, doe resist it, the Sea will not acording to the saying:—

Naturam expellas furca: licit usqui recurrit.[2]

Also, they observed vpon this Coast that, lying a Hull, 25 leagues off Land, the wind blowing trade,[3] they were brought to within two leagues thereof, contrary to expectation.[4]

This part of the Country, they hold to be more populous and fruitfull then any discovered before, and better stored

---

[1] This doth argue that this Streight doth goe through *Groenland* into *Fretum Davis.*—F. [As has already been explained (p. 38), it does nothing of the kind.—C.]

[2] This proverb, taken from Horace, may be freely translated: "Natural forces cannot be changed."—C.

[3] The "trade wind" blows N.E.—C.

[4] The 15 leag. was to the E. from Land; and this hulling must be neere some indraft. If so, I am perswaded the S. part of *Groenland* is all Ilands.—F.

with grasse, fowle, and wild Beasts. And heere they see greater Boates then before, and of the contents of 20 persons. And they thinke that, being 60 leag. vp the foresaid Straight, they saw Land on Larboard.[1]

To returne backe againe the same way out of this Streight, along the supposed backeside of the Continent of *America*: at the Queenes *Forland*, he espies a Gut to goe through in *Frobrisher's Straights*; sends the *Gabriell* through, who meetes againe in the Streight, so the Queenes *Forland* proved an Iland.

They anchor in the Streight, at a Land they named *Hatton's Head-land*, where they met 7 of their ships, and staies for the rest. The 26, they had a cruell storme of winde and snow, which disperced their Fleete, and were most cruelly weather-beaten.

The 2 of *August*, all the Fleete arrived except 4, and harbours neere Mount *Oxford*. The 6 day, they got vp as high as *Leicester* Point. Then they hold a Consultation for inhabiting, but doth not. One ship they had lost then, as they thought, but she came home without doing anything. The rest searches for Mynes, and findes one; cals it *Best's Blessing*,[2] after his owne name that found it; but, in bringing their ship thereto, she grounded vpon a Rocke &

---

[1] There can be no doubt, from the description given, that the strait up which Frobisher had thus sailed for sixty leagues in 1578 was Hudson's Strait, and that he therefore preceded Hudson's "discovery" of it by over thirty years. If he had sailed along the northern shore of the strait for the distance named he probably arrived somewhere near the Upper Savage Isles, and it is possible that on his way there his men may have seen the opposite or southern shore of the strait. It is not to be greatly wondered at that Frobisher did not follow up this discovery, important though it must have seemed to him, for it is very clear that his voyage was undertaken much more with a view of extending commerce than geographical discovery pure and simple. It is clear, too, that he had, as he says, to think of the safety of the rest of his scattered fleet.—C.

[2] Captain Best was in command of the *Anne Francis*, one of Frobisher's ships. He wrote one of the existing accounts of the expedition.—C.

halfe dryed, so as they were forced to vnderprop her with their mayne-yard, and thereby escaped the danger which they might otherwise have fallen into.

Now, the Fleete being all laden and ready, they furnish vp a little house with Bells, Babies, Pictures of men and women, Glasses, Whissles, Pipes, with an oven with baked bread, & left it to the Natives. And, vpon the last of *August*, the whole Fleete was ready to depart, but a cruell storme tooke them: some at Sea, some at anchor in Sounds. The *Busse* was faine to seek a new way to Sea through a Sound; she rid in vpon the backside of *Beare* sound, and got to Sea to the N. of *Frobrisher's Straights*; but the Generall came home in the *Gabriell*, and could not get aboard his owne ship the *Aide*.

The *Busse*, comming home, found an Iland in 57^{d} and a halfe[1]; sailed a long [it] 3 dayes, and saith it is a fruitfull Champion country, and wooddy.[2]

The *Busse* fell on the South of *Freezeland*, the 8 of September. They steered from thence S.E. and by S. untill the 12, when they discryed land 5 leagues off. The S.W. part bore S.E. by E.; the Northermost, N.N.E. or N.E.; the Master accounted the S.E. point of *Freezland* was then from him N.W. by N. 50 leag.; he accounts this Iland to be 25 leagues long, S.E. and N.W.; the S. side is in 57½. They had sight of it 28 houres, they opened 2 harbours therein. The Master did account himselfe 50 leagues S.E. by S. from *Freezeland*, when he first discovered this land.

There dyed about 40 persons upon this voyage.

---

[1] This island has no existence. Its alleged discovery led to much confusion, as it was marked in the position indicated on several early maps. There was no vessel named the *Busse* in Frobisher's fleet but the *Emanuel* of Bridgewater was a "buss", that is to say, a small, strongly-built, two-masted, Dutch fishing-boat, generally of from fifty to sixty tons burden.—C.

[2] If this Iland were found againe, there is great store of Fish about it.—F.

It is to be observed in these Voyages,[1] that these are but Histories, and that they did not know whether they saw *Asia* and *America* or no, as I am sure they did not, nor [do I] know in what land they were; yet for certaine they were at *Meta Incognita*, otherwise *Groneland.* I thinke they meant to have kept this golden Country to themselves; for the courses, distance, latitude, longitude, variation, and other marine observations herein is none, only one, Latitude of 63. 8 min. [for] the entrance of *Frobisher's* Straights, and *Freezeland* they have placed in 61 deg., while this new Iland the *Busse* [discovered] is in 57° and a halfe, Northerne Latitude.

There is Beares, Hares, Foxes, and innumerable of Sea-Fowle, whereof his men kild in one day 15 hundred. He found of *Ginnie* beanes[2] in their tents, of colour Red. The Inhabitants are good markemen with their darts; for the most part, they will strike a Ducke in the eye, but altogether in the head.

The first Voyage of Captaine Iohn Davis *of* Sandrige *in* Devonshire, 1585, *to the North-West.*[3]

HEe departed from *Dar[t]mouth* the 7 day of *Iune*, with 2 Barques, *viz.* the *Sunshine*, of 50 tonnes, 23 persons [including four musicians]; and the *Moonshine* of 35 tonnes,

---

[1] The following is not a portion of the original narrative, but a comment by Foxe.—C.

[2] This may be "Guinea-pepper", or *capsicum* (*vide* Ash's *Dictionary*).—C.

[3] Foxe has abridged this narrative from that given by *Hakluyt* (1589 ed., pp. 776-780), which was "written by John Janes, Marchant, seruant to the Worshipful Mr. William Sanderson". The fullest and best information concerning this and the two later Arctic voyages of Captain John Davis may be found in Captain A. H. Markham's *Voyages and Works of John Davis, the Navigator* (London, Hakluyt Society, 1878), and in Mr. Clements R. Markham's *Life of John Davis* (The World's Great Explorers Series, London, 1889).—C.

17 persons. He put into *Falmouth* the 8, and remained there untill the 13; he went by the West.[1]

June 14. He puts into *Silley*, and had contrary windes untill the 28. In his course N. Westward, he see many Whales and Porposes. They kild a Darly-head or Pork-fish, which eate as sweete as any Mutton.

July the 19, they fall into a great whirling or brisling of a tyde setting to Northwards, and they heard a mighty roaring of the Sea, as if it had beene the breach of some Shore. The weather was fogge and mist. They lanch a boate to sound, but findes no ground at 300 fathomes, and found the roaring to be the Sea and Ice beating together.

The 20, they descryed land, the most deformed that ever was seene. It seemed like the forme of Suger loafes, over-topping the Clouds and covered over with snow, the shore belaid with Ice, a league off; he names this Land *Desolation*.

The 21, they perceived themselves imbaid very deepe, and great store of Ice to the N.N.E., W., and S. They cleared themselves by running S.S.W. along the shoare.

The Captaine attempts to land, but could not for Ice. They try for fish, but could get none; the water was black and thick, like to a standing poole.[2] Here were many Seales.

They see woods on land, like to those on *New-found-land*. They had great store of float-wood upon the Coast; they tooke up one Tree 60 foote long and 14 hands about. They bend their course to South with intent to double the Land.

The 23, they coast the land, which did lye E.N.E. and W.S.W.

---

[1] Thought to find it [*i.e.*, the Passage] betwixt Frobrisher's, 62°, & Cabots, 58°, as I suppose.—F.

[2] This water came from land for certaine.—F.

The 24, they coast the Land lying E. and W.; not able to come neere shore for Ice; the weather something thicke and colde; the allowance of victuall was encreased, so as 5 men had every morning ½ pound of bread and one can of beere to breakefast. The weather [was] like *Aprill* in *England;* but, when the winde blew from land or Ice, it was colde; but, when it came off the Sea, it was very hot. They depart from this land; sailes N. Westward above 4 days.

29. They descry land in 63. 15., bearing N.E. The ayre and sea cleare and temperate. He stands with land; espies many faire sounds and harbours and many Inlets into the land. He Iudges this land to be a number of Ilands. He anchors, goes on land, findes where the Inhabitants had beene. He findes also a Shooe, pieces of Leather sowed with seames, and pieces of furre, and wooll like to Beaver.

The Country people come to him. He causes his Musitians to play, wherein they tooke great delight, and fals a dauncing. At night he comes a board; they all depart.

The 30, in the morning, came 30 Canoes by the Ship, cals them to land, and they both make protestation by clapping on their brests and pointing to the Sunne. They become familiar; they will sell their cloaths from their backs, with the buskins, hose, and gloves, made of Seale-skins and bird-skins, the leather well dressed and artificially sowed. They had one paire of Buskins full of wooll like Beaver. They are very tractable, voide of subtilty, and easie to be brought to civility. He thinkes they worship the Sunne.

During their stay here, they found a reasonable quantity of wood, as Furre, Spruce, and Iunupir, which had floated thither. They saw abundance of Seales in Shoales, as it had all beene fish; the cliffes was such as Sir *Martin*

Frobisher brought from *Meta Incognita.* There was divers flowers, Slud or *Muscovia-glasse.* They found an hearbe growing upon the Rockes whose fruite was sweete, full of red Iuyce. The ripe ones were like Currans[1]; they iudge the people to have store of Furres; they make shew, after they see he would have skins and furrs, that they would goe into the Country and fetch such things as they had, but the winde comming faire he came away.

The first of *August,* they proceeded N.N.W. for the discoverie.

The 6, he descried land in 66. 40 min., voyde of Ice. He anchors in a bay neare a faire Mount, the Cliffes thereof as orient as Gold. He names it Mount *Raleigh*[2]; the road, *Totnes*-road; the Sound encompassing the road, *Exeter's* Sound; the N. For-land, *Dyer's* Cape; the S. For-land or cheekes of the Sound, Cape *Walsingham.* He espies 4 white Beares, and kills one first and two afterwards.

The next day they kill a Beare; the fore-paw was 14 inches over. This Coast was Mountainous, without wood or any thing growing thereon. The aire was very temperate.

The 8, he sets saile from Mount *Raleigh,* and coasts along S.S.W.

The 9, his men complained their allowance was too small. It was augmented to 5 men 4 lb. of bread a day, 12 quarts of beere, 6 *New-land* fishes, and on the flesh

---

[1] No doubt some species of the genus *Ribes. R. prostatum* is a native of Newfoundland; and *R. recurvatum,* bearing a black berry resembling a grape, is found on the shores of Hudson's Bay.—C.

[2] "Mount Raleigh of Davis, . . . which is the easternmost on this side of Davis' Strait, is of pyramidal form and exceedingly high; our observation makes it in lat. 66° 37′ N., and long. 61° 14′ W. Cape Walsingham being in lat. 66° 37′ N., and long. 60° 50′ W. (var. 67° W.), is the easternmost land; and consequently the breadth of Davis' Strait, at its narrowest part, is about one hundred and sixty miles." (Ross's *Voyage to Baffin's Bay,* London, 1819, pp. 215-216.)—C.

day, one gill of Pease more, but Butter and Cheese was restrained from them.

The 11, he comes to the South-most Cape of this Land; he named it the *Cape of God's mercy;* the weather foggie. He coasts the N. side; and, at the fogges vp-breaking, he was entered into a faire passage, in some places 20 leag. broad; tolerable weather; voyde of Ice; the Sea of the nature, colour, and quality of the maine Ocean.

Heare he hath great hope of a passage; sailes 60 leag. N.N.W.; discovers certaine Ilands in the midst, but passage on both sides. He divides both the Ships; the one sailes on the N. side, the other on the S., where they stayed 5 daies with S.E. winde, fogge, and foule weather.

14. They goe on Land; find signes of people, and tame doggs with collers about their neckes, a bone in their Pizels, and are vsed to traile sleddes, which they found, like ours; one made of Furre, spruce, and oken boards; the other of Whalebone. They had hung vpon the top of these sleds, the heads of 3 Beasts they had killed. They found other trifles nothing worth, but onely to show that the people had lately been there, but they find this place all Ilands, with great Sounds passing betwixt them. And here they found whales comming from the Westerne Sea, and to the East-ward they had not seene one.

As they were rowing into a Sound lying S.W., soddenly there came a violent Counter-checke of a tide from S.W. against the flood which they came with; but, sounding, they could have no ground heere at 300 fathoms. It flowed uppe and downe 6 or 7 fathoms, and they could not perceive from whence it was maintayned. *If care had beene taken at their entrance into this Streight, they might easily have resolved themselves.*[1]

[1] This is, of course, a comment by Foxe.—C.

The 21, they coast the S. shore; they see many Sounds.

The 23, they enter into a faire Sound at the S. entrance of this Streight in 25 fathoms, greene Ose.

The 26, they depart from the sight of the N. land of this entrance, directing their course homewards.

The 10, he had sight of *Desolation.*

13. Hee departs from sight thereof.

The 27, he had sight of *England;* and the 30, he came into *Dartmouth.*

The Observation.[1]

He set forth from *England*, Iune 28; his furthest was 66d. 40 N. latitude; he sayles then N.W. into a passage 60 leag. upon the *America* side, as was then supposed, and found no hinderance. Yet he returnes homewards the 21 of *August.* Hee was the first wee know of, that ever was on the West-side of *Groenland*, or sayled so farre West in that paralell.[2]

He discovered upon *Groenland*-side, from *Desolation* to 64° 15 min., and, on the West-side, from 66. 40 to the South-side of his new Entrance, and returned home safely.

Captaine Iohn Davis, *his second Voyage.* 1587.[3]

May 7. HE departed from *Dartmouth* with 4 ships, *viz.* The *Mermayd*, 100 Tonnes, the *Sunshine*, 60 tonnes, the *Mooneshine*, 35 tonnes, the *North-starre*, 10 tonnes.

---

[1] This is obviously a summary by Foxe of the main results of the voyage. He appends similar critical summaries to most of the later voyages of which he inserts abstracts.—C.

[2] He certainly was not the first, as Frobisher had been there before him. The origin of this erroneous idea of Foxe's has already been explained (see note to p. 38).—C.

[3] Foxe is again careless enough to give the date wrongly. He has it 1587: it should be 1586. Foxe has taken this abstract from the original narrative in Hakluyt (1589 ed., pp. 781-789), as in the case of the preceding voyage.—C.

June the 15, discovers land in 66 deg. and in longitude from the Meridian of *London* 47; mightily pestered with Ice and snow. From land the Ice lay in some places 10, 20, 50 leagues; hee was constrained to beare backe into 57 deg. to acquit the Ice.

The 29, he meets land in 64° and in longitude from *London* Meridian 58° 30; for divers reasons, he beares into this known harbor, and to set up his Pinnace. He findes many goodly harbors, with high land little troubled with snow, and sea altogether voyd of Ice. He sends his boats to search before the ships for shoale-water to anchor in. The Countrey people come to them with cries and shouts; but, after they espied some of the Company whom they knew before, they came to their boates and hung vpon them with great joy. The Captaine, with divers others, goeth on Land. The people come to him with dauncing and leaping, and made signes they knew all those that had beene there the year before. At this present, there were 18 of them, to whom he gaue to every one a knife. They offered him skinnes, but he shewed them that he bestowed them in curtesie, and so dismist them with signes that they should returne in 4 houres.

The people repaire to him the next day and brought with them the skinnes of Seales, Stagges, white Hares, Seale-fish, Salmon-peale, small Codde, dry Caplin, with other fish, and Birds.

He sent to search their habitation, with command that no hurt should be offered. They find Tents framed vpon wood, covered with Seale-skinnes; they find therein dry Caplin, bags of Traine oyle, and Seale-skins in tan-tubbs.

He mans his Boate, attended with 50 Canoes, intending to view the Country. The people very carefully helpe him up and downe the steepe Rockes. In leaping, our men outstript them; in wrastling, they cast our best wrastler that had both skill and strength.

In a certaine Iland, they found a grave, wherin men lay buried, covered over with Seale-skins, and a Crosse laid over them.[1] They be people of good stature, broad-faced; every time they come, they make new truce by pointing to the Sunne, and crying *Eleout*,[2] striking vpon their brest. He takes them to be Idolaters and witches. They are simple in all their conversation, but very theevish in stealing of Iron, of which they make great account. They in the end began to shew their Nature in cutting of Cables, their Boate from their sterne; and their Cloathes where they laid to aire.

They also stole their Oares, a Calliver,[3] a Boare-speare, a sword, wherevpon they[4] brake the Peace by shooting off a Musket and a Faulcon,[5] at which noise they all departed with great feare. They returned againe within 10 houres, to intreat peace, which was immediately granted, they brought Seale-skinnes and Salmon-peale, but seeing Iron they could not forbeare to steale. They eate their meate raw, drinke salt water, and eate grasse and yce with delight.

Their weapons are for the most darts; but some have Bowes and Arrowes, and Slinges, with their Nettes made of whale fynne, which they doe artificially catch fish with. They have warre with some other Nation or Inland people, for many of them are wounded. He[6] had amongst them Copper Ore, blacke Copper, and red Copper. Thinking to search the habitation of this Country in his Pinnace,

---

[1] Mr. C. R. Markham observes: "It is possible that this may have been a relic of the Norsemen."—C.

[2] According to both Davis and Baffin, the word *Eleout*, *Ilyout*, or *Yliaout* may be translated: "I mean no harm."—C.

[3] See p. 49.—C.

[4] Clearly this refers to Davis's men, not to the savages.—C.

[5] A small kind of cannon, now disused. (See p. 42.)—C.

[6] Presumably this should be *they*.—C.

he entred a large River and went on Land to discover, but the high Mountaines hindred his prospect. He gathers Muscles for his supper and tooke harbour for that night vnder the Rockes: where he see a mighty whirlewind taking vp the water in great abundance for the space of 3 houres without any intermission. To conclude, he found this not to be firme Land: but mighty Rivers and Sounds, and Throughlets betweene vast and desert Ilands: with passage betweene Sea and Sea[1]; he returnes to his ship.

In his absence, the people had stolne an Anchor, and with slings had thrown stones into the ship, of a half a pound weight; he seemed to the Inhabitants to take no notice of the injurie done him; hee tills[2] them on land; gives them bracelets and other toyes, and intices 7 or 8 on board; some of them goes into the maine top.

After Sunset, they begin againe to assault them with stones in slings into the *Mooneshine*, and with one stone strucke the Boatswaine, that he overthrew him.

The 11, they came to make a new truce; the Ringleader of the mischief was one. The truce made, they take one prisoner, who pointed to his fellowes to bring the things that were stolne, and he should be enlarged unto them.

The wind within an houre came faire; they brought the fellow away. One of his consorts came and followed, talking to him; at length they tooke leave, making great lamentation. The prisoner spake 4 or 5 words to the other, clapping his hands vpon his face; the other doing the like, they depart. This prisoner in few dayes grew a pleasant Companion, trimmed vp his darts and fishing tooles, made Okum, and would lay his hand vpon a Roape to haile; his meate was first dry Caplin they had taken

---

[1] Meaning the E. Sea, and *Fretum Davis*.—F.

[2] In Hakluyt, the word is *tolled*. Nares gives *to toll*: to draw on, to entice, or to pull.—C.

there in their Tents; when it was done he eate poore *John.*[1]

Iuly. The 14 of this Moneth, one man dyed; the rest were in good Health. The 17, in the Latit. of 63. 8 min., he fell with a huge Iland of Ice in one entire Masse, so big as they could not draw the limits, with Bay and Capes, and like huge Cliffes [so] as he tooke it to be Land at first. And in this place he had stickle and strong Currents. *No other but what the Ice made, being forced through the water by the windes, and drawing so much water as they bee eyther on ground or neere. Also, as his motion doth trouble and alter the water's true course, which causeth the Tides to edy, being neere it; as Ilands in the Sea, standing in the Flood or Ebbes way, will doe the like.* He coasts to S. off this Ice vntill the 30 of *Iuly*, and saith, it was such a Barre to his proceedings, as all his hopes were banished.

The 24, all his Ropes were frozen by a grosse fogge; his men begin to grow sicke and feeble, and told him he ought in Conscience to regard the safetie of his owne life and preservation of others: and not, through his over-boldnesse, to leave their Widdowes and Fatherlesse children to give him bitter curses; *(leave these excuses and come home* Davis, *come home:)* besides, the great Ship was too great and unwieldy to discover withall; besides her charge was 100 pound a moneth. So, with divers other excuses, he sends her homewards, and, with the *Mooneshine*, made shift to steere E.S.E. from the Ice to seeke the next Land.

The first of *August*, he sees land 66. 33 Longitude, from *London* 70 deg. Here he graues the *Moonelight* (that had beene forth but 3 moneths) in a very goode roade. He findes this land to be all Ilands, with Sea on E. on W. on N., but a *Musketo* stung him grievously; the people here

[1] Hake or stock fish (sometimes cod) salted and dried: cheap, coarse food. Shakespeare mentions it in *The Tempest*, Act ii, Scene 2. —C.

sends him a Seale driving with the tide, which they had boyd up with bladders.

The people trade with him for skins, as the others did, and are in all things alike, but in pronunciation of language more plaine, and not hollow in the throate. Their Salvage kept him close, and made signes to them to get him a Companion.

Here he left the *Mermaid* at Anchor the 12 day and sailes W. above 50 leagues; sees land in 66. 19; this land is 70 leagues from the other. He anchors by an Iland of Ice from clock 9 to 3 in the morning.

The 15, he departs this land to the South; sailes untill the 18, and then he sees land N.W.; a faire Promontory in 65, and no land to southward. Heere he had great hope of a Passage: He sayles still southwards and sees Land S.W. and by S. The 17, by observation, he was in 64. 20 m. He had sailed by Chart and precise account 15 leagues S. by W., yet, upon observation, he found it S.W., so as he saith it was by a Westerne Current.

August 19, it fell snow and foule weather; they lie at hull all Night within 5 leagues of land.

The 20, the weather breakes up. They beare in with land and got into a harbour, close for all weathers; they goe on land and can discerne it to be all Ilands; they come away in the afternoone with a N.E. winde, faire weather; shapes their course to the South, whereby they may discover the passage.

They coast the land untill the 28, finding it still to continue to the S. from 67 to 57°. He sees marvailous store of Sea fowle, as Guls and others; he tries for fish; in one glasse, kills an 100 Codde, although he was but badly provided; he, doubting the weather, steps into harbor in 56 d.; sailes 10 leag. up a River 2 leagues broad; very faire Woods on both sides; stayes here untill the first of *Sept.*; had 2 great stormes; he went 6 miles on land. The woods were Furre, Pyne-apple,[1] Elder, Ewe, Withe, and Birch; he sees a black Beare, and here were store of land and river fowle, as Goose, Ducks, Black-birdes, Iayes, Thrush, and of Partridge and Feasant; he kils great store with Bowe and arrowes. At the harbour's mouth were great store of Cod-fish.

The first of *September*, he set saile, and with faire weather, coasts along to the Southward.

The 3 day, being calme, lets fall a Cadger[2] to prove for fish. In which place there was such abundance as the hooke was no sooner over-board but it was taken. It was the largest and best-fed fish that ever he see, and some of his men, which were Fishermen, said they never saw a bigger Skull[3] of fish in their lives.

The 4, he anchors in a good roade among Ilands; the Country low-land, pleasant, and full of woods. To the N. of this place, 8 leag., finding a mighty great Sea between 2 lands to the W. (the S. land to his judgement is nothing but Iles), he greatly desired to have gone into this Sea, but winde or something was against him; he anchored in 4 fadome, fine sand. In this place is fish and fowle mighty store; he had left on land some fish to drie; he sent 5 men on land for them. The Countrey people lay lurking in the wood, and on a sudden assaulted them. They slew 2, and greatly wounded other 2; one escaped by swimming, with an arrow shot through his arme.

---

[1] This is a misprint. In Hakluyt, it reads "firre, pine, apple", etc. —C.

[2] The kedge, or kedger, is a small anchor used to keep a ship steady and clear of her bower-anchor while she rides in harbour, especially at the turn of the tide. Smyth (*Sailors' Word-Book*, p. 152) says it may be a corruption of *cadger* (a carrier or something carried, from the old word *cadge*, "to carry"), because it was easily carriable.—C.

[3] A form of *school*, or *scull*, a term applied to a shoal of many kinds of fish and cetaceans.—C.

Great store of fish and fowle; they made a hook of a crooked Spike; before the baite was changed, they tooke more then 40 great Cods, the fish swimming so abundantly thicke about the hooke as is incredible to be reported. This *Davis* in his Hydrographicall[1] doth describe.

The 10, he shapes his course for homeward; he arrived in *England* the beginning of *October*.

The Observation.[2]

He departed from *England* the 7 of *May*; 15 of *June* he fell with Land in 66^{d} 70 d. Longitude from *London*.[3] He was troubled with Ice. He findes that there is Copper Ore; the people may be brought to trade, but are thievish and treacherous; he stayed in harbours 24 dayes; went no further to the Northwards then 66^{d}. 17m. in Long. from *London* 70^{d}. He returns the beginning of *August*, and upon the coast of *America* in 55^{d}. finds great store of great Cod-fish. Having not done so much as he did in his first voyage, he arrived very safely in *England* the beginning of *October*.

You shall understand that the Sun-shine *and the* North-starre, *were to seeke the Passage betwixt* Iseland *and* Groenland *by the appointement of Captaine*

---

[1] This refers to *The Worlde's Hydrographicall Description*, a curious and now exceedingly rare work, published by Davis in 1595. It has been reprinted in some of the later editions of Hakluyt, and also in Capt. Markham's *Voyages and Works of John Davis* (Hakluyt Society, 1880). The passage given above (which seems to have been misplaced, either by Foxe or his printer) will be found on p. 208 of the last-named reprint.—C.

[2] See p. 65.—C.

[3] This is another of Foxe's careless errors. In the original, it reads: "discovered land in the latitude of 60 degrees and in longitude from the meridian of London westward 47 degrees." Captain Markham considers this was probably Cape Farewell (*Voyages of Davis*, p. 15).—C.

Iohn Davis, *of which their Voyage the relation is writ by* Henry Morgan, *servant to Master* William Saunderson, *Merchant, of* London, *a worthy and principall Adventurer in the Voyages of* Davis, *as followeth.*

From Captaine *Davis.*[1]

THey departed in the Latitude of 60 deg. the 7 of *June*, and sailed into 66, and tooke harbour in *Iseland* the 12, and staies there untill the 3 of *July;* then he comes forth and being a little troubled with Ice, shapes his course for *Groynland.* The 7, he sees it, and ranges along untill the moneths end.

The 3 of *August*, he comes to *Gilbert's* Sound in 64 deg. 14 min., *Davis* his Randevow[2]; and it seemeth was so appoynted to meete *Davis* and his fleete, who departed from thence the 11 of *July.* The people came and traded with them, and in the end sought to betray them; they plaid at football and our men cast them.

They depart from thence 5 or 6 leag. Southwards; the same people come to them againe. They see Foxes and Dogs runne upon the Ilands, as they observe them to be; they found the hornes and footing of Stags, but see none.

The 30 of *August*, they depart for *England.* The winde takes them contrary, so as they take another harbour; there come [some] of the people, and bring them Seale skinnes.

The Master would have changed one of the boates he

---

[1] This rather vague passage (which is, of course, an insertion by Foxe) is intended to explain that Davis had thought fit to divide his fleet of four vessels. He himself, with the *Mermaid* and the *Moonshine*, sailed north-west, while the *Sunshine* and the *North Star* proceeded to explore along the east side of Greenland, with instructions to meet again later on at a certain *rendezvous.*—C.

[2] This was the spot Davis had appointed beforehand as that at which his divided fleet was to re-unite.--C.

had formerly bought, which they would have taken from them by violence; they fall to combat with our men, and throwing their darts, struck one of our men; another of our men shot one of them in the brest with an arrow. The fight continued; our men tooke one of them into our boate, his boate and all; our men kild three of them, two of them were shot with arrowes, and the other hurt with a sword; he with his boate taken was shot with an arrow, our men cast him over-board, and his consort tooke him up, and conveyed him away; they departed and would not come to us as before.

The 31 of *August* we departed from *Gilbert* Sound; the 3 of *September*, they lost sight of the North Starre[1]; the 30 they enter our Channell. They brought home 500 Seale skinnes, 140 halfe skinnes. What bone it was they brought home, they know not.

Captaine Davis *his* 3 *Voyage North-West*, 1587.[2]

May 12. [HE sailed] from *Dartmouth* with 3 Shippes, the *Elizabeth*, the *Sunneshine*, and a Clincker[3] called the *Hellen*, of *London*.

The 12 of *Iune*, there fell difference betwixt the Master of the *Sunneshine* and the Marriners, for that the Seamen would goe on the Voyage a fishing, the Master would not untill he had the company of the *Elizabeth*, but the matter was reconciled, and all were content to goe to the place of Randevow.

---

1 This ship was never heard of again.—C.

2 As in the case of the two previous voyages, Foxe has taken this abstract from the original, as it appears in *Hakluyt* (1589 ed., pp. 789-792).—C.

3 A vessel built of clincher-work, in which the edges of the planks overlap, as in most small boats: the contrary of carvel-work, in which the edges of the planks are laid together flush with one another.—C.

The 14 of *Iune*, they descry land high and mountainous, but did imagine themselves to be 16 or 17 leagues off.

The 16, they anchored in harbour. The people came according to their olde order with crying *Eliout*, and shewed them Seale skinnes.

The 17, they make way to set up their Pinnace they brought from *London*.

The 18, hee passed about the Iland. Hee found blacke Pumice-stones, and salt kerned upon the Rockes, white and glistering. This day he tooke one of the people; a strong fellow.

The 20, the Salvages came to the Iland where the Pinnace was set up and made ready to be lancht, and tore the two upper stroaks away from her for love of the Iron (I doubt in revenge of their prisoners, and for harme done the last yeare); but, being thus much made unserviceable, it was agreed the *Elizabeth* should have her to fish.

Now, as they were ready to depart, newes was brought the Captaine that the ships that they were to venter their lives in (I thinke for the discovery) had at one time 300 stroakes,[1] yet they agree to commit themselves to God's mercy in her, rather than returne with disgrace. So they stand North-wards along the Land which they call the land of their Merchants, because the people come and trafique with them; but here they were in doubt of their Ship. (I take it the *Elizabeth* and *Sunneshine* are gone a fishing home.) Himselfe went N.ward into 67.40, where they had great store of Whales and fowle, which they call *Cortenous*.[2] Two Canoes came to them at Sea; they cry *Eliout*, and gives Birds for bracelets; one had a dart with

---

[1] This means that the ship had sprung a leak, and needed three hundred strokes of the pump to clear her.—C.

[2] Query *cormorants*. Capt. Markham regards the word as a misprint.—C.

a peece of Vnicorne horne; the Salvage made stay thereof vntill he saw a knife, and then he truckt.[1] They went along with them 3 houres.

The 25, came 30 Canoes 10 leagues off Land, and brought Salmon-peeles, Birds, and Caplyn; they give them Pinnes, Needles, Bracelets, Nailes, Knives, Bells, looking-Glasses, and other trifles. For a Knife, a Naile or a Bracelet, they will sell their coates or any thing they have.

They brought not above 20 skinnes, but made signs that if they would goe ashore they should have more. *Chichesanege* skinnes[2] I thinke.

Iune the 30, he was in 72 deg. 12 min. at midnight; the Compasse set the variation 28 deg. West-ward; he Coasted this Land, which he called *London* Coast,[3] from the 21 to the 30, the Sea all open to the West and North-ward, the Land on Starboard E. from him; the wind shifted to the North.

Then he left that shore, and named the N.most part he did discover *Hope Sanderson;* and, shaping his Course West, runne 40 leag. and better, without sight of any Land.

Iuly the 2, he meetes with a mighty banke of Ice, West from him. He would faine have quit it by the North-wards, but the wind would not; which, if he had, he would have runne W. vntill he had seene Land, and have beene resolved. The 6, being faire weather, he puts the Barke amongst the Ice, but could not prevaile; the 7, 8, 9, 10, he coasts the Ice; the 11, was fogge and calme.

The 13, he determining to goe againe to the shore, and

---

[1] This is an old word signifying *barter* or *exchange.* (See Captain Markham's *Voyages of Davis*, p. 45.)—C.

[2] I am unable to learn what kind of skins these are. The name occurs in Hakluyt.—C.

[3] This was the West side of *Groynland.*—F.

harbor for 5 or 6 dayes, hoping in that time the extreame heate of the Sunne and beating of the Sea would have made way with the Ice; but, when he was nigh Land, he durst not anchor, for depth of water. The Salvages came oft and truct for skinnes. Darts[1] they had for old and new knives, and they would gladly have had him to the Land, but he beares away.

The 15, he finds himselfe driven 6 points west beyond his Course. He layes the fault either in the Ship or Current.

The 16, he falls with the banke of Ice againe.

The 17, he had sight of Mount *Raleigh;* at 12 at Night, he was thwart of his old hole againe. He sailes 60 leagues N.W. vp the streights. The 23, he anchors in the bottome of the Gulfe; and calls the Iles, *Cumberland* Iles.[2]

Whilst he was at anchor, a Whale passed vp by him; here the Compasse set at 30 d. variation. This day also he departs shaping his course S.E. and seeketh to recover the Sea.

The 25, he was becalmed in the bottome of the Gulfe; the aire extreame hot. *Bruton* the Master goes a Land to course Dogges. They find many graves and Trane spilt; the Salvage dogges was so fat they could scarce goe.

The 26, was a pretty storme at S.E.; 27, 28, 29, faire weather. He had coasted the South-side shore of *Cumberland's* sound, and was got cleare out into 62 deg. betwixt which and 63 deg. he espies an opening, and names it *Lumley's* Inlet,[3] and tells of great falls and Gulfes of water.[4]

The 31, he see a Head-land he names *Warwick's* Forland.

---

[1] It is to be vnderstood, that these Darts were headed with Vnicorne Horne.—F.

[2] All this time he ranged South-ward.—F.

[3] Frobisher's Strait, or Bay. (See p. 38.)—C.

[4] See p. 56, *note*.—C.

August the 1, he falls with the South-west Cape of the Gulfe, and names it *Chidley's* Cape, in 61 deg. 10 min.

From the first to the 12, he trents along the South-land. Sees 5 Deere on the top of an Iland he calls *Darcyes* Iland. They take to another Iland; his Boate was too little to carry his men and chase the Deere; though it were in the water, one of them was as big as a pretty Cowe, and very fat, their feete as broad as Oxe feete.

The 13, in 54 d. Latitude. Heere he struck vpon a Rocke. He stops his leake and Coasts along into 52 deg., not finding his fishing ships as was appointed for them to stay and fish in Latit. betweene 54 and 55 deg. vntill the fine of this moneth, but in 16 dayes they were fisht and gone home. Himselfe arrives at *Dartmouth* the 15 of *September.*

The Copie of Davis *his Letter to Mr.* Saunderson.

G*Ood Mr.* Saunderson, *with God's great mercy, I have made my safere turne in health with all my company; and have sailed* 60 *leag. further then my determination at my departure. I have beene in* 73 *deg. finding the Sea all open, and* 40 *leag. betweene Land and Land. The passage most probable; the Execution easie, as at my comming you shall fully know.*

[*Yesterday, the* 15 *of* September, *I landed all weary; therefore I pray you pardon my shortnesse.*]

[Sandridge.
this 16 *of* September,
Anno 1587.[1]]

*Yours equall as mine owne
which by triall you shall best know*
John Davis.

[1] The conclusion of this letter, which is omitted by Foxe, is taken from Hakluyt (1589 ed., p. 792).—C.

The Marine Observation.

THat he Coasted the West side of *Groynland* farther then before, from 65 deg. odde min. to 72 deg. odde minutes, naming it *London*-Coast, and on the West side was as farre vp his former streights as before ; onely he then forgot to name the Earle of *Cumberland's* Iles, which now he hath done. And, besides, he hath, in his returne home, seene and named *Lumley's* Inlet, and passed by *Fretum Hudson*, vnknowne : Yet he hath named Cape *Warwicke*, which is the East part of *Resolution*, & *Chidley's* Cape, the South bounds thereof, now called *Button's* Iles ; But vntruely, these two things are both ; although in his Letter writ to Mr. *Saunderson* at his arrivall the 2 Voyage, he doth assure the Passage to be in one of the 4 places, vpon perill of his Life. But I thinke he durst not venture it. He went forth the 7 of *May*, and returnes homewards the 23 of *July*. His greatest Variation West was 30 deg. And the 15 of *Iuly* he was driven 5 points W. of beyond his Course, by what accident he knoweth not. For, to vse his owne words, speaking of *Warwicke's* Head-land : "This Cape, as it was the most S. limit of the Gulfe, wee "passed over the 30 of this Moneth : So was it the N. "promontorie or first beginning of a very great Inlet, whose "South limit at this present wee see not : Which Inlet or "Gulfe this afternoone and in the Night, wee passed over "to our great admiration, for the waters fall."

These abstracts are more at large to be seene in the first and third Volumes of Mr. Hackluit's *Voyages.*

The Voyage of Captaine George Waymouth, *with two Fly-boates,*[1] *one of* 70, *th'other of* 60 *Tonnes;* 35 *men, victuailed for* 18 *monethes; set forth by* the Muscovia *and* Turkie *Companies.*[2]

HE set forth the second of *May*, 1602. Hee went by the North; made the *Start*, or one of the Westmost Iles of *Orkney*, it being low land, bearing West, in

---

[1] A fly-boat was a large, flat-bottomed, Dutch vessel, generally of from 300 to 600 tons' burden, and distinguished by a remarkably high stern and very broad buttocks.—C.

[2] Comes to *Hudson's* Bay, the 23 of *May*.—F. The meaning of this remark of Foxe's is not clear. A mistake of some kind seems inevitable. There is no such note in the original narrative of Weymouth's voyage, as given by Purchas (*his Pilgrimes*, vol. iii, lib. IIII, pp. 809-814). Foxe has copied from Purchas (*loc. cit.*) the totally erroneous statement, for which Weymouth himself seems to be responsible, that the credit of promoting this expedition belongs to the Muscovy and Turkey Companies. It was, in reality, carried out under the sole patronage and direction of the "Wor'll Fellowship of the M'rch'nts of London trading to the East Indies". A detailed narrative of the preparations for, and the events following upon, the voyage are given by Rundall (*Voyages North-West*, pp. 51-64). The account of the voyage itself, as given here by Foxe, is very much condensed by him from Weymouth's own narrative, as it appears in Purchas. The results of the voyage were very small from the geographical point of view, and the chief interest of the undertaking lies in the fact that he entered, and sailed some hundred leagues up, Hudson's Strait. This appears clear from the narrative, though Foxe (see p. 84) contemptuously ejaculates: "No such matter", while Sir John Barrow (*Arctic Voyages*, p. 168) pronounces it impossible. Professor Asher says of Weymouth (*Henry Hudson*, p. cxix) that on "the 25th of July, he arrived at Hatton's Headland, in 61° 40', the northern [side of the] entrance to Hudson's Bay [? Strait]. According to his own words, he sailed 'an hundred leagues west and by south' into the Strait. There must be either a slight exaggeration in the distance, or the statement as regards the course must be slightly incorrect. The latter is, indeed, the case: this the journal clearly shows. But there is no reason to pass on Weymouth the severe verdict, *that he pretends to have done a thing which is impossible*—a verdict first pronounced by Foxe, whose acquaintance with the

latitude 59 deg. 30 min.; he shapes a course betwixt N. and West, until hee brings the *Start* North; sayles away W. and by N. in 59 deg. 40 min. Then steeres away W.S.W. in 57 deg. 55 min. and there had no variation at all. He hailed away Westward, and had some fogge, much raine, but warme as in *England*.

16. This day at noone he was 57 deg. 35 min. and had not seene the Sunne or Moone in 76 houres before, and now had variation 11 deg.

18. He saw a great Iland of Ice. In the afternoon he got sight also of the Southmost part of *Groenland;* hee coasts this Ice to the North, comming sometime into black water, and presently the Sea would be cleare againe. He could not gaine ground in 120 Fathoms, neither could hee discerne any Current, at which [place] he reckons Cape *Desolation* N.N.E., 24 leagues off him.

22. He was in 60 deg. 37 min. latitude.

27. The weather warme as in *England*. He had great store of sea-Guls.

28. He directs his course Westwards, and sees the land of *America* in 62 deg. 30 min. Makes it to be *Warwicke's*

---

south [side] of Hudson's Strait was very imperfect; then confirmed by Sir John Barrow, who probably did not take the trouble to look into a map; and then repeated by others. That Weymouth really sailed a considerable distance into Hudson's Strait does not allow of a doubt." In this statement I fully agree. Weymouth's narrative of his voyage is so vague and confused that it is impossible to follow the greater part of his course on a chart with anything approaching certainty. The chief interest connected with his proceedings lies in the fact that Hudson's fourth known voyage in 1610, when Hudson's Bay was practically discovered, was undertaken with the idea of following up Weymouth's rediscovery of the Strait now known as Hudson's. On this occasion, Hudson sailed in one of Weymouth's vessels, the *Discovery*, and had with him, apparently, at the start the William Cobreath or Colbert, who was master of one of Weymouth's ships (Rundall's *Voyages*, p. 238), and who signed Weymouth's narrative as a witness to its veracity (*Purchas his Pilgrimes*, pt. iii, lib. IIII, p. 814).—C.

foreland. It was high land and covered with snow, with other small Ilands. There was great store of Ice upon the forelands East side, but the sea was altogether cleare; the land lay N. by E. about 6 leagues in length.

29. He was beaten to the Southwards by N.E. winds, and finds *Warwicks* foreland to be an Iland. Hee discovers *Lumley's* Inlet[1]; a great Current setteth to the West, the greatest hope of a passage this way; the weather was fogge and snow; he had a great whirling of a Current in latitude 61 deg., 12 leagues from the Coast of *America.*[2]

The 1 of *Julie* was fogge and snow; the ayre very cold; hee traverst to and againe in many overfals, but by his course he could not discerne which way the Current set, but most like to the West. He travers'd therein 16 or 17 dayes, and could find no ground in 120 fathom.

Iuly 2. He discernes a maine banke of Ice in 60 deg.; faire weather. He lancht his boate and loaded her twice therewith, to dissolve to fresh water. Hee sets into many overfals alongst this coast of *America*, which coast here he conceives to be broken land.

3. The [wind] S.W. He stands in with the coast of *America*, and meets with Ice 10 leagues off; the water blacke and thick as puddle.

8. He had beate it to the Northward, and descries the land of *America*, in 60 deg. 53 min., being very high. It bore S.W., covered with snow; he was 5 leagues off, but could not come neare it for Ice.[3]

9. A storme began at N.E. He cleares himselfe of the land and Ice by standing to Southwards. The storme

---

[1] These were formerly found and named by *Davis*.—F.

[2] Now was hee in the entrance of the passage, *Chidlies* Isles W. by S. halfe South.—F.

[3] This was upon the South side of *Davis* his entrance into *Cumberlands* Isles.—F.

continued, so that he stood to the Southwards in fore-course.

17. From the 9 to the 17, he was in traverse, and heere he heard a fearefull noyse of Ice. He had thicke weather, his roapes and sailes all frozen.

18. The wind N.E., extreame cold and frost, the ayre very cleere. His roapes were froze, and it froze so extreamely that it was a maine barre to his proceedings and destruction to his men.

19. He stood to Eastward with wind N.E., and the same night his men conspired to beare up the helme, and keep him in Cabbin. They shew the reason for so doing in writing; by good chance, he understands thereof and prevents them.

The Reasons [given by the men].

That, although it were granted that we might winter betweene 60 and 70 degrees of latitude with safety of lives and vessels, yet it will be *May* next before we can dismure them to lanch out into the Sea; and, therefore, if the Merchants should have purpose to proceed on the discovery of the N.W. parts of *America* the next yeare, you may be in the foresaid latitude from *England* by the first of *May*, and so be furnished better with men and victuals, to passe and proceed in the foresaid action.

Seeing, then, that you cannot assure us of a safe harbour to the Northward, wee purpose to beare up the helme for *England*; yet with this limitation: that, if in your wisedome you shall thinke good to make any discovery (it seemeth there were some with him understood more then himselfe[1]), either in 60 or 57 degrees, with this Northwest winde, wee will yeeld our lives with your selfe to encounter any danger. Thus much wee thought needfull to signifie, as a matter builded upon reason, and not proceeding upon feare or cowardice.

Then, being in latitude 68 and 55 min.,[2] there was no meanes to perswade them, but they would beare up the helme, whereupon he came out of his Cabbin to enquire

[1] This, of course, is a comment by Foxe.—C.

[2] This cannot bee.—F.

who was the cause; they answered, one and all, hoysing up sayles, and directing the course South by West.

22. Hee sent for the chiefest of the Mutineeres and punished them severely.[1] This day, hee came by an Iland of Ice; both ships launcht their boates to fetch some to make fresh-water; this Iland crackt two or three times, as though it had beene thunder-clappes, and it brake in sunder to the great danger of his boates, the one being halfe laden with Ice.

25. It blew hard, the course West by South, with fogge, and [he] was in latitude 61 degrees and 40 minutes, and findes an Inlet in this latitude.

27. The South-south-East wind blew very hard, with fog and raine; his course West.

30. The wind came in a shower to West, Northwest, and blew hard; and, because the yeare was farre spent, and many men sicke in both Ships, he thought good to returne, with great hope of this Inlet to be a passage of more probability then *Davis his Straights*, because he found it not pestred with Ice, and a straight of 40 leagues broad. He saith, he sayled 100 leagues West by South, into the Inlet.[2] He saith also, he found the variation to be 35 degrees Westward, and the needle to decline, or rather incline.

Observe.

83 Degrees and a Halfe: the 5 of *Julie*,[3] he was cleere off the Inlet. The 6, South-east wind and fogge; the 7, 8, and 9th, hee passed by many great Ilands of Ice, & discovered an Iland upon the coast of *America*, latitude 55

---

[1] This doth not appeare [reasonable] that he could punish, and yet suffer them to carry the ship backe.—F.

[2] No such matter.—F. However, see note on p. 80. The inlet can only have been that now known as Hudson's Strait.—C.

[3] This should be August.—C.

degrees, 30 minutes.[1] 14. He stood off and on this coast from the 5 untill the 14, had some foule weather, and made some Ilands; he stands into an Inlet in 56 degrees, and had good hope of a passage for divers probable reasons.

I finde nothing more of note, but that, upon the Coast ot *America*, betwixt 55 degrees 30 and 50 minutes, he observed two variations, the one of 17 deg. 15 min., the other 18 degr. 12 min. The coast was voyd of Ice, unlesse some great Ilands drive from the North, and that the ship had like to have perished for want of spare decks[2]; he saw one whirlewind upon this Coast take up the Sea into the Ayre extreamely; he entred 30 leagues within one Inlet, latitude 56 degrees, where, if the wind had come Northerly, South, or East, but one day, he had perished.

August 4. He had sight of the Iland of *Silly*. The next day, hee came into *Dartmouth*.

The generall Observation.

He set forth the 2 of *May*, and returned homewards the 30 of *Iuly*. His greatest latitude wee can be certaine of was 63 deg. 53 min., and, passing the *Grand Meridian*, betwixt *Orkney* and *Desolation*, hee had no variation; his greatest [? longitude] was 35 degrees Westward. Hee neither discovered nor named any thing more then *Davis*, nor had any sight of *Groenland*, nor was not so farre North; nor can I conceive hee hath added any thing more to this designe; yet these two, *Davis* and he, did (I conceive) light *Hudson* into his *Straights*[3]; nor did he try in so long

---

[1] Here was *Davis*.—F.

[2] The *spar-deck* is the entire upper deck of a double-banked vessel, but is also at times a temporary deck.—C.

[3] In this opinion Foxe is unquestionably correct. His remark, however, would have been still truer had he said that Frobisher, Davis, and Waymouth all three "lighted Hudson into his Straights". —C.

time, being to and againe, upon the Coast of *America* for *Davis* his fishing, having such abundance of the largest and best fed Cod-fish that he saw. His ships were never separated, which shewed that they were not greatly distrest.

Master Iames Hall *of* Kingston *upon* Hull, *Pilot Major of three Ships, set forth by the King of* Denmark *for the discovery of* Groenland. 1605.
His first Voyage Abstracted.[1]

FRom *Denmarke*, he set forth the 2 of *May*, and saith hee found the Compasse varie Eastward at the *Naes* of *Norway*, 7 deg. 10 min.

He saith that one league to the Northward of *Faire Isle*, he found the race of a Tyde setting so strongly Northwestward, as if it had bin in the race of *Portland*.

Faire Isle, bearing E.S.E. foure leagues off; *Swinborne* head, N.E. by N. eight leagues off; the Ile of *Foule*, N.E., the Compasse was varied to the Eastward of true North, 63 deg. 10 minutes, and he thinketh that the Island *Busse*, discovered by the *Busse* of *Bridgewater* in *Frobrisher's* last Voyage, is not truly placed in the *Marine Charts*.[2]

At his falling with *Groenland*, hee named a headland Cape *Christianus*, after the King of *Denmarke*, in latitude 59 deg. 50 minutes, and he found it due, for that none other before him hath named it.[3] S.W. by W. five leagues from thence, hee had 12 deg. 15 min. variation Westwards.

---

[1] The account of this voyage is abstracted by Foxe from *Purchas his Pilgrimes*, part iii, lib. IIII, pp. 814-821, including his topographical description of Greenland, which follows. Hall was, like Luke Foxe, a native of Hull. The object of the voyage was to discover the lost colony of Greenland.—C.

[2] See p. 59, where it is explained that this island had no real existence.—C.

[3] It can be no other but Cape *Farwell*.—F

Standing from thence to Seawards, he sayled three houres in blacke water as thicke as puddle.

He found Cape *Christianus* and *Desolation* to lie W. by N. 50 leagues distance, and a Current S.S.W. set him violently into the Ice; he also findes the Current upon the side of *America* to set to the North; but, contrary, on the *Groenland* coast, to the South.

He findes a harbour upon the Coast of *Groenland*, and sailes 6 leagues up a great Inlet, or river, before he could find 16 fathomes to anchor in; the land on both sides was steepe and mountainous.

He goes on land and findes houses, or rather Tents, covered with Seale-skinnes; the people came to him crying *Eliout*, holding up their hands. Their boates were covered all over with Seale skinnes. About their Tents, was great abundance of the flesh of Seales (to drie), with Caplin, and Pilchards innumerable, of which, with other fishes, their rivers are full. Their dogs were very fat; they found in their Tents Foxe and Seale skins, very well drest, also certaine coates of Seale and Fowle skins, with the feather-side inwards. They also found a certain vessell boyling upon a lampe, the vessell made after the mãner of a little pan, the bottome of stone, the sides of Whale's Gils; therein was Seales flesh, boyling in Seale oyle; and, in another, a dogge's head boyled. By those Tents lay two great boates, with which he supposed they transported themselves from one place to another, this not being the place of their continuall habits. The boats were open, with 8 or 10 thoughts,[1] and 20 foote in length at least; for a saile, they have the guts of some beast, well drest, and neatly sowed together.

[*June*] 13. After this, the people came to them in their boates, and bartered Seale skins, and their Coats even, for

[1] An old form of *thwarts*.—C.

olde nailes, or for a knife; they will sell coate and boate, Vnicorne horne, Mors teeth, or Whale finne, with which they head their darts and weapons. The Latitude of this Harbour's mouth is 66 deg. 30 min. An E. and W. Moone makes a full Sea; it floweth 3 fathomes and an halfe up and downe, he had made about a barrell and ½ of oyle, and, leaving it on land all night, the Salvages let it forth.

18. The Salvages came the next day and bartered, and going on land upon a sodaine, without violence done them (which shewes their wicked condition), they assailed them with stones out of slings in most violent manner; at the shooting of a Falcon-gun[1] they all fled.

19. The next day againe they repaire to the number of sixty, making new truce by crying *Eliout*, but perceiving they had bags full of stones by them, at the report of a Pistoll they all departed; and, after that, they came to the same Cliffe againe, and violently assaults them that no man could stand upon the hatches, so as hee was glad to shield himselfe by loosing his Bonnets,[2] and lacing them about his ship, and, at the firing of a Musket, they would ducke downe behind a Rocke; the report gone, they would afresh assault them. He departs from hence and came to an anchor in an excellent haven on the S. side of a high hill, which he named Mount *Coningham;* this Sound, for the goodnesse thereof, he named *Denmarke-*haven.

20. He loosed from this harbor; the *Salvages* came againe to the number of 73, beating and making a hideous noise; they enter into barter and throw Shels and toyes into his Boate; he, causing his boy to fetch them, they shoote him through both buttockes with a Dart; there were at this time mustered upon the Ilands, to the number of 300 people.

[1] See p. 42.—C. [2] See p. 55.—C.

Now followeth Mr. Iames Hall *his Topographicall Description of the Land, as hee discovered the same.*

THe land of *Groenland* is a very high, ragged, and mountainous Countrey, having many good Rivers, Harbours, and Bayes, into 5 of which hee sayled 10 or 12 English leagues, [they] being very navigable, with abundance of Fish of sundry sorts. The Land, in all places where [he] came, seemed to bee fertile, according to the Climate wherein it lyeth, for betweene the Mountaines was most pleasant Plaines and Vallies, insomuch as if he had not seene the same, hee would not have beleeved that such a fertile land in shew[1] could have bin in those Northerne Regions. There is also great store of Fowle, as Ravens, Crowes, Partridges, Pheasants, Seamewes, Gulls, with other sorts. Of Beasts, he hath not seene any, except blacke Foxes, of which there are very many. Hee doth suppose there are also many Deere, for, about their Tents, they found many Harts-hornes, with the bones of other beasts; also, within the land, he saw the footing and dung of divers other beasts; he found the footing of one beast to be 8 inches over. In the rivers, were Fishes, as Seales, Whales and Salmon, with divers other sorts of fishes. The coast is a very good and faire land; for, 3 leag. off, he found 15 fathomes, and, as he approached the same, 13, 12, 10 fathomes, very faire sandy ground. The people are a kind of *Samoid*,[2] or wandering nation, removing from one place

---

[1] *I.e.*, in appearance.—C.

[2] The Samoyeds are a wandering race of people, belonging to the Ural-Altaic race, often mentioned by the early English and Dutch travellers to the North-East, and still inhabiting the most northerly portions of Siberia. They dwell in tents, or *chooms*, made of reindeer-skins, which they take with them from place to place. The first meeting between West Europeans and Samoyeds was during Bur-

unto another. They are people of a reasonable stature, browne of colour, very like the people of the East and W. *India;* they are active and warlike, vsing their darts and slings very nimbly. They eate their meate raw, or little parboild, with blood, oyle or water. They apparell themselves in skinnes of such beasts as they kill, but especially with Seales and fowles, which they can dresse very soft and smooth, in Summer turning the haire and feather side outwards, in Winter inwards; their weapons are slings, bowes, [and] darts headed with bone or yron. He supposeth them to be Idolatrous, worshipping the Sun. He met, all the coast along, much drift wood, but from whence it came he knew not. He coasted this Coast along from 66 deg. to 69 deg. and found many good sounds and harbours, and returning towards his Shippe, which he found in a harbour by espying certaine Worlockes[1] which the Captaine had caused to be set as Beacons, for to give him knowledge of their being.

In this his absence from the Admirall,[2] the Saluages had done them much violence; the Captaine had taken three of them, whom he kindly intreated, others of them he slew.

[*July* 10.] This evening he takes in his provision of fresh water.

He sets on land one young man to be left in the Countrey to his cruell fortune; and this was done by

---

rough's voyage, in 1556 (*Voyage of the Vega*, vol. i, p. 223). Purchas (*his Pilgrimes*, part iii, lib. III, p. 555) gives a lengthy account of their habits, mode of life, etc. Professor Nordenskiöld (*Voyage of the Vega*, vol. i, pp. 73-104) gives some account of his visit to this tribe, and Mr. Henry Seebohm (*Siberia in Asia*, p. 176) mentions being presented to their chief.—C.

[1] In Purchas the expression used is "*warlockes of stones*". Probably cairns or beacons are meant.—C.

[2] See p. 49. Foxe has omitted to mention that Hall had left his ship on June 16th, in order to explore up the coast in his pinnace.—C.

expresse command of the *State-holder* of *Denmarke* before his coming forth; they also in the Pinnace set another on land, both being malefactors, giving of them small necessaries. (*It may be those people lived a long time after, and may bee yet living, if the Salvages have not devoured them.*) He sets sailes and comes to Sea, where he found much drift Ice with a high Sea, which he thought to be a current setting through *Fretum Davis* to the Southward, as by experience he proved; for, by observation on this day at noone, he was in Latitude 62 deg. 40 m., whereas, the day before, he was in Latitude 66 deg. 10 min., having made by account a S. by W. way, about 10 leagues; this current he did find to set along the Coast of *Groenland*, South by East.

15. This day he was in Latitude 57 degrees. The 16 day, close weather, he meetes with a mighty skull[1] of Whales, amongst drift Ice, and meetes also a great current setting West North-west over for *America*. *This is the current found by* Frobrisher, *comming from the East, and butting upon the East-side of Groenland, and doth strike along the Land to Cape Christianus, otherwise called Cape Farewell.*

August 1. Hee met with a skull[1] of Herrings, so that he knew himselfe not farre from *Orkney;* he was in Latitude 58 deg. 40 m., and, sounding, had 42 fathomes, very sandy ground, with some blacke dents[2]; when, shortly after, and the same day in the evening, he sounded againe, and had but 20 fathomes, dented ground, he was neere the shore before he saw it, for it was thicke weather.

10. He came to *Elsenore* Rode in *Denmarke.*

---

[1] See p. 71.—C.

[2] Ogilvie gives *dent* = indented: impressed with little hollows. The word appears in Purchas.—C.

The second Voyage of Master Iames Hall *from* Denmarke *to the further discovery of* Groenland, *with five Shippes, (observed* 1606).[1]

May 27. HEe set forth from *Copemanhaven*, and went betwixt *Orkney* and *Shotland* the 7 of *Iune*, which day one of the *Groenlanders* dyed; it seemed it was one of them he had brought from thence the yeare before.

14. He accounteth himselfe to be 19 deg. 45 min. from the Meridian of the *Naes* of *Norway*. He steereth away W., the wind S.E. and thick weather. He imagineth himselfe in 58 deg. 10 min. latit., at which time, by reason of a Northerly Current, contrary to his expectation, he had made a West way Southerly, 22 leagues, and then, as he supposeth, the Compasse varied West-ward one point.

July 1. He sees land, being 8 leag. off, with a great banke of Ice lying off the S.W. end thereof; he supposeth it to bee *Busse* Iland, and that it lyeth more to the West-ward then it is placed in the Marine charts.[2]

Steering away W. by N., he was in a great Current, setting S.S.W., the which he did suppose did set betwixt *Iseland* and *Busse* Ile, over with *America*; from hence, he steeres away W.N.W.

6. He found himselfe to be in 58 deg. 50 min., whereby, contrary to his expectation, he did plainly see the Southerne current to be the cause. This Evening, he found the Compasse to be varied 12 deg. 5 min. West-ward. This Night, their Pinnace and Vice-admirall come foule of one another.[3]

---

[1] This voyage had the same object and the same promoter as the last. The narrative of it is taken from *Purchas his Pilgrimes*, pt. iii, lib. IIII, pp. 821-826.—C.

[2] *Busse* Ile again discovered.—F. But see pp. 59 and 86.—C.

[3] A caveat for Commanders in Fleets.—F.

8. He was in 59 deg. 30 min. and findes still the Current and variation to carry him to the South-ward of West.

10. He sees the Coast of *America* in Latit. 60 deg. 16 min., about 9 leagues off, and findes the needle varied 23 deg. W. The hill-tops were covered with snow; the shore to the N. full of Ice. He had a Current set West into the shore, and indangers him, had not a gale fresh at S. West brought him off.

18. Vntill this day, he passed many Mountaines of Ice. At Noone, was in 63 deg. 45 min. Latit.

19. Being amongst much Ice, and plying to get cleare, [he] saith he seeth the Land of *America* in 64 Latit. It lay S. and N., very high and ragged, covered with Snow. He findes still a strong Current to the West. From the Latit. 51, his Compasse was placed $\frac{3}{4}$ of a point to Eastward of N., and [he] was carried almost 4 points to Westwards beyond his judgment. He found this Current to set W.N.W.[1] The Compasse varied 23 degrees.

From the 20 vntill the 25, he passeth and traverseth over from the West side for *Groyneland*, and had sight of Queene *Anne's* Cape, 10 leagues off.

27. He seeth the Capes he named the last yeare, as Cape *Anne*, Cape *Sophy*, the Foords also, as *Rumell's* foord, *Christianus* foord, and puts into *Coningham's* foord, where, he saith, the Silver was. The *Salvages* come on board and barter with them for Iron, with Seale skinnes and Whale fynne. He searcht vp the Foord and findes it to be but a Bay, with many greene and pleasant Ilands. The people, to the number of 25, followes them with their Boates; some of his men on Land travailes vp the Mountaines, and sees raine Deere.

August 6. There fell some small difference amongst them about choosing of an anchoring place, the water being

---

[1] *Davis* doth write of the same not farre from this place.—F.

deepe, and they removing from their first Road-sted further vp the River; where it floweth S.E. and N.W. in Latitude 66 deg. 25 minutes.

9. The Captaine went vp the River with his Boate, where they came to see their winter Houses, which were builded with Whales' bones; the baulkes thereof were of Whale's ribbes, and covered with Earth. They had certaine vaults or roomes vnderground, 4 square, two yards deepe in the Earth. The towne consisteth of about 40 houses; they found the buriall of their dead, the Corps wrapped in Seale-skinnes, and stones laid in the manner of a Coffin over them.

This day, they take 5 of the inhabitants to bring into *Denmarke*, to be informed of their Country, which they call *Secanunga*, and say that, within the Land, they have a great King, who is carried vpon men's shoulders.

10. They come forth of *Rumell's* foord, and in great danger got to Sea, being inforced betweene certaine Ilands which lye off Cape *Sophie* 4 leagues into the Sea. The last yeare, he named them *Knight's* Ilands. Being got there, he came into 66 deg. 50 min. Latit., W.N.W. from Cape *Sophy*, 15 leag. off.

18. He goes within sight of shore, being all high land like Ilands, and great store of Ice betwixt him and the same.

28. He coasted to and againe amongst Ice vntill this day, having had a storme, and meetes one of his Fleete which had beene separated from him; he fell [in] with two bankes of Ice, and is 8 leag. off *Desolation* by account, but could not see it by reason of fogge.

31. He meetes Ice, which he marvailes at, being that the South-most part of *Groynland* described in the Marine Chart is in the Latit. of 60 deg., he being then in 59 deg. 10 min., *Desolation* bearing W.N.W. halfe N., 64 leag. off; and Cape *Christianus*, the next knowne part of *Groynland*,

N.W. by W. Westerly, 38 leag. He holds an E.S.E. Course, and sees Land the same day, being very high, lying along E.S.E. 16 leag.; the W. part seemed to fall away E. by N. The land was very high, and covered with snow; he was not certaine whether it was the Maine or an Iland. He named it *Frost* Iland. (I cannot conceive but that, by his Latit., it must be the same he had formerly named Cape *Christianus*.)

September 1. He comes homewards the 8 day, and is in 58 d. 36 m., variation 2 d. 45 m. W. The 10 day, he was in 59 deg. 10 min., variation 1 deg. 4 min.

18. He espies the Iland of *Farrey*, Latit. 62 d. 3 m.; the streame sets vnder the Iles next hand E. and W.

October 4. He arrived at *Copeman* Haven. His conclusion is that *Bredarenses* Foord is the most Northerly, *Coningham's* Foord is next in 67 d. and odde m.; the Foord where they see the Towne is 2 leag. to S. of *Coningham's;* the towne stands 10 leag. vp the Foord; the *King's* Foord is in 66 d. ½ (but he did not find the Siluer Myne).

[Hall's *Third Voyage*, 1607.[1]]

Master Iames Hall *was imployed in a third Voyage from* Denmarke, 1607; *but, after that he had made the Land, the* Danes *mutinied, and in fine forced the ships backe againe for* Iseland. *There being then nothing done towards the edifying of this worke, it is needlesse to make any further search after the same.*

---

[1] The information relating to Hall's third voyage is taken from *Purchas his Pilgrimes*, pt. iii, lib. IIII, p. 827, and is nearly all the information Purchas gives, except that he says the voyage was again to Greenland, but that, as it produced no fresh discoveries, he omitted the account of it, though he had a narrative of it, "written, and with representations of land-sights curiously delineated, by Josias Hubert of Hull." Hubert had also sailed with Hall in the previous year, and we shall find that he afterwards sailed with Button in 1612-13.—C.

The 4th *and last Voyage of Master* Iames Hall, *from* Kingstone *vpon* Hull, *with* 2 *Ships, viz. the* Patience, *and the* Hearts-Ease.[1] [1612.] *Written by* William Baffin.

HE saith that, in the Morning, perceiving the Sunne and Moone to shine very cleare, he purposed to find out the Longitude, and this day he spent in finding out the true Meridian Line vpon an Iland, by hanging at the extreames thereof 2 plummets in threed, instead of an Index or sight.

Iuly 9. He observed the Moone iust vpon the Meridian, at which very instant the Sunnes almicanter[2] was 8 deg. 53 min. N., the Poles elevation 65 deg. 20 min., by which working, according to the doctrine of *Sphæricall* triangles having 3 sides given, to wit, the Complement of the Poles elevation, the Complement of the *Almicanter*, and the complement of the Sunnes declination, to find out the quantity of the Angle at the Pole; by which working, he findes the time to be clocke 4, 17 min. and 24 sec.; he

---

[1] This voyage, the date of which Foxe omits, was undertaken in 1612. The narrative of it here given, together with the description of Greenland, is taken from *Purchas his Pilgrimes*, pt. iii, lib. IIII, pp. 831-836; but Purchas' narrative is a mere fragment, commencing on July 8th, whereas Hall sailed on April 10th. Another account of the voyage, written by one John Gatonby of Hull, is to be found in Churchill's *Collection of Voyages and Travels* (1732 ed., vol. vi, pp. 241-251). The first portion of this narrative, down to July 8th, when the other account begins, has been reprinted by Mr. C. R. Markham (*Voyages of Baffin*, pp. 1-19). The narrative in Purchas was written by William Baffin, pilot of the *Patience*, of whom we shall hear again. The object of the voyage was to make search in Greenland for mineral ores, and the "adventurers" were Sir Thomas Smith, Sir James Lancaster, Sir William Cockayne, and Mr. Richard Ball.—C.

[2] Almacantars are imaginary circles parallel to the horizon, and supposed to pass through every degree of the meridian. The word is Arabic, and is synonymous with parallels of altitude.—C.

findes likewise, by the Ephemerides,[1] that the Moone came to the Meridian of *London* at 4 a'clocke 25 m. 34 sec., which 17 m. and 24 sec. substracted, leaveth 8 min. 10 sec. of time, for the difference of the Moones passing betwixt both the Meridians; the Moones motion that day was 12 degrees, seaven minutes, which converted into Minutes of time were 48 minutes 29 sec. This, wrought by the rule of proportion, is thus; if 48 min. 29 sec., the time that the Moone commeth sooner to the Meridian then she did the day before give 360 deg., the Earths Circumference; what shall 8 min. 10 sec., to wit, 60 deg. 30 min. the difference betwixt the Meridian of *London* and *Cockings* Sound in *Groenland*.[2]

This he affirmeth to be without any great error, the

[1] This term, which is now applied to almanacs and other periodical publications, formerly meant merely the astronomical calculations, which took their names from their projectors, as *Searle's Ephemeris*, *Origanus' Ephemeris*, and which were all the early navigators had to work with. David Origanus was the author of an *Ephemeris* for the years 1595 to 1650, which was published in 1599 at Frankfort, where he held a Professorship in Greek, as well as in Mathematics. His meridian was Wittenberg. "*John Searle* received his licence to practise chirurgery in 1607, and published, in 1609, *An Ephemeris from* 1609 *to* 1617, *whereunto is annexed three succinct Treatises of the use of an Ephemeris of the fixed Starres, and four Sections of Astrologie* (4to., London). The book contains, among other tables, a correction of time in respect of difference of meridians; a list of places, with latitude and longitude in time; a table for converting degrees and minutes into time; eclipses; and a table of the inequality of days, and the equation or correction of them. The copy of Searle's *Ephemeris* at the British Museum wants the title-page; that at the Bodleian is a perfect copy" (Markham's *Voyages of Baffin*, p. 124). In 1766, was published the first *Nautical Almanac*, a tabular statement of the geocentric planetary positions, which may be said to have created a new era in voyaging. This book, with certain alterations, was in force up to 1830, when a commission of the Royal Society, with certain astronomers, established the present *Ephemeris*, a much valued work published annually, but computed four years in advance (*Encyc. Brit.*).—C.

[2] Longitude wrought in *Cockings* Sound Latit. 65 deg. 20 min.—F.

rules truely observed and Ephemerides exact, for that no instrument can be exact enough to find out the true Houre, Minute, and second, the losse of one Minute being the losse of 7 deg. in Longitude.

[*July*] 22. Master *Hall* was slaine with a Dart, throwne into his body by one of the *Gröenlanders*, before whose death, and since the ninth day, little was done worthy note; but the supposed Myne [was] sought for; and in that search many brave Rivers and Harbours were found, with the footing of some great Deere or Elke, as bigge as an Oxe. It may be supposed that the cause of this Blow was for the losse of the Brother, or other the Friends of the Homicide, which might be some of those five the *Danes* carried away the last yeare (and but one before); for that in all trading, both before and after his Death, they did never offer any violence vnto any of the Company, but before, would oftentimes be pointing and ayming, as it were at him, calling him Captaine one to another.

They search further for the Myne, and find many places where the *Danes* had digged, with stone of Orient couler, but, when it was refined, it proved drosse, having no mettall at all therein, but was like to *Muscovia* sludde. They also found a pleasant Vallie.

They find not the Myne, as also the people forbeare to trade with them as before. They made way againe out of this River to come to their Admirall, in which way & in many Ilands they find where many of their winter houses had bin, & some of their Tents were but lately carried away, in which place they also found of their long Boates made of Wood, and bound together with skinnes of Whale-fyn, and covered with Seale skinnes, being 33 foote long, and 5 broad, having in them 10 thoughts or Seates. This day they depart *Rumell's* Foord, in 67 degrees Latitude, and 24 degrees 16 minutes variation, it being one of the fairest Rivers he see in that Country; it lyeth in

East, and E. by South, and that Night hee came to the Admirall, then in the *King's* Foord.

26. They consult about comming home, for the Captain was slaine, and the people refuse to trade with them as they were wont, I thinke for feare; for the same Voyage, one *Iames Pullie* was also slaine by a Salvage, for offering to take one of their people out of his Boate by violence.

They depart homewards, Master Andrew Barker, *of* Hull, *in the Admirall,* William Huntris, *Master in the Pinnace.*

August 18. HE is in 58 deg. 50 min.; he findes the variation to be 13 deg. 22 min., contrary to observation of other men in this place, and because this was the first Sea-Voyage of this young Arts-man, I will trace his owne words. The 18, the Sunnes declination, saith he, was 9 d. 58 min. for the Meridian of *London;* but, being almost 4 houres of time to the West-wards, there is $\frac{1}{3}$ to be abated from the rest, for his declination was 9 deg. 55 min., his altitude 24 deg. 40 min. in Latitude 59 deg., his distance from the South by the Compasse 81 deg. Latitude 57 deg., variation 11 deg. 10 m.

September 6. Nothing of note but one storme vntill this day. The latitude was 61 deg. 18 min., the variation 6 deg. East; the deepe 68 Fathomes.

8. He came to the Ile of *Orkney,* where, anchoring, the people came and gave them Hennes, geese, and sheepe, for old cloathes or shoes.

11. He came to *Kingstone* upon *Hull.*

The Observation he made of Groenland.

That it is an exceeding high land, and Mountaines which are very high within the land. They are of stone, some of one colour, some of another, all glistering, but nothing worth. If there be any Mettall, it lyeth low in the earth, and cannot well be come by. There are some rockes purer then Alablaster. The Northside of the Mountaines are continually covered with snow; there are few trees but in one place, 40 miles within the land. In a river called *Ball's*-river, upon the Southside of a Mountaine, there is a little Grove of Wood, about 6 or 7 foot high, like a Coppice in *England*, it being of Willow, Iuniper, and such like. They found much *Angelica*,[1] and he thinkes the Countrey people do eat thereof, for he saw many of those rootes in their boates.

There are store of Foxes in the Maine and Ilands, of sundry colours; they are as white as snow, and long furred; there is divers deere, but they be farre up within the land; for the Inhabitants doe hunt them sore that come towards the See, where themselves live; he see 7 at one time; his men had divers Darts & horns of Deere; he see the foot of one beast bigger than the foot of an Oxe, their Dogs and Foxe Pizzels have a bone within them. The people all the Sommer time use nothing but fishing, drying their fish and Seale's flesh upon the rocks for their winter's provision. Every one, both man and woman, have a boat covered with Seale's skinnes, close sowed, that no water can enter them. Some of them are 20 foot long, and not above 2 foot broad, shaped like a Weaver's shuttle, so light that a man may carry many of them at once. They use but one oare, with a washe at both ends; it is incre-

[1] *Angelica* is a genus of plants belonging to the order *Umbelliferæ*. Most of its species are found in the temperate or colder regions of the Northern Hemisphere. Several species are natives of North America. *Archangelica officinalis*, the garden angelica, is a native of Northern Europe. Its roots and stalks, when young, are eaten raw, and are brittl eand sweet, like celery. It was formerly much grown in gardens, among other herbs, and was believed to be a sovereign remedy against the plague and all poisons. The leaves were held to be good against sorcery and enchantment: hence its name.—C.

dible to see how swiftly they rowe, no ship being able to sayle so fast; they sit in the midst of their boates, and holds their oare in the middle. In these boates, they catch their fish, as Seales, Salmon, Morses, and others; they strike some with Darts, and angle others; their coard is made of Whale-bone, [&] their hooke of a bone, with which lines and hooks wee have catched very much fish.

I could not learne of their rites and Ceremonies, but generally they worship the Sun. They remove from place to place as their fishing doth serve. They live in tents in the Summer, and in winter in houses somewhat within the ground. When they approach you, they will hold up their hand to the Sunne, crying *Eliout*, which, we answering in like fashion, they dare boldly come to us. They make a Cave with stones, wherein they bury their dead, according to the bignesse of the Corpes, defending them strongly from the prey of Foxes, or Ravenous beasts; they make another neare the former, wherein they bury his bowe, arrowes, and darts, and other his provision, and he is buried in his apparell; the coldnesse of the clime keepes the body from putrifaction.

They eate their meate rawe, yet they use fire; they drink Salt-water by the Ship side.

Some of our men conceived them to be Man-eaters, but he thinks not, because they might have killed 3 of our men at one time filling water in an Iland far from our Ship and without any weapon, at which time a great company of them came to them, and, searching their boate for Iron, they gave them all they had, with their chest keyes, whereupon they depart without doing them any harme; but let others take heed.

Since that I have entred thus farre into Groenland *by the way of Capt.* Davis *and Mast.* Iames Hall, *heare the report of* Dethmar Plef-kins,[1] *a Minister, sent into* Iseland *from* Hamburgh, 1563.

G*Roenland* was first so named in the yeare of CHRIST, 900. And although I purposed, saith he, to passe over *Groenland* with silence, yet, seeing I touched upon the land, and observed some few things, I thought it not impertinent to make mention of them. There was, in a Monastery in *Iseland*, called *Helgafiel*, a certaine blind Monke who lived miserably there. He was borne in *Groenland*, of a darke complexion, and broad face. The Governour commanded him to be brought unto him, that he might know some part of the State of *Groenland;* he said there was a Monastery of St. *Thomas* in *Groenland*,

[1] Dethmar Plefkins or Dithmar Blefken was probably a Dane. Purchas tells us (*his Pilgrimes*, pt. iii, lib. III, p. 643) that, in 1563, he was staying at Hamburg, awaiting his library from Rostock, when two Hamburg merchant-ships, bound on an expedition to Iceland, requested Doctor Paulus of Etgen, who was then Superintendent of the church of Hamburg, to provide them with a minister. "This office [Plefkins says] was bestowed upon me, which I undertook the more willingly, because I had a longing desire to know strange things and divers countries." The ships sailed from Hamburg on the 10th of April, and reached Iceland on the 15th of June. There they remained until March following, when they proceeded to Greenland and explored the country, returning to Iceland in June. Plefkins then proceeded to explore the country, and particularly Mount Hecla, which he had sufficient opportunity to do, for the Hamburgher, despairing of his return from the interior, set sail for home without him. He remained in Iceland three years, and then took passage in a Portuguese ship and arrived safely at Lisbon. Whilst in Iceland, he became well acquainted with Arngrim Jonas (see *ante*, p. 17). His Description of Iceland, to which he added Histories of Iceland and Greenland, was published in Latin at Leyden in 1607. There are one or two later editions. Foxe has abstracted the following from the translation given by Purchas (*loc. cit.*).—C.

into the which his parents thrust him when he was but young, and after[wards] he was taken out by the Bishop of *Groenland*, when he was 30 yeeres of age, to sayle with him into *Norway*, to the Archbishop of *Nidrosia* (or *Drunton*)[1] to whom the Iland Bishops are subject. On his returne, hee was left in a Monastery by the Bishop, whose country *Groenland* was. This was done as he said, in 1546. He said that Iland was called *Groenland*, *antiphrastically*[2]: for that it seldome or never waxeth greene, and that there is so great cold there throughout the whole yeere (except *Iune*, *Iuly*, and *August*) that being clothed and covered with Furres, they could scarce be warme, and that they had at home certaine round pieces of wood, which being moved with their feet, kept their feet warme. He saith that it aboundeth, as *Island* doth, with fishes, and that they had Beares and white Foxes, nay, Pigmies, and Vnicornes, and that the *Day* did not appeare untill the Sun had run through *Pisces*.

This Monke told us marvellous strange things; that there was in the Monastery of S. *Thomas* (where he lived) a Fountaine which sent forth burning and flaming water, that this water was conveyed through Pipes of stone, to the severall Cels of the Monks, and that it made them warme as stoves do with us, and all kind of meats might be boyled in this Fountain and fiery water, & no otherwise than if it had bin on a fire indeed; he advertised, moreover, that the wals of the Monastery were made with Pumice stones out of a certaine mountain not farre from the Monastery: like to *Hecla* in *Iseland*, for if you powre this water upon the Pumice stone, there will follow a slymie matter, which insteed of lyme they use for morter.

---

[1] Drontheim.—C.

[2] Antiphrasis is a figure of speech by which words are used ironically to convey the opposite of what they mean.—C.

After the Governor's conference with him, I came privately to demand certaine particulars touching the Pigmies and other things. He had a little skill in the Latine tongue; he understood me speaking Latine, but answered me by an Interpreter. He said the Pigmies represent the most perfect shape of man; that they were hayrie to the outermost joynts of the finger; and that the males have beards down to the knees, but, although they have the shape of man, yet they have little sense or understanding, or distinct speech, but make shew of a kind of hissing, after the manner of Geese; that his Abbot kept two of them in his Monasterie, male and female, but they lived not long, and that they were unreasonable creatures, and live in perpetuall darkenesse; that some say, they have Warre with the Cranes; but that, he knew not.

He affirmed that the same manner of food was in *Groenland* as in *Iseland;* to wit, of Fish, but not of Cattell, because they have no Cattell; and that the Countrey is not populous. Forthwith from *Iseland* begins the *Hiperborion Sea,* which beates upon *Groenland,* and the Countrey of the Pigmies, which at this day is called *Nova Zembla,* and there the frozen Sea hath a Bay, which is called the White Sea, or *Mare album;* and there are certaine passages whereby they saile into the *Scythian* Sea, if they can for Ice. And the Governour had a ship of the King of *Denmarkes* furnished with all necessaries, but when he heard (by the Monke) of the short cut into the Kingdome of China, by the *Tartarian* Sea, which had often been attempted by others, held it but in vaine.

The last of *March,* 1564, he commanded that ship to sayle to those places, and me also together with them, enjoyning me diligently to marke well the scituation of the places and whatsoever we met with worthy of sight or report. We were in the ship 3 score and 4 men, as well *Danes* as *Iselanders,* and the 20 day of *Aprill,* we arrived in a certaine *Promontorie* of *Groenland;* and when we found no harbour to the which we may safely commit our selves, letting downe the lead, we found the depth of the Sea, and it was such as we could not anchor there, and the abundance of Ice was so great that it was not possible to saile near the rocks. 24 of us armed, with great labor and danger, went on shore in our Skiffe, among whom I was, to try whether I could find a harbour or no, and what kind of men *Groenland* had. In the meane time, the ship floated in the Sea and Ice in a great Calme; halfe our Company abode on the shore (to keepe the Skiffe); another part (and I with them) ranne abroad to discover; they that were left on shore to keepe the Skiffe, going hither and thither, found a little man dead, with a long beard, with a little boat, and a crooked hooke of the bone of a fish, and a leather Cord; foure fish bladders were bound unto the boat (as is supposed) that it should not be drowned, whereof 3 were sunke and fallen flat; this boat (because it was very unlike ours) the Governour sent to the King of *Denmarke.*

Wee wandred in the meane season in a land unknowne unto us, which was covered with snow and Ice; we found neither footing of men or any inhabitants, nor fit place, but the Sea was closed and fensed on every side with craggy Rocks; yet we met with a great white Beare, which neither feared us, nor could be driven away with our cries, but came full upon us, as to his certaine prey, and when he came neare unto us, being twice shot through with a gun, he stood bolt upright, with his fore-feete like as a man standeth, untill he was shot through the 3 time, and so fell downe dead. His skinne was sent to the King of *Denmark.* Wee agreed amongst ourselves before we went on shore, if we found a fit harbour or else had need of their helpe, that we should plucke up our *Stander,* which we carried out with us for that purpose; and that, if they would call us backe, should signifie the same with their Ordnance. A tempest arising in the meane while, the Master of the Ship gives us a signe to returne, and calls us backe unto the Ship, all of us after great labor, in 3 dayes returned unto the Ship with the Beares skin; we sailed therefore to the other side of the Island to the North, to the Country of the *Pigmies* or *Nova Zembla*,[1] that by the mouth of the White Sea, wee might come to the *Scythian* or *Tartarian* Sea; from thence, they say, there is a passage to the kingdome of *China* and *Cathay;* but being hindred by the Ice, wee could not passe the mouth of that Sea, wherefore, without doing anything, we returned into *Iseland* the 16 of *June*.

The Voyage of Master Iohn Knight, *imployed into* Groynland, *as Captaine, the yeare before by the King of* Denmark, *but now out of* England *to search the N. W. passage*, 1606.[2]

HEe was set forth at the cost and charge of the *Muscovia* Company, and the *East India* Merchants.

He set sayle from *Gravesend* the 18 of *Aprill*, and arrived in *Orkney* the 26; hee stayed there 14 dayes with contrary windes, and in the meane time hee entertaines 2 men of the Country, to goe along with him on the Voyage.

---

[1] A great mistake. The *Russ.* were the first that named this land *Nova Zembla*, so *Stephen Burrowes*, 1556. But this cannot be *Pigmies*-land, for that must ioyne or be neare to *Groenland*.—F.

[2] The narrative of this unfortunate voyage is condensed from that given by Purchas (*his Pilgrimes*, pt. iii, lib. IIII, pp. 827-831). Knight had sailed under Capt. Hall from Denmark to Greenland in the previous year.—C.

May 12. He set forth from St. *Margaret's* Sound in *Orkney* and steered away W. by S.

16. This day at noone, holding still his course Westward of S., the Latit. was 58 d. 19 m., his magneticall declination 8 deg.; also this mornings Sun, being 10 d. above the Horizon, was distant from E. Northward 22 degrees.

17. This day he was in 58 d. 10 m.; this morning the Sun did rise 50 d. to the N. of East.

21. He was in Latit. 57 d. 50 m. continuing his course still Southwards of W.; from the last day, much wind at N.N.E. and fog, and he had a current setting to the North.

22. He saw many Sea guls and Rock weede.

23. He iudged his way to be made S.W. by W., but it proved W. by S. or rather W., 20 leag. by reason of a current (there he sees an Owle).

28. He had Lat. 57 d. 57 m., the variation 14 d. 30 m. Westward; this day he had black water streames, leaches and fals of currents, it seemed to the N. and some to the West.

29. Latit. 58 degrees he was in a tyde-gate,[1] which hee judged to set N. and S., or that it was the Eddie of the currents which he saw the other day. He sees white Fowles chirping like sparrowes, and also many dead Cowes adrift (*I think rather Crowes*). His Latit. was 58 d. 3 m., the variation 24 d. to the N. Westwards; this night he observed the Sun to set 21 d. W. from N.

June 1. He was in Latit. 57 d. 35 m. and saw many black fowles, like Willockes[2] flying in flocks together.

4. He was in Latit. 54 d. 40 min.

5. He was in 56 d. and had the variation by the Scale 20 d., by his other Instruments 24 d. W.; the Sun was

---

[1] A place where the tide runs strongly (Smith's *Sailors' Word-Book*).—C.

[2] No doubt Black Guillemots or Dovekies (*Lomvia grylle*).—C.

22 d. ½ high; and to the North off W. 30 d. by the Instruments, and 26 d. by the skale.

11. At noone, he had Latit. 58; at night, the Sun set 14 d. W. from N.; at morning, it riseth 50 d. to the E. of North.

13. He had sight of land like Ilands in Latit. 57 d. 25 min., and had some quantity of Ice driving to the South.

On Friday, he makes fast to the Ice, but the wind calming, he rowes betwixt Ice and Ice, towards the land, but in a fog he makes fast againe to the Ice; when it cleered, he loosed againe and Rowes in, and is exceedingly pestered with Ice, suffering a mighty storme (with thick and foggie weather) which so bruised his Shippe betwixt the Ilands of Ice that he was in danger to be crushed to pieces, though [he] imployed all his industry to the uttermost of his power.

19. He again discovers the Iland of *America*, bearing N. and by W. about 15 leag. off, the variation 25 d. W., Latit. 56 d. 48 m. This coast sheweth like broken Ilands, and the Tyde of flood commeth from the North.

24. He had a violent storme Northerly, and such a suffe[1] of the Sea came in that his roapes broke that were fast on shoare; his Rudder was driven from his sterne by the force of the mighty Ilands of Ice, so that he was forct to hale close into the bottome of a cove, to save his Cloathes, furniture and victuals; but, before he had done, the Ship was halfe full of water, and he tooke little rest for that night (which may very well be beleeved).

25. The ship on ground, he goeth about to set the water out of her, and to stop so many of her leakes as he could come by, and some went to building the shallop. He caused his boate to be lanched over the Ice, and sent his mate

[1] Nares mentions the use of the word *sough* in a somewhat similar connection.—C.

Edward Gorill, with 3 more, to seeke for a better place where to bring his ship on ground, if it were possible to mend her againe. They returned without any certainty, by reason of the abundance of Ice, which choaked every place, *yet they found Wood growing on the shoare.*

Here Master Iohn Knight *ended writing his Iournall with his life; another proceedes as followeth.*

June 26. ON Thursday, in the morning, he caused some of his men to goe on board to save what things they could; himselfe, *Edw. Gorrill* his mate, and 3 more of his company, tooke the boate, having with them 4 pistols, 3 muskets, 5 swords, and 2 halfe pikes, for to goe over to a great Iland, not above a mile from the ship, to see for Harbour to mend his ship in. They tooke with them an Æquinoctiall Diall, a paper to draw the land when they were passed over. The Mr., his Brother, and Mate, with one more, went on shore, leaving 2 in the Boate, from 10 of the clocke in the morning untill 11 at night, who heard no newes of them after they departed from the top of the hill; then did the Trumpeter, being one of them left in the Boate, sound two or 3 times, and the other did discharge his Musket two or 3 times more, and so they came away to the W. side with the Boate, where the ship was, who were watching for their comming; but, when they see these two come and no more, they marvelled where the rest were; and, when they were on land, the others inquired for the Master and the rest, but they could tell no newes of them after their departure from the boate, but that they did see them goe on to the top of the Iland, which report did strike all the men into great feare, to thinke in what extremity they were, because they wanted their Master & 3 of their men, their ship sunk, and they nothing to trust too but their shallop, which was at that time but ½ furnished. This night, lying on shoare in their Tent, which was betwixt 2 Rockes, they kept very good watch, for feare of any peoples suddaine assault, or if the Master and his company had travailed so farre as they could not come home againe that night, if they should shoote a musket, they might heare them; but they came not at all.

27. They consulted that 7 of them should goe over with the boat to try if they could see or learne any newes of their Master or of their men, supposing that they were either surprised by the *Salvages* of the Country or else devoured by the wild beasts; so they tooke with them 7 muskets, swords and Targets,[1] and such provision as they had in the ship, and went downe to the Sea side, but they could not get over for Ice; at length they returned, with much adoe to come to shoare, and went to the ship to save such things as they could get out of her.

28. Faire weather; they make cleare the ship, and helpe to save and mend all things; she lay upon the Rocks, therefore they make her as light as they could, for beating and bruising of her Hull. That night it rained very sore, and about clock 1, in the boat-swain and Steward's watch (it being almost out), the Steward goes on board the ship to pumpe, leaving the boate-swaine at watch some muskets' shot from their Tent. Now, while he was pumping, there came over the Rockes a great sort of the Countrey people, to the place where the boate-swaine was at watch, who, when they saw him, they shot arrowes at him, running to him as fast as they could, whereupon he discharged his musket at them, and so fled backe to the Tent as fast as he could, thinking they had beset it, they were so many. The Steward, hearing the Musket goe off, came forth of

[1] A target, or targe, was a kind of buckler, or shield, worn on the arm.—C.

the Ship, & when he was comming, saw the *Salvages* approaching their Shallop, and cryed out to them in the Tent to save the boate and the Shallop, who made what hast they could ; but, when they came at their boat and see so many of the *Salv.* in the Shallop, they were then afraid of being betraied. At this time it rained sore, yet calling their wits together, they sent 2 of their men back to the tent, the rest made towards the *Salvages*, and shot at them some 3 or 4 muskets, who, when they perceived it, they stood in the shallop and held up their hands to them, calling one to another; they, thinking it were better to dye in their owne defence in pursuing the *Salvages* then they them, for it was in the night, and they were still in sight; thus, recovering their Shallop, they sent more men to keepe the Tent. The *Salvages* were but 8 men, and a dog, yet the rest followed, but they were got into other boates, before they overtooke them. The ice was so thick that they stucke fast in them; they came so neere them as they could, and shot a doozen shot at them, before they could get cleare ; the shot caused them to cry out one to another very sore, for their boates were full of men; as far as they could iudge, they are little people, tawny coloured, thick haired, little or no beard, flat nosed, and are man-eaters.

June 29. They carry all their provision aboard their Ship for feare of the *Salvages'* second assault, the ship lying betweene two Rocks, and all without so full of Ice that they could not passe any way to sea, no not with a boate. This day, 2 of our men watcht, that we might have warning if any of them came againe with their boates; the Carpenter made what hast he could with the shallop, and did onely tinch,[1] but neither calkt nor pitcht her, yet they brought her to the ship.

---

[1] In Purchas, the word is *tench*. Presumably the word is a form of *clinch*, meaning that they clinched her rivets only, but Nares does not give it.—C.

30. They fall to worke with axes and pick-axes to cut the Ice, for all about the Iland was nothing else, and no place to ride free in. That night, it pleased God that they rowed her away with their oares, but she was exceeding leaky and the shallop also, and, which was worse, they had never a Rudder to steere withall, yet they rowed all night amongst Ice.

July 1 and 2. They rowed up and down amongst the driving Ice, with little hopes to recover their Countrey.

3. The wind at N., they had a great current set to Southward; they make fast to a peece of Ice and went to worke to stow their things snug and close downe within board to make her stiff, for they had no ballast; the Carpenter makes what shift he could to hang their Rudder; having nothing to make Gudgeons nor Pintels,[1] they were faine to breake open the Master's Chest, to take the Iron bands, to make fast 2 pickaxes for 2 pintels; this Night they hang the Rudder with 2 pintels, and a Cable through the middle of it, to keepe it too with two tackes; now were they in good hopes to get cleare, for before their ship being leaky, with her stem sore beaten with Ice and Rockes, themselves, with pumping and rowing, were both sore and wearie.

Iuly 4. This day, the wind came W.N.W. and was faire weather (they got cleere out of the Bay the Noone-watch) and was the first they had of long time before, being all glad to watch to conduct the Ship cleare from the Ice; their ship was so leaky that, if she stood vnpumped but halfe an houre, they could not dry her with 1000 stroakes;

[1] The rudder is hung on to a ship by pintles and braces or gudgeons. The braces are secured firmly to the stern-post by jaws, which spread, and are bolted on each side. The pintles are hooks which enter the braces.—C.

therefore they were inforced to Rumige,[1] and found many leakes, but not that which caused them to pumpe so sore; at last they find it, close abaft the fore-foote, where the keele was split in 2 or 3 places. They could not come to stop it, for it was vnder a timber, in at which the Sea came so fast as it was not possible to keepe her free with both pumpes; then did they take their maine bonnet and basted it with Okum, and put it over-board right against their leake, which eased them 4 or 500 stroakes in an houre. They all this day consult to shape their course for *New-found-land*, hoping to meete with some *English* or *French*, and to mend their Ship. At this time, one of their men was very sicke; another had his hand splinted, and most of them all were so sore with rowing and pumping as they were not able to stirre, but that they must perforce.

5. They shape their Course for *New-found-land.*

21. They fall with land, being nothing but Ilands; Latit. 49 degrees, 30 minutes.

22. Faire weather; they stand in among Ilands, and a great Current set from Iland to Iland; no ground at 100 Fathoms, they keepe too and fro all this night in great danger, being among broken Rockes, with thicke weather.

23. They espie a dozen shallops fishing; they make towards them; they tooke harbour and remained in this Bay of *Fogo* untill the 22 of *August*, repairing their ship and refreshing themselves; the 24 of *September*, they arrive at *Dartmouth.*

This journal, from the death of Master Iohn Knight, *was writ by* Oliver Browne,[2] *one of the Company.*

---

[1] In former days, to rummage a ship was to thoroughly overhaul both her and her contents. Nowadays, the term is usually confined to the search by Custom-house officers for smuggled goods.—C.

[2] This is probably a misprint for Brownel. What Mr. Markham regards as the upper part of a broken letter *l* appears in Purchas after the word Browne, though it seems to me more like a comma which has

An Abstract of the Voyage of Master Henry Hudson, *to the Northwest, begun the* 17 *of Aprill*,[1] 1610, *and ended with his life, being treacherously exposed by some of his Company.*[2]

April 22. IN the Road of *Lee*, in the River of *Thames*,[3] he caused Master *Coolbrand* to be set in a Pinke,[4] to bee carried backe againe to *London*. This

---

slipped out of its place. Oliver Brunel or Brownel was a well-known Dutch navigator, who seems to have been with Knight on this voyage (see *The Voyages of Barentz*, second edition, edited by Lieut. Koolemans Beynen; Hakluyt Society, 1876). Further information as to Knight's voyage may be found in *The Voyages of Sir James Lancaster, Kt., . . . and the Voyage of Captain John Knight* (1606) *to Seek the North-West Passage*, edited by Mr. Clements R. Markham (Hakluyt Society, 1877), wherein (p. 278) is printed a narrative of the voyage slightly different from that given by Purchas, taken from a MS. preserved in the India Office.—C.

[1] Bylot's deposition says the 18th.—C.

[2] The following account is an abstract of that given in *Purchas his Pilgrimes*, pt. iii, lib. III, pp. 596-597. For further information concerning Hudson, the reader cannot do better than consult Dr. Asher's admirable work, *Henry Hudson, the Navigator* (Hakluyt Society, 1860), wherein will be found practically everything known of Hudson and his voyages until the recent discovery, at the Trinity House, of some very interesting records relating to his last voyage. These records (which consist mainly of extracts from the log and depositions of the survivors taken before the Corporation of Trinity House, together with the opinions of the Corporation upon them), I have pleasure in being now able to print for the first time (see Appendix). They have been already described in the Introduction, and their bearing upon the narrative given above is shown in greater detail in many footnotes hereafter. Bylot's deposition states that Hudson's ship was victualled for eight months, but Prickett (see p. 131) says six.—C.

[3] By this is no doubt meant Hadleigh Ray, a creek running between Canvey Island and the north (or Essex) shore of the Thames. It was, no doubt, so called from the ancient port and fishing-station of Leigh, which overlooks the creek.—C.

[4] A ship with a very narrow stern, having a small square part above. The shape is of old date, but continued in use, especially among the Danes, for the advantage of the quarter guns.—C.

Coolbrand was every way held to be a better man than himselfe, being put in by the Adventurers as his assistant, who, envying the same (*he having the command in his owne hands*), devised this course to send himselfe the same way, though in a farre worse place, *as hereafter followeth.*[1]

May 5. He came to the Isles of *Orkney*, and here he set the N. end of the Needle, and the North end of the Fly all one.

6. He was in latitude 59 d. 23 m., and there he perceived that the N. end of *Scotland*, *Orkney*, and *Shotland* (*for hee visited them all*), as he saith, are not so Northerly as is commonly set downe in the Charts.

8. He saw *Farre Ilands* in Lat. 62. 24 m., but he staid not there.

11. He fell [in] with the E. part of *Iseland*, then plyed up along the S. part of the nd and came to the West-

---

[1] No definite information seems to be forthcoming as to who this man was, or as to the reason for Hudson's action regarding him. Even his precise name seems doubtful; for, while Hudson calls him Master *Coleburne*, Prickett calls him Master *Colbert*, and Foxe here speaks of him as Master *Coolbrand*. It may, however, be noted that a certain *William Cobreth* was master of the *Discovery* (the very vessel Hudson sailed in) when she sailed under Captain Weymouth in 1602 (see p. 80); and knowing, as we do, how exceedingly variable was the spelling of surnames in those days, it is very likely that this was the same individual. If so, it seems not at all an unreasonable conclusion that the "adventurers" had wished that he should accompany Hudson (who, we know, was setting out on this voyage, largely at least, to follow up Weymouth's discovery), but that Hudson, for some reason, objected, and took this means to rid himself of a companion of whom he was jealous. The estimate of Coolbrand's ability above expressed is, it should be noted, that of Foxe. Its correctness has been hotly repudiated by Dr. Asher (*Henry Hudson*, pp. xliii and cxcvi); but, if my surmise as to the identity of this man with Weymouth's Captain William Cobreth be correct, the discrepancy is to some extent explained. Unfortunately, the Trinity House depositions contain no reference to this man. Rundall gives a facsimile of his autograph (*Voyages towards the North-West*, p. 238).—C.

most,[1] and the 15 he still plyed up untill the last of *May*, and got some fowles of divers sorts.

Iune 1. He put to Sea out of a harbour in the Westmost part of *Iseland*, (and according as hee writeth) plyed to the Westward in Latit. 66 d. 34 m.

2. He was in Latit. 65 d. 57 m. Small wind Easterly.

4. He saw *Groenland* perfectly over the Ice. This night, Sun set at N. and rise N.N.E.

The 5 he plyed in 65 d., still incombered with Ice, which hang upon the coast of *Groenland*.

9. He was off *Frobrisher's straits*, and plide Southward untill the 15, and then he was in Latit. 59 d. 27 m. and had sight of *Desolation*, and finds the errour of the former lying downe of the land.[2] Running to the Northward, as he saith, untill this day [20th] in 60 d. 42 m., he saw much Ice, many riplings and overfallings, and a strong streame setting West Northwest.

23. In sight of much Ice. Wind variable, and in latitude 62 degrees, 19 minutes.

25. About midnight, he saw the land North, but [it] was suddenly lost[3]; yet he runne still Westward in 62 deg. 19 min., and he plyed upon the South side seeking the shore; he was troubled with much Ice in latitude 62 deg. 16 min.

Iuly 8. Hee plyed off the shore againe until this day. The *Poles* elevation 60 d. 0 m. He saw the land from

---

[1] A misprint for *Westmony*, meaning the Vestmannaeyjar (Westmen's or Irishmen's) Islands, a little south of Iceland.—C.

[2] Davis his error, but not beleeved.—F. See *post*, p. 122.—C.

[3] The course from the 25th to the 30th of June is described in the depositions, wherein it is stated that they entered Hudson's Strait on the 26th, and on the 27th sailed westward, apparently along the north shore. On the 30th, they anchored to the ice and found a set of tide to the westward. From this date to nearly the end of July, the course is described both in Purchas and in the depositions, with some slight discrepancies.—C.

N.W. by W. $\frac{1}{2}$ N. unto the S.W. by W., covered with snow, a Champion land, and cals it *Desire provoked.*

11. He plyed still to Westward and, fearing a storme, he anchored by 3 ragged Ilands, in uncertain deepes, betweene 8 and 9 fathomes. He findes the harbour unsufficient, by reason of sunken rockes, one of the which was the next morning 2 fathomes above water (which he had gone over); hee calls them the *Isles of God's mercy.* It floweth here better than 4 fathomes; the flood came from North, flowing 8 a Clock the Change day. The latitude in this place is 62 d. 9 min.

16. Plying to Southward untill this day, he was in 58 deg. 50 min. There he was Imbayed with land, and had much Ice.

19. Vntil this day he plyed Westward, and found his latitude 61 deg. 24 min. where he see a Bay in the Southland, which he named *Hold with hope.*

21. Hence he plies to the Northward, had variable winds, and findes the Sea more growne then he had any time since he left *England.*

23. The *Poles* height was 61 deg. 33 min.

25. He saw the Southland, and named it *Magna Britania.*

26. He was in latitude 62 deg. 44 min.

28. He plied Southward off the Westward, and was in 63 deg. 10 min.

31. Plying Southerly, he found himselfe in 62 d. 24 min.

August 1. He had sight of the North shoare, from the N.E. by E. to the W. by S., the N. [part] 12 leagues off, the W. part 20 leagues. He had no ground at 180 fathomes, and he thought hee see land bearing E.N.E. on the Sun's side, but could not make it perfectly. Here he found the latitude 62 deg. 50 min.

2. This day, he had sight of a *Faire* headland, 6 leagues off, which he called *Salisburie's Foreland;* he ran from

thence W.S.W. 14 leagues, in the midst of which, he came into a great whirling Sea, whether caused by the meeting of 2 streames or overfals he knew not; thence, sayling W. by S. 7 leagues farther, he was in the mouth of a straight, and had no ground at 100 fathomes, the straight not being above 2 leagues broad in the passage in this Westerne part, which from the Eastern part of *Fretum Davis* is distant 250 leagues.

3. This day he put thorough this passage (which was narrow) after his men that did observe had been on land. The flood-tide did come from N., flowing by the shore 5 fathoms.[1] He names the Cape on the Star-board, C. *Digges*,[2] being

---

[1] On this date, Hudson's own detailed narrative ceases, though the paragraph following (which either Foxe or his printer misplaced, as may be seen by comparing his original narrative with that in Purchas) continues the account in a vague fashion, and without dates. Unfortunately, the narrative given in the depositions ceases on the same date. However, it recommences on the 12th and continues very briefly until the 22nd, thus giving us the only account we have, in anything like detail, of Hudson's course southward along the eastern shore of the Bay.—C.

[2] Dr. Robert Bell, of the Geological Survey of Canada, says (*Report of Progress*, 1885, p. 11, DD.):—"The south-eastern part of Inner Digges Island presents a high and nearly-vertical cliff, facing the still higher bluffs of Cape Wolstenholme, of which an outline-sketch was given in my report for 1880. From the Cape, these bluffs continue southward for some miles, diminishing in height and merging into the rounded hills of the coast further down". Like Cape Dudley Digges in Baffin's Bay (discovered and named by Baffin in 1616), this island was named after Sir Dudley Digges, one of the principal promoters of the early voyages of exploration to the north-west. He was a shareholder in the East India Company, and was employed on various embassies to Russia, Holland, and elsewhere, in connection with the Muscovy and the East India Companies. He was Member of Parliament for Tewkesbury, and attacked the system of farming the customs, and monopolies altogether, vigorously asserting the privileges of the Commons, both in and out of Parliament. He was appointed Master of the Rolls in 1636, and died in 1638-9. To him and to Hakluyt (who died in 1616), Purchas says he was indebted for much of his information, and especially for that relating to Hudson's voyages. Pur-

[on] an Iland; that on the Larboard stands upon the maine, named by him Cape *Worstenholme*.[1]

chas's extravagant eulogy of Digges may be found in *His Pilgrimage* (4th ed., 1626, p. 817). He seems to have been the author of a small and now-exceedingly-rare anonymous work entitled *Of the Circumference of the Earth, or a Treatise of the North-East* [a misprint for *West*] *Passage*, which was "imprinted at London by W. W., for John Barnes, 1611." A second edition (apparently identical, except for the omission of a Latin quotation) appeared in the following year. It is a small pott octavo tract of twenty-six pages. No recognised bibliographer has, so far as I know, attributed this publication to Digges; but among the State Papers, Domestic, Jas. I, vol. lxvii, No. 82, there is a letter (printed in *The Court and Times of James I*, 2 vols., London, 8vo, 1848, vol. i, p. 153) from John Chamberlain, Esq., to Sir Dudley Carleton, dated London, Dec. 4th, 1611 (a few months after the return of Hudson's crew), in which he says: "Sr Dudley Digges is in consideration of this new discouerie of this north-west passage (wherein he is a great undertaker). [This matter] will [not] give him leave to thincke of anything els, for yt possesseth him wholy, and they are preparing shipps against the spring, as yf there were no doubt nor difficultie at all in the matter, and the Prince is become patron and protector of this new discouerie." This of course, refers to Button's intended expedition, which was then in preparation. On March 21st, 1612, the same writer, in another letter to Sir Dudley Carleton, says: "There is a litle treatise of the North-West Passage, written by S. D. Digges; but I may say *beatus qui intelligit*, especially the first periode, wch is but a bad beginning to stumble at the threshold. Some of his goode frends say he had better have geven five hundred pound then published such a pamphlet. But he is wonderfully possessed wth the opinion and hopes of that passage" (see State Papers, Domestic Series, Jas. I, vol. lxviii, No. 78; also *loc. cit.*, vol. i, p. 139). The entire pamphlet was reprinted by John Petherham in his *Bibliographical Miscellany* (No. 2, Dec. 15, 1853, pp. 9-16, and No. 3, Jan. 20, 1854, pp. 17-18). Of the first edition, he says: "The only copy of this with which I am acquainted is in the possession of Dr. Bliss, of Oxford, who kindly allowed me to make a transcript for publication." Of the second edition, he says: "I can only refer at present to three copies—one in

[1] This Hudson named after Sir John Wolstenholme, one of his "adventurers", and afterwards one of the chief promoters of Foxe's expedition in 1631. A biographical notice of him is given in the Introduction.—C.

After he had sailed West by South 10 leagues, the land fell to the Southwards, and the Iles to the West left him.[1] He observed, and found himselfe in 61 d. 20 m. and a Sea from the Westwards.

Here is all I finde extant of his owne writing, although he lived untill *Iuly*[2] following, before he was exposed.

A large discourse of the said Voyage, and the success thereof, written by Abacuk Pricket, *who lived to come home.*[3]

May. They came to the Iles of *Orkney*, and from thence to *Farre.*[4] One day, being calme, they fell to fishing, and kild

the Bodleian Library, a second in the British Musuem [there are now two], and a third in Mr. Bright's library, sold in 1846." If Digges was the author and had, as Chamberlaine thought, cause to be ashamed of the fact, the rarity of the two editions may be accounted for by the supposition that he called in all the copies he could. Mr. Petherham, who has compared the two editions, testifies that, although the differences between them are very slight, they are distinct impressions. The second edition was registered by John Barnes, on June 27th, 1612 (Arber's *Transcript of the Registers of the Stationers' Company*, vol. ii, p. 488), but I can find no record of the first edition, which Petherham states was published under the title of *Fata mihi totum mea sunt agitanda per orbem.* The word *East* on the title-page should clearly be *West*, as the contents show. Although the references to Hudson's expedition in the pamphlet are very slight, there seems to be no doubt that its issue was, largely, at least, a result of that expedition, and two extracts from it, quoted hereafter, have undoubted reference to Hudson's voyage. Good biographical notices of Digges may be found in the *Dictionary of National Biography* and in the Introduction to Markham's *Voyages of William Baffin*, pp. x-xvi.—C.

[1] It is not clear which isles are meant. It seems not improbable that by this Hudson may have meant Mansel Island, which he could have seen from the point he had now reached, and which is clearly marked on his chart, though we are not told that he visited it.—C.

[2] He may have done so, but he was "exposed" in the month of *June* following, not July.—C.

[3] But little dependence can be placed on this narrative, and as a guide to discoveries made it is of little value. Prickett scarcely

[4] The Faroe Isles.—C.

good store of Cod and Ling. From thence, they raise the Iles of *Westmony*,[1] in one of which the King of *Denmarke* hath a fortresse, by which they passed to raise the *Snow-hill-foote*,[2] a mountaine so called, on the N.W. part of the land; and in that course, they see the famous Mount *Hecla*, which cast out much fire, a signe of foule weather to come. They leave *Island* on Sterne, and met a maine of

---

gives a date, distance, or latitude, and his account of the mutiny must be read with doubt, from his connection with the mutineers, and his being permitted to remain with them in the ship. Surprise has often been expressed that no punishment should have been inflicted upon the mutineers for their atrocious crime, and there certainly does not seem to be any very satisfactory record of their having suffered any. But, in the four editions of Hessel Gerritz's *Descriptio ac delineatio Geographica Detectionis Freti*, published at Amsterdam in 1612-13 (for translations see Dr. Asher's *Henry Hudson*, pp. 181-194; also Muller's reprint—Amsterdam, 4to, 1878), it is stated that the mutineers returned home by the way they had gone, arriving in England in September 1611 (a date Prickett omits to give), and were at once thrown into prison, the intention being to keep them there until their fellows should have been found and brought home. Moreover, it is to be remarked that Hudson's journal terminates on the 3rd August 1610, but Hudson was not sent adrift in the boat until about the 21st June 1611. Yet no inquiry seems to have been made for the master's journal during this long period, although Prickett admits that he took charge of the master's cabin and that Greene gave him the key of the master's chest. We can only suppose that both Robert Bylot and Abacuk Prickett were able to substantiate their innocence, as they were both afterwards engaged to proceed on the same voyage under Captain (afterwards Sir) Thomas Button, and as Bylot sailed as captain both in 1615 and in 1616. We gather incidentally from Purchas (*his Pilgrimage*, 4th ed., 1626, p. 818) that Prickett had been "a servant of Sir Dudley Digges", and that the mutineers had saved him alive in the hope that he might, through his master's influence, procure for them a pardon for their misdeeds. Prickett's narrative is very little condensed or altered by Foxe from its original form as it appears in *Purchas his Pilgrimes* (pt. iii, lib. III, pp. 597-608).—C.

[1] See p. 116.—C.

[2] Snæfell-Jökull, a mountain on the west coast of Iceland, in Westland, district of Sneefieldness, 4,500 feet high. John Barrow, who ascended it in 1834, gives a good description of it (*A Visit to Iceland*, London, 1835).—C.

Ice lying upon the North part thereof, which, when they see, they stood backe for a Harbour in *Island*, called *Derefer*, where he kild good store of Fowle; from thence they put to Sea againe, but the wind not serving, he puts back againe into another harbour, called *Lowsie Bay;* heare they found a Bath so hot as would scald a Fowle.

June 1. He put forth for *Groenland*, and thought he saw land, but it proved a fog banke.

He raised *Groenland*, but great store of Ice hung upon the land. This land is Mountainous, with hils like Suger loaves, covered with Snow.

He coasted along betwixt W. and N.W. untill he saw *Dissolation*, which is a great Iland on the N.W. part of *Groenland*.[1] Here he saw great store of Whales; some came about and under the ship, but did no harme. From hence, he directed his course N.W. and would have hailed more Northerly if the Winde would have suffered him. In this course, he see the first mountainous Ice. About the last of *June*, he raised the land N. of him; he tooke the same to be that Iland which Capt. *Davis* set in his Chart on the W. side of this straight; he would have gone to N. thereof, but the winde would not suffer him, so he fell on the South side into a great ripling or overfall of a current; into this current he went, and made his way Northward of W. untill he met with Ice which hung on this Island; wherefore he cast about to the South, and cleared himselfe, and then stood to the W. amongst some store of floting Ice, and upon the Ice store of Seales. He

---

[1] Desolation Island was a name applied to the southern part of Greenland, which was supposed to be an island divided from the more northerly part, or mainland, by Frobisher's Strait, as will be seen represented on ancient maps. The Strait in question was, in those days, quite wrongly placed, as has been already pointed out (p. 38). Foxe speaks of Desolation being on the *north-west* part of Greenland, but Prickett says on the *west*, and *south* would have been more correct.—C.

still made way N.W. and meetes sometimes with Ice, and then againe cleare. Thus, proceeding betwixt Ice and Ice, he sees a great Iland of Ice overtumble, which was a good warning to him not to come neere them. The next day he had a great storme, and was driven to put in amongst the Ice, and there to lye. Some of his men fell sicke. He will not say for feare, though he see no other signe of griefe.

The storme ceasing, he put from the Ice, standing W.S.W. and N.W. and as the cleare Sea and Ice would suffer him. Seeing the sea thus continued, he would seeke to the S., but the more he strove, the worse he was, for he was fast inclosed and began to dispaire (as he after told this Writer).[1] Hee thought he should never have got out of this Ice, but there haue perished.

Wherfore he shewed him his Card, that he was 100 leag. further than ever any *English* man was,[2] and referred to their choyse to proceed further or no. Whereupon some wisht them[selves] at home, others any where so from amongst the Ice; but, in this distraction, there passed words which was thought upon long time after.

There was one told the Master that, if he had 100 pound, he would give 90 therof to be at home; but the Carpenter made answere and said, if he had 100, he would not give 10 upon any such condition, but would hold it as good money as ever he had any, and by God's leave to bring it as well home. Well, to worke they goe, and with labour get the ship cleare, the sea being also cleere a league or 2

---

[1] This means that Hudson told Prickett.—C.

[2] See Weymouth.—F. This refers to Waymouth's statement that in 1602 he sailed 100 leagues into the Strait, which Foxe ridicules in a side-note (see p. 84), though without good cause. There is, however, undoubtedly an error here, probably due to Prickett's vague recollections; for Hudson had not yet sailed as far into his Strait as either Frobisher in 1576 or Weymouth in 1602. For Dr. Asher's remarks on this point, see *Henry Hudson*, p. 101.—C.

off. The course now is N.N.W. In the end, he raised high land at S.W. (covered with snow) and names it *Desire provokes.*[1] He heard the noyse of a great overfall of a tide that came out of the Land, and now he could well discerne that formerly he had beene imbayed, and that Time made his acquaintance so well knowne to the Ice that, when wet fogge or foule weather came, he should make fast to the broad peece of Ice, and run and sport, and fill sweete fresh water upon the same. After he had brought this land to beare S. of him, he had a current or tyde to open the Ice, being carried first one way and then another, but in bayes they lay as in a pond, without mooving. In the Bay wherein he was thus troubled, he see of those Icie mountaines on ground; here was 120 and 140 fathomes of water, and here he sees a Beare upon the Ice; his men gave chase to him, but lost him; by the tydes carrying away the Ice, the Beare was gone out of their reach.

He continued his course towards the North-West and raised land to the North, and Ice, where there was a Beare which came from Ice to Ice, untill she came neere them, and then she runne away. Hee stood along on the South side of this land, meeting with Ice, which seeing he stood into the shore, and found a harbour in the West end of this Iland. He went over one Rock of two fathomes and a halfe, and at low water it was so much bare; he names it the Iles of *God's mercy.*[2] The Master sent this Author[3] over to discover this Iland to the North and North-West; he met with a covey of Partridges, and kild onely the old one. It was barren land, having nothing thereon but water plashes,[4] and torne Rockes, as though it had

---

[1] No doubt Akpatok Island.—C.

[2] This name still appears on our charts.—C.

[3] This means that Hudson sent Prickett.—C.

[4] Shallow *pools of water* are meant (see Nares's *Glossary*).—C.

beene subiect to earthquakes ; to the North of this Iland, there is a great bay or Sea ; they know not what it may prove. There was a great Iland of Ice on ground, which with the spring tide was set on float and carried to the N.W. but came not backe within sight. Here they found some drift wood on shoare.

From thence he stood to S.W., to double the land to the W. of him, through many peeces of Ice ; at length, he found a cleare Sea and raised land to the N.W., whereupon he stood more to S. then before (and fell amongst Ice), which he would have doubled to the N. but could not ; then he stood to the S.W., amongst the Ice, and strove to get to the land, but could not for the Ice. From out of this bay, he stood to the North, and was soone cleare of the Ice ; then he stood to S.W. and W., where he was inclosed with land and Ice, having land on the one side from the S. to the N.W. and on the other side he see land from E. to W., but the land that lay to the N. and lay E. and W. was but an Iland. He went so farre as he could (and made fast to the Ice the tyde brought upon him), but the ebbe did open and made way, and 7 or 8 houres he was cleare, and with the great Ilands of Ice, was carried to the N.W.

He stands to the W. along the S. shoare and raised 3 Capes or headlands one above another ; the middlemost is an Iland with a Bay, which he thinkes will prove a good harbour ; he names them Prince *Henries Forland.* When he had laid these, he raised another which was the extreame part of the land looking towards the N. ; upon it are two hills (but one above the rest) like an humlocke, he names it King *Iames his Cape.* To the N. of this lyes certaine Ilands ; he names it Queen *Anne's Forland.* He still followed the N. shoare beyond the *King's Forland.* There is a Bay wherein lyeth some broken land and close to the maine, but he passed by in the night.

From thence he stood to the N. and W., to double this land againe, and fell with land that stretched from the maine, like a shoare from S. to N. and from N. to W. and downe to S. again. Here a storme takes him and he stands to N. and raises land; then he stood to S. againe, for he was loath at any time to see the N. shoare. The storme continuing, and he comming to the S., he found himselfe shot to the W. a great way, which he mervailed at (considering his leaward way). To S.W-ward of this land, there is a hill he named *Mount Charles;* to the N. and beyond this is an Iland; that to the East hath a fore head-land, and beyond it to the W. other broken land, where he thinks may be found a good harbour. He names this *Cape Salisbury.*

He then left those lands to the North-East, and fell into a rippling or overfall of a current, which he thought to be shoald water, but had no ground[1]; he put on still in sight of the South-land, and raised land 2 leag. from the maine. He tooke it to be the North maine, but it proved an Iland, having a very faire Head-land to the West, which he named Cape *Digges.* On the other side, to the East, was another Cape or head-land, which he called Cape *Worstenholme*[2]; betwixt which two he sailed South.[3]

He sends the Boate on shore to *Digges* his Iland, and in her going she was overtaken by a storme of Raine, Thunder, and Lightning. They came to the N.-East side, being high-land, but with much adoe, going to the highest part, they find some plaine ground, and Deere, 4 or 5, and after[wards] 16 in one Heard, but could not come nigh them with[in] Musket shot.

---

1 This means that he sounded but was unable to reach the bottom as is made clear in Prickett's own narrative.—C.

2 See p. 119.—C.

3 It should be noted that Hudson's own journal ends at this point. What follows is on the uncorroborated authority of Prickett.—C.

Thus, going from place to place, they see (to the West) a Hill higher then all the rest. It was steepe, and they could not get vp to it but on the South-west side; and on that side was a great water-pond, from whence, on the South, runneth a streame of water, as much as would drive a Mill, falling into the Sea; and in this place bred great store of Fowle, and the best grasse grew there they had seene since their comming from *England*. They found Sorrell[1] and Scurvey-grasse[2] in great aboundance; they found likewise Hills,[3] made like to haycockes, within which were great store of Fowle, hanged by the neckes; they tooke many of them, and downe a valley carried them to their Boate.

In this time, the Master had brought in the Ship betweene the two lands, & shot off some peeces to call the Boat on board, for it was a fogge. They perswaded him to stay heere for refreshing, but he would not, but sailed downe to the S. The land beares E. from him, and the same Maine that he had all the time followed. He looseth

---

[1] *Rumex* is a large genus of plants, containing the Docks and Sorrels, and extending over the greater part of the world. The particular species Hudson met with was probably *Rumex* (*Oxyria*) *digynus* which extends far into the Arctic regions, and appears as an alpine on all the great mountain ranges of Europe and Asia. It is common on the Scotch mountains. Like many allied species (some long under cultivation), it is an excellent pot-herb and anti-scorbutic.—C.

[2] *Cochlearia* is a genus of cruciferous plants, having no affinity to the grasses, and chiefly inhabiting northern countries. *C. officinalis*, the common scurvy grass, is by far the best known species, and was unquestionably that which Hudson met with. It is mainly a littoral plant, growing commonly on coasts all round the Arctic circle, and is not uncommon on the British shores, especially on those of Scotland. Its leaves possess valuable anti-scorbutic properties, which the early navigators frequently made use of as a remedy for scurvy.—C.

[3] It appears from the original that stone huts are meant. Probably the natives hung up their fowls in this manner to preserve them from foxes and other predacious animals.—C.

the sight thereof, because it fell away to the East, and after he had sailed 25 or 30 leag. he came to shallow water, broken ground, and Rockes, which he passed to the S., and in a storme of winde, the water still shoalding, he came to anchor in 15 fathomes.

After this he wayed and stood S.E., for so the Land laid, and came to have land on both sides; then he anchored and sent the Boate on land. The Land on the W. was a very narrow point, and to the S. there was a large Sea. He stood to the South betweene these two Lands[1] in this place, not above two leagues, and in the sight of the East-shore. In the end he lost sight thereof, and came into the bottome of a bay, into 6 or 7 fathom water; then he stands vp againe to the N. by the West-shore, vntill he came to an Iland in 53 deg., where he tooke in water and ballast.[2]

From thence, he passed to the N., but some 2 or 3 dayes after, there fell some reasoning concerning their comming into this Bay and going out; the Master takes occasion to reviue old matters and displaces his Mate, *Robert Ivet*, and also his Boate-swaine, for words spoken in the great Bay of Ice. He places *Robert Bylot*, his Mate, and *William Wilson*, Boate-swaine,[3] and then stands vp to the N. vntill he raised land, and then downe to the S.; then vp againe to the N. and then downe to the S., and on

---

[1] Dr. Asher surmises (*Henry Hudson*, p. 108) that Hudson was now between the East Main and Charlton Island; but in this he is unquestionably mistaken. There can, I think, be no doubt that he was between Cape Smith and Smith Island, in 60° 50′ N., for the narrative distinctly states that the passage between the two lands was "not two leagues broad". There is no other point on the eastern side of the Bay answering so well to the description.—C.

[2] This must have been Agoomska Island in James's Bay.—C.

[3] A fuller account of these changes, which took place on September 10th, will be found in Woodhouse's "note" (see p. 160).—C.

Michaelmas day he came in and went out from certaine Ilands which he sets downe for *Michalmas* Bay.[1] From thence, he stood to the N. and came into shole water; the weather thicke and foule.

He anchored in 6 or 7 fadomes, and lay there 8 dayes, before [the end of] which time he could not have one houre to get vp his anchor; the winde then ceasing, the Master would have vp the anchor, against the mind of all that knew what belonged therevnto; but, when he had his anchor on peake, the Ship tooke one sea and threw them all from Capstone, and diuers were hurt. He left his anchor and saved most of his Cable, for the Carpenter had laid his axe ready to cut the Cable if occasion should so fall out.

From hence he stands to S.W., through a cleare Sea of divers soundings, and came to a Sea of two colours, one blacke, the other white, and 16 or 17 fadome water, betweene which he went 5 or 6 leag. The night coming on, he goes in his maine and fore-saile and came into 5 or 6 fathomes. He saw no land, for it was darke, so that he stood to the East, and had deeper water, and then stood to the S. and S.W. and came to the Westermost bay of all,[2] (*in this bay wintred Captaine* Iames *of* Bristow).[3] In this bay, neerest to the N. shore, he anchored and sent on Land his boate; he found the land flat, and his men saw the footing of a man & a ducke in the snowy Rockes, and found good store of wood. Heere he saw a ledge of Rockes, lying S. and N., to the South-ward of him, and flowed over at full sea, and a strong tide set in there. He weyed anchor at

---

[1] So named because entered on (old) Michaelmas Day (Oct. 11th). It was probably Hannah Bay, as Dr. Asher suggests, but it is impossible to be certain. It is not named on Hudson's Chart.—C.

[2] The westmost of Hudson's Bayes.—F. The meaning of this expression is given on page 131.—C.

[3] This is an addition by Foxe. Captain James wintered in this bay in 1631-32, or twenty years after Hudson.—C.

midnight and thought to stand forth as he came in, but it fortuned that hee runne upon these Rockes and sate there for 12 houres, but by God's mercy he got off againe unhurt, though not unafrighted.

He then stood up to the East, and raised 3 hils lying North and South. He went to the furthermost, and left it to the North. He came into a Bay and anchored, and sent the Carpenter and this writer[1] to looke for a place to winter, it being the last of *October*, the nights long and cold, the earth all covered with Snow, themselves wearied, having spent 3 moneths in a labyrinth in this Bay. They went downe to the East, to the bottome there[of], but returned not with that they went for.

November. The next day, he went to the South and S. West, where he found a place unto which hee brought his ship, and hailed her on ground,[2] and on the 10 day she was froze in; now hee lookes to the lengthning of his provision.[3]

---

[1] Prickett, not Foxe, is here meant.—C.

[2] Purchas, in *his Pilgrimage* (4th ed., 1626, p. 818), says it was the 3rd, not the 1st, of November when they laid up.—C.

[3] Prickett's narrative is so obscure that one cannot gain from it any precise idea of the route sailed over during the two months or more immediately preceding the laying up for the winter on November 1st. It is, however, quite clear that during the whole of this time they had been cruising aimlessly about among the bays and islands in what is now known as James' Bay. Prickett (who was not a sailor) was probably quite unable to give any precise idea of its geographical configuration. The only geographical point which he makes tolerably clear is that Hudson wintered at the extreme southernmost arm of Rupert's Bay (which is nameless on the latest Admiralty Charts), at the south-east corner of James' Bay, in lat. 51° 10′ N. In this opinion, Dr. Asher seems to share, although the footnote in which he says so seems to be wrongly numbered, and to have been inserted on p. 108 instead of on p. 110, of his admirable work, *Henry Hudson*. The chart of Hudson's discoveries which Gerritz published in 1612, was no doubt founded on Hudson's own "card", which we know the mutineers brought home with them (see p. 149). It was probably altered somewhat from its original form, in accord-

He was victualled for 6 moneths with good provision, and might have had more from home if he would; now

---

ance with additional information supplied by either Bylot or Prickett, or both ; but it cannot be said to be a very accurate representation of the configuration of Hudson's Bay. It (and, of course, all the charts which were drawn from it for many years after) show James' Bay as if it was divided into *two* very deep, narrow bays (known respectively as Hudson's East and Hudson's West Bay) by a long, narrow promontory of land. This may be seen on the maps of both Foxe and James, and on some others of later date. It is not easy to say what led to this misconception. It is probable enough that Hudson, during the period above mentioned, had sighted and sailed along the entire western coast of James' Bay, but there is nothing to prove that he rounded, or even reached, its north-western extremity, Cape Henrietta Maria, as it was afterwards named by Capt. James in 1631. It is true that Hessel Gerritz states, in the first edition of his work, that Hudson sailed up the western shore of the Bay as high as lat. 60°, and in a later edition as high as 62° or 63° N. (for translations, see Asher's *Henry Hudson*, pp. 182, 188, and 192) ; but there is every reason to suppose these statements erroneous. Gerritz's statements upon this point were probably founded upon the fact that on his chart of Hudson's discoveries the land on the west side of Hudson's Bay is shown to extend as far north as 60° exactly (see a facsimile of it in Asher's *Henry Hudson*) ; but the narrative of Hudson's voyage, vague though it is, shows it to be impossible that either he on the voyage out or the survivors on their return voyage can have reached so far north in this direction. It may be that the latter imagined themselves to have done so, or that they purposely drew the chart incorrectly in this respect in order to make it appear that they had made more extensive discoveries than they really had. On the whole, it seems most probable that the most northerly land on the western side of the Bay shown on Gerritz's chart may be identified with what is now known as Cape Henrietta Maria, though that cape lies, not in 60° N., but in 55° N. This identification is rendered more probable by Gerritz's statement (*loc. cit.*) that, at the most northerly point Hudson reached on the western side of James's Bay (though it was certainly not in any of the latitudes mentioned by Gerritz), he met with a "wide sea, agitated by mighty tides from the north-west", which was no doubt Hudson's Bay of modern maps. If this surmise be correct (and it was evidently shared by Luke Fox—see p. 225, orig. ed.), it will be seen that Hudson's expedition explored the whole coast-line of what is now known as James' Bay. Although the narrative nowhere clearly tells us that they did so, it contains nothing to show that they did not.

hee must pinch, for that he knew of no supply untill he came the next yeere to Cape *Digs*,[1] where the Fowle breed, for there was all his hopes; wherefore hee propounded reward to him that killed either beast, fish, or fowle. His Gunner[2] dyed about the middle of this moneth; he blameth the Master's uncharitable dealing with this man, but note what followed.

The Master kept in his house at *London* a young man named *Henry Greene*, borne in *Kent*, of worthy Parents; but, by his life and conversation, he had lost the love of all his friends, and spent all that he had; but, by the means of one Master *Venson*, his mother parted with 4l. to buy him cloathes, which money he was not trusted with the disposing of himselfe. This *Henry Greene* was not knowne to the Adventurers, nor had any wages, but came onely on board at *Graves-end*, and at *Harwich*, would have gone into the field with one *Wilkinson*[3] of our Company. At

---

The islands shown on the chart certainly bear some resemblance to those which actually exist in James' Bay; but there is nothing to explain why it was that the explorers supposed the bay they were exploring to be divided into *two bays* by a long narrow tongue of land, as stated above. On some later maps—for instance, on those of Foxe and James—this non-existent tongue of land is shown to be even longer than on Hudson's chart itself.—C.

[1] See *ante*, pp. 118 and 126.—C.

[2] Prickett gives this man's name as John Williams. The depositions also give his name, but in them it is stated that he "dyed in 9 Octob."—C.

[3] Presumably this means that he would have *fought* with Wilkinson. If so, his quarrelsome disposition is well shown, for the distance between Gravesend and Harwich is not more than a day's sail. Forster, however (*Voyages and Discoveries in the North*, p. 337), seems to think the passage means that he would have *deserted* with Wilkinson. It is not very clear who Wilkinson was. The depositions show that there was no one of this name among Hudson's crew. The name may be a misprint for either Wilson or Williams. In Prickett's narrative, he is spoken of as "one Wilkinson", the words "of our company" having apparently been added by Foxe, though he could not have meant his *own* company, as he had with him no men who had been on Arctic voyages before.—C.

Island he fell out with the Chirurgion in *Dutch*, and he beat him on shore in *English*, which set all the Company in a rage, so that they had much adoe to get the Chirurgion on board againe. This Author told the Master of it, but he bade him let the matter alone; for (said he) the Chirurgion hath a tongue that would wrong the best friend he had; but *Robert Iuet*, the Master's Mate, would needs put his fingers in the Embers, and told the Carpenter a long tale (when he was drunke) that the Master had brought in *Greene* to cracke his credit that should displease him, which, when the Master heard of (being forty leagues from *Island*), he would have gone backe to *Island* to have sent *Juet* his Mate home in a Fisherman, but, being otherwise perswaded, all was well, and *Greene* stood upright and was very inward with the Master, and was a serviceable man every way for manhood; but for Religion he would say he was white Paper, whereon he might write what he would. Now the Gunner was dead, and then (as order is in such cases) if the Company stand in neede of any thing belonged to the man that is deceassed, then is it brought to the maine Mast, and there sold to them that will give the most for it. This Gunner had a gray cloth Gowne, which *Greene* prayed the Master to befriend him so much as to let him have it, paying for it as [much as] another would give. The Master said he should, and therefore answered some who sought for it that *Greene* should have it and none else.

Now, out of time and season, the Master called the Carpenter to goe in hand with a house on shore, which at the beginning hee would not heare of, when it might have been done. The Carpenter told him that the Snow and Frost was such as hee neither could nor would goe in hand with such worke, which when he [*i.e.*, *Hudson*] heard, he feretted him out of his cabine and struck him, calling him by many foule words and threatened to hang him.

The Carpenter tolde him that he knew what belonged to his place better then he did, and that he was no house Carpenter; yet the house was made with much labour [but] to no end. The next day after the Master and the Carpenter fell out, the Carpenter tooke his Peece[1] and *Henry Greene* with him, for it was ordered that none should goe out alone, but one with a Peece, another with a Pike. This did move the Master so much more against *Greene*, that *Robert Bylot*, his mate, must have the Gowne, and had it delivered unto him; which *Henry Greene* seeing, he charged the Mr. with his promise, but the Mr. did so raile on *Greene*, & with so many words of disgrace, telling him that all his friends durst not trust him with 20 shillings, & therefore why should he, and as for wages he was to have none, nor should if he did not please him, yet the Mr. had promised him as good wages as any man in the ship, and to have him [made] one of the Princes *Guard* at his home coming; but you shall see how the Divell so wrought out of this that *Greene* did the Master what mischiefe he could, in seeking to discredit him, and to thrust him and other honest men out of the ship. To speake of all the troubles, and of this cold Winter, would be too tedious.[2]

Now he sheweth how mercifully God dealt with them in this time, for in the space of three moneths he had such store of one kinde of Fowle, which were Partridge,[3] as white as Milke, of which he killed at least one hundred

---

[1] An old name for a *gun*: hence our modern "fowling-piece".—C.

[2] The depositions throw no light on the events of the wintering.—C.

[3] These were the Willow Ptarmigan (*Lagopus lagopus*), an abundant species in the forest region south of Hudson's Bay. It serves as an important article of food during winter to the servants of the Hudson's Bay Company, at the Company's posts in the region this bird inhabits. As many as 10,000 are said to have been killed at some of the posts in one year.—C.

dozen, of sundry sorts, for all was fish that came to the net. At the Spring, this fowle left them, and in their places came other of divers sorts, as Swannes, Goose, Ducke, and Teale, but hard to come by. He thought that they would have bred there in those broken grounds, but they doe not, but came from the S. and flew to the N., further then he was this Voyage; yet, if they had beene taken short with N. or N.E. winds, they stay there against the winds' returne, and then take their flight to the N. ward, and in short time none of them are to be seene there.

Then he searched the woody hills and vallies for all things that had any show of substance (for food), how vile soever; the mosse of the ground, and the frog in his engendring time was not spared; but amongst divers sorts of buds, it pleased God that *Thomas Woodhouse* brought one of a tree[1] that was full of Turpentine substance; of this, the Surgion made a decoction to drinke, and applyed the buds hot to them that were troubled with ache in any part, from whence they received present ease.

Now, about that time the Ice began to breake out of the bayes, there came a *Salvage* to their Ship, as it were to see and to be seene, and was the first that they had seen in all that time. He intreats him well, and used him kindly, promising to himselfe great matters by this meanes, and therefore called for all the Knives and hatchets which every man had to his private vse, and to this *Salvage* he gave a knife, a looking-glasse, and buttons, who received them thankefully, and made signes that after he

---

[1] Probably the buds were those of some conifer, perhaps, as Forster suggests (*Voyages and Discovery in the North*, p. 338), of the Tamarac (*Larix americana*), or of the Spruce (*Abies nigra* and *A. alba*), both of which trees extend much further north. In *his Pilgrimage* (4th ed., 1626, p. 818), Purchas gives fuller information about the benefits derived from this tree, but he is certainly in error in saying that it *blossomed* in December.—C.

had slept he would come againe; which he did, and brought with him a sled, which he drew after him, and upon it two Deere skins and two Beaver skinnes. He had a scrip under his arme, out of which he drew those things the Master had given him, and layes the knife upon the Beaver skinnes, and the glasses and buttons upon the other, and so gave them to the Master, who received them; and the *Salvage* tooke those things which the Master had given him, and put them into his scrip againe, then the Master showed him a hatchet, for which he would have given him one of his Deere skins, but the Master would have both, and so he had, although not willingly. After many signes of people to the N. and to the S., and that after so many sleepes he would come again, he went his way, but came no more.

Now, the Ice being broke out of the Sounds [so] that a boate might passe, the M^r^. appointed *William Wilson*, *Henry Greene*, *Michael Peirce*, *John Thomas*, *Andrew Motter*,[1] *Bennet Mathewes*, and *Arnold Lodlo*,[2] to go on fishing. These men, the first day they went, caught 500 fish as bigge as good Herrings, and some Trouts. Here was good hope to have their want supplyed, but those were the most that ever they got in one day; for many dayes they got not a quarter so many. In this time of their fishing, *Henry Greene* and *William Wilson*, with some others, plotted to take the net & Shallop,[3] which the Carpenter

---

[1] In the depositions, and also in Woodhouse's "note" (see p. 161), this man's name is given as *Adrian* Motter.—C.

[2] This man's name appears as *Ladley* in the depositions. In Woodhouse's "note" (*Purchas his Pilgrimes*, pt. iii, p. 609), he is spoken of as "Ladlie Arnold".—C.

[3] It is impossible to say exactly what is meant by "shallop" in this case, as (according to Smythe) the name has been applied to various kinds of fishing and other boats. The word often occurs in accounts of early voyages, and is derived from the French *chaloupe*.—C.

had new set up, and so to shift for themselves[1]; but, the Shallop being ready, the Master would goe in it himselfe to the S. and S.W. to see if he could meete with people, for to that end it was set up; and that way he could see the woods set on firè by them. The Master takes the net and shallop, and so much victuals as would serve for nine dayes, and went to the Southward; they that remained on board were appointed to take in water, wood, and ballast, and to make the ship ready against his returne; but he could set no time, for hee was perswaded that if he met with the Salvages, hee should have refreshing of fresh meat, and that good store; but, in conclusion, he returned worse then he went, and though the Inhabitants set the woods on fire before him, yet they would not come to him.

Being now returned, he makes ready for his home coming, and first he delivers all the bread in the fore roome, wch came to a pound a peece for every man's share, and delivered unto every man a bill of returne, willing them to have that to shew, if it pleased God they came home, and he wept when he gave it unto them. But, to help themselves to some reliefe, the boat went to fish from Friday morning until Sunday noone, and brought but 80 small fish for 18 hungry bellies[2]; whereupon he waighed and came away from his wintering place into the Sea, where, his bread being gone, that store of cheese[s] which he had must stop the gap, being but five; the Company grudged, for they made account of nine; but those five were equally divided by the Master, some counselling him to the contrary, for there were some who, having it, would

---

[1] Prickett's deposition corroborates this statement.—C.

[2] It seems as if this number should be twenty-two; for we know that Hudson's company numbered twenty-three, and only one had died up to this time. The number of Hudson's crew is stated hereafter on the authority of Woodhouse, and is confirmed by the depositions, wherein a list of the men is given.—C.

make haste to be rid thereof, because they could not gouerne it. I know (saith the writer) when *Henry Greene* gave halfe his bread which he had for 14 dayes, to one to keepe, and prayed him not to let him have any untill the next Munday, but before Wednesday at night, he never left untill he had it againe, having eaten up his first week's bread before. So *Wilson*, the boatswaine, had eat in one day his 14 dayes bread, and hath laid in bed 2 or 3 dayes for his labour. The cause why the Master delivered all the cheese[s] was because they were not all of one goodnesse, and therefore they should see that they had no wrong done them, but that every man had the best and worst together, which was 3 pound and one halfe for 7 dayes.[1]

The wind serving, he wayed and stands to the N.W., and on Munday night, the 18 of *June*, he fell into the Ice, and the next day, with W. wind, he lay fast within sight of land untill Sunday following. Now, being here, the Master told *Nicholas Simmes*[2] that there would be a breaking up of Chests, and a search for bread, and willed him if he had any to bring it to him, which he did, and delivered the Master 30 Cakes in a bag. This deed of the Mr. (if it be true) hath made me marvell what should be the reason why hee did not stop the breach in the beginning, but let it grow to that height as it overthrew himselfe and many other honest men ; but there are many devices in the heart of man, but *the counsell of the Lord should stand.*

---

[1] No wonder they were short of provisions ; for they had been out fourteen months already ; and we know, on Bylot's authority, that they had been victualled for only eight, while Prickett says for six only. That Hudson had not dealt quite fairly with his crew respecting the provisions, seems likely enough ; for all the survivors, in their depositions, affirm that he had not. They accuse him of having wasted the provisions, and with having fed his favourites at the expense of the others.—C.

[2] The depositions show that Nicholas Syms was a boy.—C.

Being thus in the Ice, on Saturday, the one and twentieth of *Iune*,[1] at night, *Wilson*, the Boatswaine, and *Henry Greene* came to this writer lying lame in his Cabbin,[2] and told him that they and the rest of their associates would shift the Company, and turne the Master and all the sicke men into the Shallop, and let them shift for themselves; for there was not 14 dayes victuall left for all the Company, at that poore allowance they were at; and that there they lay, the Master not caring to go one way or other; and that they had not eaten anything this three dayes, and therefore were resolute either to mend or end, and what they had begun, they would go through therewith, or die. When he heard this, he told them he marvelled to heare so much from them, considering that they were married men and had wives and children, and that for their sakes they should commit so foule a thing in the sight of God and man as that would be; for why should they banish themselves from their native countrey? *Henry Greene* bad him hold his peace, for he knew the worst of it; which was to be hanged when he came at home, and therefore of the two he would rather be hanged at home then starved abroad; and for the good-will they bore him, they would have him to stay in the ship. He gave them thankes, and told them that hee came into the ship, not to forsake her, nor yet to hurt himselfe and others by any such deed. *Henry Greene* told him then that hee must take his fortune in the Shallop; if there bee no remedy, quoth he, the will of God be done.

Away goes *Greene* in a rage, swearing to cut his throat that went about to disturbe them, and left *Wilson* by him,[3]

---

[1] There is an error here. The 18th of June, 1611, was a *Tuesday* (not a Monday, as stated) while the 21st was a *Friday*, not a Saturday. —C.

[2] This means Prickett, not Foxe. Foxe's narrative here follows Prickett's, as it appears in Purchas, almost word for word.—C.

[3] This still refers to Prickett.—C.

with whom he had some conference, but to no good, for hee was perswaded to goe on with the action whilst it was hot, lest their party should faile them and the mischiefe they intended to others should fall upon their owne shoulders. *Greene* comes againe and demaunded what he said. *Wilson* answered and said, He is in his old song, still patient. Then he spake to *Greene* to stay 3 dayes, in which time hee would so deale with the M^r^. as all should bee well. But, being denied, he dealt with him but for 2 dayes nay, for 12 houres. There is no way then, say they, but, out of hand: then he told them that, if they would stay while[1] Munday, he would joine with them to share all the victuals in the ship, and would justifie it when he came at home; but this would not serve, wherefore he told them it was some worse matter they had in hand then they made shew of, and that it was blood and revenge he sought, or else he would not undertake such a deed at such a time of night. *Greene* with that tooke his Bible, which lay before him, and sware that he would doe no harme, and what he did, it was for the good of the Voyage, and for nothing else, and that all the rest should do the like. The like did *Wilson* sweare.

Greene went his way, and presently comes *Ivet*, who because hee was an ancient man, he[2] hoped to have found some reason in him, but he was worse then *Greene*, for hee swore plainely that he would justifie this deed at home-comming. After him came *Iohn Thomas* and *Michael Pierce*, as birds of one feather, but because they died, as hereafter shall be shewed, he let them passe. Then came *Motter* and *Bennet*, of whom he demaunded if they were well advised what they had taken in hand. They answered, they were, and therefore came to take their oath.

Now, saith this writer, because he was much condemned

[1] A misprint for *till*.—C.

[2] This still refers to Prickett.—C.

for this oath, as one that plotted with them, and that by an oath he should binde them together to performe what they had begun; he thought good to set downe [in writing] to the view of all men, to see how well their oath and deeds agreed, and thus it was; *You shall be true to God, your Prince, and Countrey; you shall doe nothing but to the glory of God, and to the good of the action in hand, and harme to no man.* This was the oath, without adding or diminishing; he looked for more of those companions, although those were too many, but there came no more. It was darke, and they in readinesse to put this deed of darknesse into execution. He called to him *Greene* and *Wilson* and prayed them not to goe in hand with it in the dark, but to stay untill morning, for now he hoped every man would goe to his rest. But wickednesse sleepeth not, for *Henry Greene* keepeth the Master Company all night, and gave this writer bread which his Cabbin mate gave him, and others were as watchfull as hee. Then he[1] asked *Henry Greene* whom he would put out with the Master. He said the Carpenter, *Iohn King*, and the sick men. He said they should not doe well to part with the Carpenter, what need soever they should have; [the reason] why, the Carpenter was in no more regard amongst them then another, [was] for that he and *John King* were condemned for wrong done in the victuals; but the chiefest cause was because the Master loved him, and made him his mate upon this his returne from his wintring place, thereby displacing *Robert Bylot*, who they did grudge, because hee[2] could neither write nor reade; for therefore, said they, the Master

[1] *He* still continues to mean Prickett.—C.

[2] Apparently this means that Hudson appointed the carpenter, whose name was Philip Staff, to be his mate, thereby displacing Bylot. If so, this affords further evidence of the unsatisfactory relations between Hudson and his crew, for he had previously appointed Bylot mate in the place of Juet.—C.

and his ignorant mate will carry the ship whither the Master pleaseth, the Master having forbidden any man to keepe account or reckoning, having taken from all men whatsoever served for that purpose. Well, he obtained [leave] of *Henry Greene* and *Wilson* that the Carpenter should stay, by which meanes,[1] after that they had satisfied themselves, the Master and the rest might bee taken into the ship againe, or hee hoped that some one or other would give some notice to the Carpenter, *Iohn King*, or the Master; for so it might have come to passe, and have beene by some of them prevented, that were the most forward.

Now it cannot be amisse to shew how they were lodged, and to begin in the cookes Roome; there lay *Bennet* and the Cooper, lame. Without the Cookes roome, on the starboard side, lay *Thomas Woodhouse*[2] sicke; next to him lay *Sydrach Fenner*, lame; then [the Surgeon and John Hudson; then] lay *Wilson*, the boatswaine; and then *Arnold Lodlo* next to him; in the Gunner's roome, lay *Robert Iuet* and *John Thomas;* on the Larboard side, lay *Michaell But* and *Adrian Moore*, which was never well since they lost their anckor; next to him lay *Michael Peirce* and *Andrew Motter;* next to them, without the Gunne roome, lay *Iohn King*, and with him *Robert Bilot;* next to them himselfe,[3] and next to him *Francis Clements;* in the midship, betwixt the Capstone and the Pompes, *Henry Greene*, and *Nicholas Simmes*. This night *John King* was late up, and they thought he had beene with the Mr., but he was

---

[1] In Purchas it is "*by whose means he hoped*".—C.

[2] Thomas Woodhouse, or Wydowse, who is described as a "student in the Mathematickes", when he was deserted, left in his desk an interesting note, which was published by Purchas (*his Pilgrimes*, pt. iii, p. 609). It adds several interesting items of information to those given by Prickett. Foxe reprints the "note" almost verbatim (see *post*, p. 160).—C.

[3] That is, Prickett.—C.

with the Carpenter, who lay in the poope, and comming from him was met by his cabbine mate, as it were by chance, so they went to cabbine together. It was not long ere it was day; then came *Bennet* for water for the kettle; he went into the hold; when he was in, they shut the hatch on him, but who kept it downe he knoweth not; but upon the decke came *Bennet*.

In the meane time, went *Henry Greene* and another to the Carpenter, and held him talke until the Master came out of his Cabbin, which hee soone did. Then came *John Thomas* and *Bennet* before him, while *Wilson* bindes his armes behind him; he asked what they meant; they told him that he should know when hee was in the Shallop. Now *Iuet*, while this was doing, came to *John King* into the hold, who was provided for him, for hee had got a sword of his owne, and kept him at a Bay, and might have killed him but others came to helpe him, and so he came up to the Master. The Master called to the Carpenter and tolde him that he was bound, but he heard no answer made. Now *Arnold Lodlo* and *Michell But* railed at them, and told them their knavery would show it selfe. Then was the Shallop hailed up to the ship side, and the poore sicke and lame men were called up to get them into the Shallop. The Master called to this writer [*i.e.*, Prickett], who came out of his cabbine as well as he could to the hatch way to speake with him, where on his knees he besought them for the love of God to remember themselves, and to do as they would be done unto. They bad him keep himselfe well and get him into his cabbine, not suffering the Master to speak to him; but when he came into his cabbin againe, at the horn window[1] which gave light into his cabbine, the Mr. told him that *Juet* would over-

[1] The original has it that Hudson "called to me at the horne which gave light into my cabbin". Foxe has added the word "window", as if he feared the meaning would not be clear without it.—C.

throw them all; nay, said he, it is that villaine *Greene*, and spake it not softly.

Now was the Carpenter at liberty, and asked them if they would be hanged when they came at home; and, as for himselfe, he said, he would not stay in the Ship, unlesse they would force him. They bid him go then, for they would not stay him. I will, said he, so I may have my chest and all that is in it. They said he should, and presently put it into the Shallop; then came he to take his leave of this writer, who perswaded him to stay, which, if he would, he might so worke that all might be well, but he answered, that he did not thinke but that they would be glad to take them in againe, for he was so perswaded by the Master that there was not one in all the ship that could tell how to carry her home. But, saith he, if we must part (which we will not willingly doe, for they would follow the ship) prayed him if they came to the Cape[1] before them, that he would leave some token that he had beene there, neare to the place where the Fowles breed, and he would doe the like for us, and so with teares we parted.[2] Now were the sick men driven out of their

---

[1] In Purchas this is *Capes*, which clearly refers to Capes Wolstenholme and Digges.—C.

[2] It is only fair to the memory of the man who thus nobly remained faithful to his master to point out that his name was Philip Staffe. This is the more necessary because there are in the text several passages which, if not carefully read, lead to the conclusion that the carpenter's name was John King. Several writers have fallen into this error. Thus, Prof. Henry Morley, in his introduction to *Voyages in Search of the North-West Passage* (Cassell's National Library, 1886), says: "There was one stout man, John King, the carpenter, who stepped into the boat, abjuring his companions, and chose rather to die than even passively be partaker in so foul a crime. John King, we who live after will remember you!" Purchas, in a passage in *his Pilgrimage* (4th ed., 1626, p. 818), which is omitted from *his Pilgrimes*, speaks of him as, "one Philip Staffe, an Ipswich man, who, according to his name, had been a principall staff and stay to the weaker and

cabbines into the Shallop, but *Iohn Thomas* was *Francis Clement's* friend, and *Bennet*[1] was the Cooper's, so as there was words betweene them, and *Henry Greene* saying that they should goe, and the other swearing that they should not goe, but [that] such as were in the Shallop should returne. When *Henry Greene* heard that, he was compelled to give place and to put out *Arnold Lodlo* and *Michael But*, which with much a doe they did.[2]

In the meane time, there was some that plyed their worke as though the ship had beene entred by force, and they had free leave to pillage, breaking up chests and

---

more enfeebled courages of his companions in the whole action, lightening and unlightening their drooping darkened spirits with sparks from his own resolution ; their best purveyor with his piece on shore ; and both a skilful carpenter and lusty mariner on board, [who], when he could by no persuasions, seasoned with teares, divert them from their divellish designs, notwithstanding they entreated him to stay with them, yet chose rather to commit himself to God's Mercy in the forlorne shallop then [to stay] with such villaines [and] to accept of liklier hopes."—C.

[1] Bennet Mathews is, of course, meant. In several other places Prickett similarly speaks of him by his Christian name.—C.

[2] Additional information concerning the mutiny will be found in the depositions ; but, as already stated, very little reliance must be placed upon the statements of the survivors, who had, of course, to make out the best tale they could for themselves. It is significant that none of them ventured to deny that (if they did not urge the desertion of Hudson) they were, at least, not dissenting parties ; and it is still more noticeable that all agreed in imputing the chief blame to those who had been killed and were thus unable to deny the accusation. One cannot help thinking that the deaths of Green, Wilson, Thomas, and Pierce at the hands of the Esquimaux, and of Juet through starvation, served the survivors in very good stead. As to the exact date of the mutiny: Dr. Asher (p. ccxi) assigns it to June 21st, though Prickett's narrative seems to point to the following day. This is supported by Prickett's deposition, in which he says, "about the 22nd or 23rd" of June, while Bylot says the 23rd. There is no record that four of the company lost their lives at the time of the mutiny, as erroneously stated in the *St. James's Gazette* (April 20th, 1887) and in the *Dictionary of National Biography* (vol. xxviii, p. 148).—C.

rifeling all places. One of them came to this writer and asked him what they should doe, who answered that they should make an end of what they had begun, for he see him doe nothing but sharke[1] up and downe. Now [were] all the poore men in the Shallop, as *Henry Hudson*, Master, *Iohn Hudson*, his sonne,[2] *Arnold Lodlo*, *Sidrack Fenner*, *Phillip Staffe*, the Carpenter, *Thomas Woodhouse*, *Adam Moore*, *Henry King*,[3] and *Michell But;* the Carpenter got of them a peice, and powder and shot, and some pikes, an Iron pot with some meale, and other things. They stoode out of the Ice, the Shallop being fast to the sterne of the Ship, and so when they were nigh out (for he cannot say they were cleane out), they cut the head-fest[4] from the sterne of their ship, and then out went topsailes, and stood to the E. in cleare Sea, having lost sight of the Shallop. In the end, they tooke in topsailes, righted their helme, and lay in foresaile, until they had ransacked and searched all places in the Ship. In the hold, they found one of the vessels of meale whole, and another halfe spent, for they had but two. They found also two firkins of butter, some 27 peeces of porke, and halfe a bushell of pease, but in the Mr's cabbine they found 200 of Bisket

---

[1] Swindle or plunder: an old word, now obsolete.—C.

[2] In spite of the statement by Dr. Asher (*Henry Hudson*, p. 122, *note*) that the fact of John Hudson being the captain's son was "merely a conjecture", Foxe is perfectly correct in describing him as Hudson's son, a fact which he very likely knew of his own knowledge. Purchas, however, distinctly states that he was so (*Purchas his Pilgrimage*, 4th ed., 1626, p. 818), as also does Bylot in his deposition. It appears from Purchas (*his Pilgrimes*, vol. iii, lib. III, pp. 567 and 574) that he had also sailed with his father on his first and second voyages, and probably also on his third.—C.

[3] This must be a misprint for *John* King, though Henry occurs in Purchas.—C.

[4] The original has it: "they cut her head fast from the sterne of our Ship", which means that they cut her tow-rope or painter (see Smyth's *Sailors' Word-Book*, p. 375.)—C.

Cakes, a peck of meale, [and] of Beare to the quantity of a Butt. Now, when it was said that the Shallop was again come within sight, they let fall the maine saile, and out top sailes, and fly as from an enemy.

Then he prayed them to remember themselves, but *Wilson* nor the rest would heare of no such matter. Comming nigh the E. shore, they cast about to the W. and came to an Iland,[1] where they anckred in 16 fathoms, and tries on shoare with the net for fish, but could not drive for rockes. *Michaell Peirce* killed two fowle, and heare they found good store of weed called cockle grasse,[2] [some] of which they gathered (as in their wintering place) and came on board. They lay there that night and the most of the next day, in wch time they see not the Shallop, nor ever after.

Now came *Hen. Greene* and told him [*i.e.*, Prickett] that it was the companies' will that he should go up into the Mrs. cabbine, and take charge thereof. He told him that it was more fit for *Rob. Juet;* [but] he said he [*i.e.*, Juet] should not come in it, nor meddle with the Mr's Card nor Journals; so up he [Prickett] came, and *Hen. Greene* gave [him] the key of the Mr's chest and told him that he had layde the Mr's best things together, which he would use himselfe when time did serve. The bread was also delivered this writer, by tale.

The winde serving, they stand N.E., and this was *Bylot's* course, contrary to *Juet,* who would have gone N.West. They had the E. shoare in sight, and in the night had a stiffe gale of wind and stood before it until they met with Ice, and stoode amongst the same until they were fast. It was so thicke ahead, and the winde brought it so fast on a

---

[1] As Dr. Asher remarks, Prickett's geographical statements are so vague that it is absolutely impossible even to guess what island is here referred to.—C.

[2] Scurvy grass (see p. 127).—C.

sterne, that they could not stirre backewards, and so laid there 14 dayes in worse plight then ever before they had beene, where was great store [of ice], yet it lay not so broad upon the water as this; for this floating Ice continued miles and halfe miles in compasse, having a deepe Sea and a tyde of flood setting S.E. and N.W., but *Bilot* was confident to goe through to the N.E., as he did.

At length, being cleare of the Ice, he continued his course in sight of the East shore until he had raised 4 Ilands,[1] which lay North and South, but past them 6 or 7 leag. where the wind tooke them short.[2] They stood backe to them againe, and came to anckor betweene 2 of the Westmost.[3] They found nothing on land but cockle grasse. He saith he found that, before they came to this place, that he was kept in the Ship (against *Henry Greene's* minde) because he did not favour their proceeding better than he did; for he drave him to take upon him to search for such things as himselfe had stolne, and accused him of a matter no lesse then treason amongst themselves, for that he had deceived the company of 30 bread cakes.

Now they began to talk that *England* was no safe place for them, and *Henry Greene* swore that the ship should not come in any place, but keepe the Sea still, until he had the

---

1 Prof. Asher conjectures that these were not far from Portland Promontory or Cape Dufferin, 58° 50′ N., 79° W.—C.

2 We may presume that they reached this point about the 15th of July, on which date the extract from the log preserved at Trinity House commences. The narrative, which is very brief, continues until the 16th of August, when they were clear of Hudson's Strait. The entries relate chiefly to the winds, tides, dates, soundings, courses, and other matters connected with the navigation, in which respects Prickett's narrative is almost wholly wanting. The events recorded in the log, however, do not contain anything of importance not referred to by Prickett, and it is remarkable that the log contains no reference to the massacre of four of the mutineers by the Esquimaux, which Prickett hereafter narrates at such length.—C.

3 This should be *northernmost.*—C.

King's hand and seale to show for his safety. They had many devises, but *Henry Greene* was their Captaine, and so they called him. From those Ilands he stood to the N., having the Easterne land in sight, and raised those Ilands the Master had called *Romney's* Ilands,[1] between which Ilands and the shallow ground to the E. of them the Mr. stood downe into the first great bay[2] (this was entered going outward in the Master's time). They kept the East side still in sight and, comming thwart of low land, strooke once upon a Rock that lay under water, but without any harme that they saw. They continue their course and raise Land on head which stretched to the North; there they said plainely that *Robert Bilot* by his Northerne course had left the Capes to the South, and that in time they must seeke that way for reliefe, having but small store left. But *Bilot* would still follow the land to the North, saying that hee hoped in good time to finde what would relieve us that way, as soone as to the South. This Writer saith that hee told them that this Land was the Maine of *Wostenholme* Cape, and that the shallow Rockey ground was the same that the Master ran down by when he went into the great Bay. *Robert Ivet*, and all, said it was not possible, unlesse the Master had brought the Ship over land, and willed them to looke into the Master's Card, how well their course and it did agree.[3] They stood to the East, and left the

---

[1] No doubt the islands off Cape Smith, to the north of Mosquito Bay. They are not named on Hudson's Chart or referred to in the narrative. Hudson doubtless named them after Rebecca Lady Romney, who was one of his adventurers—or should one say adventur*esses* ?—C.

[2] This expression has already been explained (see pp. 128 and 131).—C.

[3] It is not much to be wondered at that Prickett, a landsman, should have been unable to give a clear account of the course as they sailed homewards, when the most competent of the seamen left on board were in the state of complete bewilderment here depicted. In their

Maine land to the North by many small Ilands into a narrow gut betweene two Lands and anchored. They went on the West side, and found a great Horne,[1] and cockle grasse on the East side; this Grasse was great reliefe to them, for without it they could not have recovered the Capes, for want of Victuall.

When they weighed Anchor, they doubled the Cape to the North, which is high land, even to the Capes, which is North and South some 25 or 30 leagues; then they stand to the North. They saw of those Fowles which breed at the Capes, and killed some; at which time with great joy they raised the Capes and, bearing for them, came to the Ilands that lay in the mouth of the Streights[2]; but, bearing in, they ran upon a Rock, and stood fast for 8 or 9 houres. It was ebbe when they grounded, but the next flood floated them off againe.[3] It was faire weather; the ebbe came

---

anxiety to reach the capes on which the fowl bred, we see them in a state of feverish fear that they had passed them and got too far to the northward.—C.

[1] Found a great horne, which they tooke to bee Sea Vnicorne.—F. The original speaks of "*the* great horne", as though it were some remarkable horn which they brought home. Foxe is responsible for the statement that they took it to be the horn of a sea-unicorn. Probably, however, it was an unusually large tusk of that animal, better known as the Narwhal (*Monodon monoceros;* see p. 45).—C.

[2] The Digges Island group (see pp. 118 and 126).—C.

[3] This grounding and release, here lightly passed over, is worth noting. In the Trinity House documents, we read: "[July] 26. They run aground on a rock in the little streight mouth [that is the narrow strait between Digges Island and the mainland, and not the wider one between Digges Island and Mansel Island] at 9 a. [? afternoon]. Fleet again at 4 in the morning." In the *Pilgrimage* (4th ed., 1626, p. 818), it is stated that, "the ship came aground at Digges Island, and so continued divers houres, till a great floud (which they, by this accident, tooke first notice of) came from the westward and set them on flote." Purchas himself regarded this as "a very probable argument of an open passage into the South Sea". In the "Grounds for a Coniecture" (Appendix), it is spoken of as "the Great Billow". Further, in Sir Dudley Digges's *Treatise on the*

from the East, and the flood from the West. Being afloate, they stood to the Eastward and anchored.

[*July* 27.] This day he sent the Boat on land to kill Fowle. They in the Ship had warning to stand as neere as they could; but, the winde being contrary, they could not fetch the place where the Fowle breed, but they found great store of Gulls upon the Cliffes, but hard to come by; but with their peeces they killed 30, and towards night returned on shipboard. Then they brought their Ship neerer the Mouth of the Streights, and anchored in 18 Fathom upon a Riffe or Shelfe; but, when they had wayed and stood to the place where the Fowle breed, they were faine to stand to and againe in the Streight's mouth under Sayle, because they could not finde ground to Anchor in, the water was so deepe.

[*July* 28.] The Boat went to *Digges* Cape, and made directly for the place where the Fowle breed, where they

---

North-West Passage (see p. 119), we find reference (p. 3) to "our streightes found in ([61° N.]), a Latitude free from feare of danger, cutting through the body of America, 200 or 300 Leagues, unto an open sea that showed a great and hollow billow, and brought a flood that rose 5 faddome." On p. 26 is a further reference to the same "flood", which clearly shows that the writer thought the north-west passage had been discovered by Hudson. He says: "And for any-thing wee yet can hear, no one Voyage to the Contrary, we See not but wee may Conclude that the Flood our People met Came from the Southerne sea; and, untill we heare more Authenticall reasons then of feare, grounded on false Cardes, [we may] beleeue that our industry, by God's grace, may this next voyage [Button's] manifest the Prophesie of Baptista Ramusius touching the North-West Passage." Taking all the evidence together, there can be no doubt that the survivors made much capital out of their observations of the strong flood at Digges Island. Moreover, there can be very little doubt that it was their report that led Button, on his return home, to try the tyde in this locality (see p. 197); and, although he says (p. 198) he found them mistaken in the set of this tide, he observed another, setting more northerly, on which, as hereafter pointed out, he himself built very great hopes.—C.

see 7 Boats came about the Easterne point towards them; but, when the Salvages saw their Boat, they draw their lesser Boats into their bigger, and when they had done, they came rowing to their Boat, and made signs to the west. Our men made ready for all essayes. The Saluages came to them, and they grew familiar one with another, so as ours tooke one of theirs into their Boate, and they tooke one of ours into theirs. Then they carried our Man to a Cove where their Tents stood to the Westward of the place where the Fowle breed. So they carried our Man into their Tents, where he remained untill our men returned theirs. In our Boat went their man to the place where the Fowle breed; and wee being desirous to know how the Salvages killed their Fowle, hee shewed them the manner how, which was thus: They tooke a long Pole with a snare at the end, which they put about the Fowles neck,[1] and so pluck them downe. When our men knew that we had a better way, and so[2] shewed the Salvages the use of our Peeces, which at one shot would kill 7 or 8. To bee short, they returned to the Cove to receive our man, and to deliver theirs. When they came, they made great joy, with dancing, leaping, and striking of their breasts; they offered divers things to our Men; but they onely tooke some Morse teeth, which they gave them for a knife and two glasse Buttons; so, receiving our Man, they came aboard, rejoycing at this chance, as if they had met with the most simple people of the World.

And *Henry Greene*, more then the rest, was so confident that by no meanes we should take care to stand upon our guarde. God blinded him so that, where he thought to receive great matters from this people, he received more than hee looked for, and that suddenly, by being made an

[1] This method of catching sea-fowl is still, or was recently, practised on the cliffs of St. Kilda's Isle.—C.

[2] To make sense, the words "and so" should be read as *they*.—C.

example for all men that make no conscience of doing evill; and that wee take heed how wee trust the Salvage people, how simple soever they seeme to be.

[*July* 29.] They made haste to be on shore, and because the Ship rid farre off, they weighed and stood as neere the place where the Fowle breed as they could; and because he, this Writer [*i.e.*, Prickett], was lame, he was to goe in the Boat to carry such things as he had in the Cabbine, of every thing some what. And so, with more hast then good speed (and not without swearing), away he went; as *Henry Greene*, *William Wilson*, *John Thomas*, *Michael Pierce*, *Andrew Motter*, and himselfe. When they came neere the shore, the people were on the hills dancing and leaping. To the Cove we came, where they had drawn up their Boats. Wee brought our Boat to the East side of the Cove, close to the Rocks; on land they goe, and make fast the Boat to a great stone on the shore. The people came and every one had something in his hand to barter; but *Henry Greene* swore that they should have nothing untill he had Venison, for that they had so promised him by signes the last day.

Now, when wee came, they made signes to their Dogs, whereof there were many, like Mongrels, as bigge as Hounds, and pointed to the Mountaines and to the Sunne, clapping their hands. Then *Henry Greene*, *John Thomas*, and *William Wilson*, stood hard by the Boat's head; *Michael Pierce*, and *Andrew Motter* were got upon the Rocks a gathering of Sorrell.[1] Not one of them had any Weapon about him, not so much as a stick, save *Henry Greene* onely, who had a piece of a Pike in his hand; nor saw he any thing they had to shoot him with. *Henry Greene* and *Wilson* had Looking glasses, Iewes-trumps,[2]

---

[1] See p. 127.—C.

[2] Jews' harps (see Latham's *Johnson's Dictionary*).—C.

and Bells, which they were shewing, the Salvages standing round about them. One of them came into the Boat's head to shew him a Bottle. This Writer made signes unto him to get him on shore, but he made as though he had not understood him; whereupon he stood up and pointed him on shore. In the meanetime, another stole behind to the sterne of the Boat; and when he saw him on shore that was on the Boat's head, hee sate downe againe, but suddenly hee sawe the leggs and feete of a man by him; wherefore hee cast up his head and sawe the Salvage, with his knife in his hand,[1] who stroke at his Brest over his head; hee casting up his arme to save his brest, the Savage wounded his arme, and stroke him into the body under his right Pap; the Salvage stroke a second blow, which he met with his left hand, and then stroke him into the right thigh, and had like to have cut off his little finger of his left hand. Now this Writer had got hold of the string of the knife, and had wound it about his left hand; he, striving with both his hands to make an end of that he had begun, found the Salvage but weake in the gripe (God enabling him), getting hold of the sleeve of his left arme he see his left side lay open to him; which when he saw, he put his sleeve of his left arme into his left hand, holding the string of the knife fast in the same hand, and having got his right hand at liberty, hee sought for somewhat wherewith to strike him, not remembering his dagger at his side, but looking downe he saw it; and therewith strooke the Salvage into the body and throat.

Whilst he was thus assaulted in the Boat, their men were set upon on the shore. *John Thomas* and *William Wilson* had their bowells cut; and *Michael Pierce* and

---

[1] Purchas, in *his Pilgrimage* (4th ed., 1626, p. 818), and also in *his Pilgrimes* (part iii, p. 610), gives particulars of this knife, which, he says, was such as they use in Java. This he regarded as further evidence of the existence of a passage.—C.

Henry Greene, being mortally wounded, came tumbling into the Boat together. When *Andrew Moter* saw this medley, hee came running downe the Rock and leaped into the Sea, and so swam to the Boat, and hung at her sterne untill *Michael Pierce* took him in, who manfully made good the Boat's head against the Salvages that pressed sore upon them. Now *Michael Pierce* had got an Hatchet, with which hee stroke one, that hee lay sprawling in the Sea. *Henry Greene* cried *coragio*, and laid about him with Trunchion. This Writer cryeth to cleere the Boat's head, and *Andrew Motter* cryeth to bee taken in. The Salvages betake them to their Bowes and Arrowes, which they sent so amongst them that *Henry Greene* was slain outright, and *Michael Pierce* received many wounds, and so did the rest. *Michael Pierce* cleareth the Boat and put it from the shore, and helpeth *Andrew Motter* in; but in the clearing of the Boat, *Pricket* received a cruel wound on his back with an Arrow. *Michael Pierce* and *Motter* rowed away the Boat, which when the Salvages sawe, they came to their Boats, which they feared they would have lanched to have followed them, but they did not. Their Ship was in the middle of the Channell, and yet could not see them all this time.

Now when they had rowed a good way from the shore, *Pierce* fainted, and could rowe no more. Then was *Motter* driven to stand in the Boat's head and wave to the Ship, which at first sawe them not; and, when they did, they could not tell what to make of them, but in they stood for them, and so tooke them up. *Greene* was throwne into the Sea; the rest was taken into the Ship, the Salvage being yet alive, but without sence. That day dyed *Wilson*, cursing and swearing in most fearefull manner. *Michael Pierce* lived two dayes and then dyed. Thus have you had the tragicall end of *Greene* and his 3 Mates, being the lustiest men in all the Ship.

The poore number that was left was to ply the Ship to and againe in the mouth of the Straits, for there was no anchoring; and besides, they were to goe in the Boat to kill Fowle to bring them home, which they did with great danger; for if the winde blew, there was an high Sea; and the Eddie of the tide would carry the Ship so neere the Rocks, as it feared[1] the Master, for so now they call Bylot.

After which great labour, and [when] on the South Cape they had killed 300 Fowle, they stood to the East; but the wind came East, and put them back againe to the Capes where they killed 100 Fowles. At length a West winde drives them homewards, for the most part along the North side of the Streight, untill he fell into broken ground about the Queene's Forland and there anchored; and from thence he came to [the Isles of] God's mercies, and from thence to those Ilands that lye in the mouth of the Straits, but not seeing the land untill they were ready to runne their Bowsprit against the Rocks in a fogge; but it cleered a little, and then they might see themselves inclosed amongst Rockie Ilands, and could finde no ground to anchor in.

They lie a trye[2] all night, and the next day the fogge continues. They seeke for ground to anchor in, but found none under 100 Fathoms. The next day, he weighed againe and stood to the East, and now they are brought to the allowance of halfe a Fowle a day, yet they had some meal left, and nothing else; and now were glad to burne of the feathers (for saving of the skins which, before, they [had] fleaed off the Fowle) because they will not pull, nor the garbidge also was not throwne away.

---

1 *Frightened:* an old expression (see Nares's *Glossary*).—C.

2 That is, they *lay-to* (see Smyth's *Sailors' Word-Book*, p. 701).—C.

He saith that, after they were cleare of those Ilands[1] which lie out with two points, one on the South East and the other on the North, making a Bay to the sight, as if there were no way through, hee continued his course East South East and South East, thinking to raise Desolation, from thence to shape his course for *Ireland*, though *Ivet* perswaded to goe for *New-found-land*, hoping there to have reliefe amongst our Countrymen. But, in Latitude 57 degrees, the winde came South West, and so it was thought fit to seek for food where some grew: *viz. Ireland*, and so the course was directed. In which time they were fain to frie their Fowles bones in Candle tallow, putting vinegar thereto, which was stirred amongst them, and every man had one pound of Candles allowed for one weeke as a great dainty. And when *Rob. Juet* said that by his reckning they were within 60 or 70 leagues of *Ireland*, they had 200 leagues thither. Their course was much longer (through evill steeridge), for their men were so weake as they were fain to sit at the Helme.

Then *Robert Juet* dyed, and the rest despaired and said they were past *Ireland*. Their last Fowle was in Stepe-tub, and the men cared not what end [of the ship] went first. The Master was glad to doe their labour and his owne, tackling going to wrack, and none regarding to helpe the same. In this extremity, it pleased God to give them sight of Land, not farre from the place where the Master said they should fall [in with it], which was the Bay of *Galloway*,[2] to the West of the Durses.[3] So they stood along to the South W. and espied a saile, which was a Boat

[1] So far as one can make out, the Islands thus alluded to must have been those now called Button's Islands.—C.

[2] Galway Bay.—C.

[3] Dursey Island, on the coast of the County Cork, and the adjacent rocky islets. They are, however, a long way from Galway Bay. Foxe, not Prickett, seems responsible for the error.—C.

of *Foye*[1] at anchor fishing. This Barke brought them into Beare Haven.[2] Here they stayed some few dayes and dealt with the Irish for Reliefe, but found none, for in that place there was neither bread, drink, nor money, or Countrymen, which were then on fishing, they found as cold in kindnesse, that they would doe nothing without present money. In the end *John Waymouth*, one of the Barke [that] brought them in, furnished them with money upon pawne of their best Anchor and Cable, wherewith they bought Bread, Beare, and Beefe.

Now, as they were beholding to *Waymouth*, so were they beholding to Captaine *Taylor* for making their Contracts, and for their men's wages, who would not goe with them home, except *Waymouth* would passe his word. Whereupon *Taylor* swore he would presse them; and if they would not goe he would hang them.

In conclusion, they agreed for 3 pound 10 shillings a man, to bring the Ship to *Plimouth* or *Phalmouth*, and to give the Pilot 5 pound; and, if they were put into *Bristow*, they were to have 4*l.* 10*s.* a Man, and the Pilot 6*l.*; and, omitting further circumstance, they came to *Plimouth*, from thence to the Downes, from thence to *Gravesend*,[3] and so to *London*; where the Master had this Writer to Sir *Thomas Smith's*.[4]

---

[1] *Fowey*: the ancient, but now decayed, Cornish sea port. Purchas, in *his Pilgrimage* (4th ed., 1626, p. 819), says it was the 6th of September when they fell in with the boat, but this date is omitted from the fuller narrative in the *Pilgrimes*.

[2] *Bear Haven*: the land-locked natural harbour in Bantry Bay, formed by the channel between Bear Island and the mainland of the County Cork.—C.

[3] Prickett does not mention the date on which he and his fellow mutineers reached Gravesend, on their return voyage in the *Discovery*, but Hessel Gerritz says that it was in the month of September (see p. 121).—C.

[4] This means that Bylot took Prickett with him to Sir Thomas Smith's, as is more clearly stated in the original in Purchas.—C.

The Printer on the behalfe of Pricket.[1]

FOr as much as this may happily be suspected by some, not so friendly to *Pricket*, who returned with that company who so cruelly had exposed *Hudson*; and therefore may seeme to lay heavy imputation, and rippe up further occasion then they will beleeve: he saith also he[2] added the report of *Thomas Woodhouse*, one of the exposed Company, who ascribed the occasions of discord to *Juet.* I take not on mee [says Purchas] to sentence, no not to examine. I present the evidence just as I had it. Let the Bench censure with both eares, that which they may see with both eyes, and there note, to which I [have] first prefixed his letter to Master *Samuel Macham.*

Master Macham, *I heartily commend me unto you. I can write unto you no newes, though I have seene much, but such as every Fisherman haunting the Coast can report better than my selfe.*[3]

We kept our Whitsunday on the N.E. end of *Iseland*, and I thinke I never fared better in *England* then we feasted here. They of the country are very poore, and live miserably; yet we found there store of dainty fresh Fish and dainty Fowle. I my selfe, in one afternoone, killed so many as feasted all our Company, being 23 persons, at one time onely with Partridges, besides Curlew,

---

[1] This heading, which is not in the original, is an addition by Foxe. The paragraph following is a comment by Purchas on Prickett's narrative which Foxe has reproduced in slightly altered form.—C.

[2] This means that Purchas says he added the report.—C.

[3] This letter appears to have been written by Woodhouse, but it may have been written by either Hudson himself or Prickett. Purchas has omitted the signature, and his remarks about it are very confused.—C.

Plover, Mallard, Teale, and Geese. I have seene two hot Bathes in *Iseland*, and have beene in one of them. We are resolved to try the utmost, and lye onely expecting a faire winde, and to refresh our selves to avoide the Ice, which now is come off the West Coast, of which we have seene whole Ilands; but God be thanked have not bene in danger of any. Thus I desire all your prayers for us.

Iseland, this 30 *of May*, 1610.

A note found in the Deske of Thomas Woodhouse,[1] *Student in the Mathematicks, and one of those exposed with* Mr. Hudson *in the Shallop.*

THE 10 of September, 1610, after dinner, our Master called all the company together, to heare and beare witnesse of the abuse of some of the company, it having beene the request of *Robert Iuet*, that the Master should redresse some abuses and slaunders, as he called them, against this *Iuet.* Which thing, after the Master had examined and heard with equitie what hee could say for himselfe, there were proved so many great abuses and mutinous matters against the Master and action by *Ivet*, that there was danger to have suffered them longer ; and it was fit time to punish and cut off further occasions of the like motives.[2]

It was first proved to his face by *Bennet Mathew*, our Trumpeter, upon our first sight of Island : and hee confest that hee suppos'd that in the action would bee man-slaughter, and prove bloody to some.

Secondly, at our comming from Island, in the hearing of the companye, he did threaten to turne the Ship['s] head

---

[1] Purchas spells the name *Wydowse.* Foxe here reprints this note in full and almost *verbatim.*—C.

[2] A misprint for *mutinies.*—C.

home from the action, which at that time was wisely pacefied by the Master, in hope of amendment.

Thirdly, it was deposed by *Phillip Staffe*, our Carpenter, and *Arnold Lodlo*, to his face upon the holy Bible, that he perswaded them to keepe Muskets charged and Swords ready in their Cabbines, for they should be charged with shot ere the voyage were over.

Fourthly, we being pestered in the Ice, he had used many words tending to mutinie, discouragement, and slaunder of the action, which easily tooke effect in those that were timorous. And, had not the Master in time prevented, it might easily have overthrowne the voiage; and now lately, being imbayed in a deepe Bay which the Master had desire to see, for some reasons to himselfe knowne, his words tended altogether to put the company into a fright[1] of extremity, by wintering in cold, jesting at our Master's hope to see *Bancum*[2] by Candlemas.

For those and divers other base slaunders against the Master, he was deposed, and *Robert Bylot*, who had shewed himselfe honestly respecting the good of the voyage, was placed Master's Mate in his stead.

Also, *Francis Clements*, the Boatswaine, at that time was put from his office, and *William Wilson*, a man thought more fit, preferred to his place; this man had basely carried himselfe to our Master and to the action.

Also, *Adrian Motter* was appointed Boatswaine's mate, and a promise from the Master, that from this day *Iuet s* wages should remaine to *Bylot*, and the Bo'sons overplus of wages should be equally devided between *Wilson* and *Iohn King*, to the owner's good liking, and one of the quarter Masters, who had very well carried themselves to the furtherance of the businesse.

---

[1] Purchas has it *fray of extremity*, which is an obsolete form of fright or alarm.—C.

[2] *Bantam* in the original.—C.

Also, the Master promised that, if the offenders yet behaved themselves honestly, he would be a meanes for their good and that hee would forget injuries, with other admonitions.

Here the Reader may observe a plaine expression of *Hudson's* good carriage, made concerning the voyage and pithily demonstrated, which makes mee ready to call *Pricket* to further question, who in all his long declaration of this voyage, hath not given *Hudson* any commendations, no not in his good parts: and yet hath taken paines enough otherwayes to make an ample expression, and to call the roague *Greene, Henry Greene.* Well, *Pricket,* I am in great doubt of thy fidelity to Master *Hudson.*[1]

The Voyage of Sir Thomas Button *with two ships, the* Resolution, *Admirall, the* Discovery, *Vice-admirall; manned and victualled for* 18 *Moneths.* 1612.[2]

COncerning this voyage, there cannot bee much expected from me, seeing that I have met with none of the Iournalls thereof. It appeareth that they have

---

[1] This paragraph (from the tone of which it is not easy to dissent) is a shrewd comment by Foxe on Prickett's narrative. It takes the place of a paragraph in the original (*Purchas his Pilgrimes*, part iii, p. 609), which Dr. Asher (*Henry Hudson*, p. 138) attributes to Woodhouse, but which is clearly a comment by Purchas himself on the documents he prints relating to Hudson's voyage. In this paragraph, Purchas tells how he had been informed that, on the homeward voyage of the mutineers, their ship ran aground at Digges Island, and that a great flood from the west floated them off (see p. 150), which he regarded as evidence of a passage into the South Sea.—C.

[2] I desire here to repeat that all the preceding voyages, as well as those of Bylot and Baffin which follow, have already been edited and fully discussed by other and more competent writers, and that I have, therefore, annotated them as briefly as possible. They would hardly have been worth reprinting had they not formed part of Foxe's

beene concealed, for what reasons I know not; but it is fitting that such things should be made extant as may

---

otherwise-valuable work. We come now, however, as already stated (see Introduction), to the most important and interesting portion of Foxe's book. The following account of Button's voyage is *original*—that is to say, it is not abstracted from the works of earlier writers, as all Foxe's preceding narratives have been. The following is, in fact, the only account we have of Button's by-no-means-unimportant voyage, with the exception of the very brief notice of it given by Purchas. In *his Pilgrimage* (4th ed., 1626, p. 819), he says that he had met with no account of the voyage, but that he had gathered a few particulars (which he gives) and had seen a chart of Button's discoveries—probably that alluded to as "Hubart's Platt", in *Purchas his Pilgrimes*, part iii, p. 848 (cf. *North-West Foxe*, orig. ed., p. 161). In *his Pilgrimes* (part iii, p. 848), Purchas also says that he had personally solicited from Button the loan of his journals, but without success. The narrative Foxe gives of Button's voyage was obtained by him from several different authorities, and is consequently very scrappy, though of fair length and fulness. In the first place, we have a very brief narrative of the events of the voyage up to the time of wintering, communicated to Foxe by Abakuck Prickett, probably by word of mouth, as we know that Foxe knew him. Next we have the report of Captain Hawkridge, which is much longer. It commences with the entry of the expedition into Hudson's Strait and continues to the time of laying up for the winter. His contribution is not of great importance, and it would be very brief were it not for the fact that he includes some lengthy replies to queries which Button drew up to employ the minds of his men during the winter. Next follows a long and detailed narrative supplied to Foxe by his patron Sir Thomas Roe, who abstracted it (Foxe says) from Button's own journal. It only commences, however, on July 15th, 1613, when Button had left his wintering-place and had returned to Hopes Checked. It continues until the expedition reached Digges Island, late in August, when it ceases. Prickett then supplies a very brief account of the return voyage. These several narratives afford a tolerably complete account of Button's voyage, but they hardly corroborate one another upon a single point, and, as a rule, they do not overlap one another in point of time. The narrative of Button's voyage commences on the first page of "signature O" of Foxe's work, and from here to the commencement of his own voyage in "signature V" (about sixty pages) has evidently been very carelessly "read" by the printer's reader. Consequently, Button's voyage is disfigured by a large number of stupid misprints. All those which

any way redound to the good of the Common-wealth ; and therefore I can but communicate what I have received from *Abacuck Pricket*,[1] who was in the same Voyage, and [from] others by Relation, and from Sir *Thomas Roe* in the last part of a Iournall of this voyage.

He departed about the beginning of May,[2] and went by

---

are pointed out in the *Errata*, have been carefully corrected in the narrative following, as well as some others which were obvious ; but many obscure passages probably still owe their want of clearness to printer's errors. The *Discovery*, which was Button's consort, was, of course, the same vessel in which Weymouth and Hudson had sailed before. Button's own ship, the *Resolution*, was probably a much larger vessel, and is several times spoken of as such, but we have no definite information about her. It is very likely that she had belonged to the Royal Navy, as a record quoted in the Introduction shows that one of the King's shipwrights was employed in her selection. It is a curious coincidence that Captain Cook, on his famous voyage to Bering's Strait in 1776-79, sailed in his old vessel the *Resolution*, while his consort was named the *Discovery*. The origin and personal history of Sir Thomas Button, the instructions under which he sailed, and many of the chief events and main results of his interesting voyage, have been already discussed at some length in the Introduction. Inasmuch, however, as this important voyage has never yet received adequate treatment, I have annotated it with considerable fulness and care.—C.

[1] This is, of course, the man, whose doubtfully-veracious account of Hudson's voyage immediately precedes this. Whether or not he was really guilty of unfaithfulness to his captain, as Foxe believes (see p. 162), he had clearly been able to free his action from blame sufficiently to be chosen to accompany Button. We also know (see Markham's *Voyages of William Baffin*, p. 111) that Robert Bylot, or Billet, who had brought the *Discovery* home after the mutiny in the previous year, was also with Button's expedition. From this source, too, we may gather that Bylot also sailed with Gibbons on his abortive voyage in 1614.—C.

[2] This date is not correct. The real date of his departure is more nearly given in the MS. *Autobiography of Phineas Pett* in the British Museum (see Introduction), on folio 66 of which occurs this passage : "The 14 day of aprill [1612], being Easter tuesday, I came to Gravesend to meet Captaine Button, who was then going away upon his voyage, & we parted together on board his Shipp, from whence I returned to Chatham."—C.

the West, and entered the passage on the south of *Resolution*,[1] and sometime was fast amongst the Ice; but at length he came to *Diggs* his Iland, where hee staied 8 dayes, and in that time set up a Pinnace he had brought from home with him in pieces. And they set from thence to the Westward, where hee discovered the Land he called *Cary's Swansnest*.[2] From thence he proceeded to the Southward of the West, falling with land in Latitude about 60 d. 40, which he named *Hopes-check*,[3] I thinke because that

[1] It seems probable that he named this island Resolution after his own ship of that name. At all events, there seems no record of its having been so named by any previous navigator.—C.

[2] A cape at the south-east end of Coats Island, 62° 10′ N.; 83° W. It is still called by this strange name, the origin of which is not clear. Possibly the nest of a wild swan was there discovered by one of his crew named Cary; or the place may have been named after the Earl of Monmouth of that day (whose family name was Carey), though he does not seem to have been one of the Company of Adventurers for discovering the North-West Passage; or it may have been named after Mr. Allwin Carey, who may have formed one of Button's companions or been appointed by the above-named Company to superintend the fitting out of the ship for the voyage. This gentleman was probably a trusted servant of the North-West Company; for we read (Markham's *Voyages of William Baffin*, p. 111) that, at the sailing of Bylot and Baffin, who were sent out by that Company in 1615, there "came abourd Mr. John Wostenholme, esquire, one of the cheefe aduenturers, and with him Mr. Allwin Carye (husband for the voyage)." In the present day a "ship's husband" is an agent appointed by all her owners to make payments, receive moneys, etc., on account of her voyages. It was after him that Baffin, on July 8th, 1616, named a group of islands which he discovered in Baffin's Bay. Or Button may have named the place after Sir Henry Carie, "Master of our Jewell House", who was one of those incorporated under the Charter of the Company for Discovering the North-west Passage. This man may have been the father of "that Honourable Gentleman, Master Thomas Carie, one of the Bed Chamber to the King", after whom, on July 2nd, 1632, Capt. James named an island in James's Bay.—C.

[3] This name has not been retained. It was merely applied to that point at which Button first encountered the west coast of Hudson's Bay in about the latitude named. It was the first time that the western shore of the Bay had been discovered, and there can be little doubt,

there his expectation was crossed; and, thereabout, enduring a grievous storme, [he] was put to the Southward and constrained to looke for harbour the 13 of August, to repaire some losses. After which time, came on the new Winter, with much stormie weather, [so] as he was constrained to winter there, in a small *Rile*[1] or Creeke on the North side of a River in Lat. 57d.10, which River he named *Port Nelson*,[2] after the name of his [Sailing] Master (whom

---

from all the evidence we have, that Button had previously believed he would have been able to sail straight on into the great Western Ocean, and that his hopes were very severely checked when he encountered the land in question. On many old maps, "the Checks" (by which name Hubart and others hereafter speak of Hopes Checked) are shown as three headlands or small islands near the shore (it is not clear which); but, as there is nothing in the narrative to explain the cause of this, conjecture is in vain.—C.

[1] This is probably a misprint for *rill*, meaning a small brook or streamlet. I cannot find that the word was ever spelled *rile*, as above.—C.

[2] Port Nelson, still so called, is formed by the estuary of the Nelson River, down which the waters of the Saskatchewan, the Red River, and other important streams, find their way into Hudson's Bay. At the mouth of Hayes River, a small branch of the Nelson, stands York Fort, the most important of the Hudson's Bay Company's posts on the Bay. We find a clue to the *exact* spot at which Button wintered in a large, interesting, and beautifully-drawn chart of Hudson's Bay and Straits "Made by John Thornton at the Sign of the Platt in the Minories: Anno 1685," and now preserved at the British Museum (Add. MSS. 5414, 20). On this chart, the wintering-place of Button and the halting-place of Foxe in Port Nelson are shown precisely. Foxe apparently laid up his ship in Heart Creek of the Admiralty Chart of the "Harbours and Anchorages in Hudson's Bay" (No. 2112), which is called "Foxe's Hole" on Thornton's chart, while Button wintered in a creek which seems to be identical with Root Creek of the Admiralty Chart, but which is described as "Sr Tho. Button's wintering-place" on Thornton's. Both creeks are on the northern side of the estuary, and the latter is a few leagues to the north-eastward of the former. Root Creek seems to lie in 57° exactly, or 10' to the south of the position above assigned to it. Hubart (see p. 173) gives the latitude as 56° 58', which is almost exactly correct. Forster (*Voyages and Discoveries in the North*, p. 345) writes as if something were known

he buried there), putting his small Ship in the foremost and Baracadoe them both (with Piles of Firre and earth) from storme of Snow, Ice, Raine, Floods, or what else might fall.

He wintered in his Ship and kept 3 fires all the Winter, but lost many men,[1] and yet was supplied with great store of white Partridges[2] and other Fowle, of which I have heard it credibly reported that this company killed 1800 dozen in the Winter season.[3]

The report of Captaine Hawkridge.[4]

HE entered in the South Channell, seeing the South shore within *Fretum Hudson*, neere *Hope's advance* twice or thrice, and once trying the tide about [the] *Savage Iles*, where it came from the South East & flowed 3 fathoms.

---

as to the origin and history of Nelson, and says he was mate in the *Resolution;* but whence he derives his information, I cannot discover.—C.

[1] Hawkridge also mentions this mortality. Moreover, Purchas (*his Pilgrimage*, 4th ed., 1626, p. 819) says he had heard that Button had "much misery of sickness in his wintering". This mortality must have begun after the commencement of the New Year; for Edward Glanville, writing apparently on December 22nd, speaks of their having been preserved from all dangers whatsoever. Purchas adds that Button "was forced to quit the great ship" (the *Resolution*), but none of those whose narratives Foxe prints allude in any way to this circumstance.—C.

[2] See p. 134.—C.

[3] Here ends the information supplied by Prickett. The following narrative by Captain Hawkridge is much more detailed.—C.

[4] He was one of Button's two volunteer companions, according to Barrow (*Voyages into the Arctic Regions*, p. 196) and Rundall (*Voyages towards the North-West*, p. 84), but I can discover no actual basis for the statement. Nothing seems to be known of his previous history or origin. In the year 1619, he made an altogether unimportant voyage in search of the North-West Passage, details of which Foxe gives hereafter. Reference to some later sufferings he underwent will be found in the Introduction.—C,

At length he came in the South Channell, betweene the Ile *Salisbury* and the South Maine, sayling in between *C. Wostenholme* and Sir *Dudley Diggs* his Ile, to the West end thereof, where he saith is a banke of *Owes*[1] to anchor upon, at 13 fathom. *C. Wostenholme* is to be brought within the West point of the Ile, which will then beare East by South. Here the *Salvages* did offer to assault his men (bound going to kill Willicks,[2] of which there is such store as in short time hee could have laded his Boat) with two Canoes, and to the number of 70 or 80 men came upon them, untill with one Musket-shot he slew one of their men and hurt more, who, much amazed with the report and execution of a Musket, retired; yet, at his comming from thence, he sending his Pinnace boat on land to take in fresh water, the Salvages were laid in ambush amongst the Rocks, and slew him 5 men dead. One escaped by swimming. It is much to bee doubted that the Salvages did slay those men in revenge for 4 of their great Canoes he tooke off the Land from this people, whereof he restored but two backe againe. And here it was where the villaines *Greene* and *Juet*[3] were slaine, after they had exposed Master *Hudson*. This is Sir *Dudley Diggs* his Ile, and there is Deare within the same.

Hee passed from hence to a Cape on the N. side of his Bayes entrance,[4] which he named *Cary's Swans nest*, and

---

[1] That is *ooze*, or soft mud.—C.

[2] Black Guillemots or Dovekies (see p. 107).—C.

[3] There is an error here. Green, Wilson, Thomas, and Pierce had lost their lives at this spot in the previous year, but Juet, it will be remembered, had died afterwards, on the voyage home, apparently of starvation (see p. 157).—C.

[4] That is, the entrance to Button's Bay; for, long after Button's voyage, a large part of the bay we now know as Hudson's Bay was known as Button's Bay, the name Hudson's Bay (or, rather, *Bays*, for it was then supposed to be two bays—see p. 131, note) being given

from thence to his *Hopes checkt*, was troubled with shift of winds and should water; took harbour the 15 of August[1] in *Port Nelson*. He endured a sharpe Winter; lost many men; [and] kild 3 Deare in the River as they were swimming from side to side. There came also to them divers Beares and Wolves, and it was the 16 of February before the River[2] was froze over, having had divers warme thawing days before. The Captain, having beene sick the whole Winter, began to mend the 24 of Ianuary. The Ice began not to cleare out of the River until the 21 of Aprill, after which they killed daily, with their Net, abundance of Fish as bigge as Mackrils.

Now, during this wintering, it appeareth that Sir *Thomas*,[3] having good time to advise and contemplate what was to be done the next yeere, drew [up] some Demands in writing, which he caused (it seemed) the most understanding men of his Company to answer. Of which, such answers as came to my hands I doe hereby freely impart for thy better understanding.

---

solely to the Bay at the bottom of which Hudson wintered, but which has been known as James' Bay since the year 1631, when Captain James wintered in it. The names may be seen applied as above described on Foxe's own map and on some others of even later date; but on James's chart, and on most others of subsequent date, we find the names "James's Bay" and "Hudson's Bay" used as they now are, though, for more than a century later, we often find the western side of the latter still called "Button's Bay". Even now, that name is retained on the latest charts for the small bay to the westward of the entrance to Port Churchill.—C.

1 Prickett says the 13th (see p. 166).—C.

2 Not above a mile broad.—F. This is probably a comment by Foxe, and is apparently not quite correct. Port Nelson (see p. 166) is at least six miles wide at its entrance, and cannot have been much narrower at the spot where Button wintered.—C.

3 He was not *Sir* Thomas at this period, not having been knighted until August 30th, 1616 (see Metcalfe's *Book of Knights*, p. 213).—C.

[BUTTON'S QUERY.]

Laus Deo, 1612, *December the* 22.

[*What is*] *the course and distance from place to place, from Cape Cleare*[1] *to this River in New Wales?*[2]

[CAPT. HAWKRIDGE'S REPLY.]

Imprimis: from Cape *Cleere* to Cape *Desolation* strait course by common Compasse North W. by W. ½.	428 Leagues. Variation 24.

The Latitude of 59 *d.* 40 *m.*

From *Desolation* to the Ile of *Resolution*, course is N.W. by W., the Latitude 61 d., the distance	170 Leagues. Variation 26 d. ½.
From *Resolution* to Sir *Dudley Diggs* his Ile, Lat. 62 d. 40 m. N.W. the distance is	142 Leagues. Variation 30 d.
From Sir *Dudley Diggs* his Ile, to the *Cheeks*, the course is W. ½ Northerly, the distance is	193 Leagues. Variation 22 d.
From the *Checks* to New *Wales*, Lat. 57, the course is S. by W.,[3] the distance [is]	90 Leagues. Variation 22 d.

---

[1] On the south-west coast of Ireland, off the coast of the County Cork.—C.

[2] Upon the whole of the western shore of "his Bay" (Hudson's Bay), Button bestowed the name of New Wales, which he unquestionably did in honour of *old* Wales, his native land. This name it retained until a very recent period, the northern part being known as New North Wales and the southern as New South Wales. The latter name appears on some maps of British North America published as late as the beginning of this century; but it was probably disused soon after, as the Australian colony which now bears that name assumed it in 1788.—C.

[3] He should have said S. by *E.* (see note on p. 173).—C.

The courses are all by the common Compasse.

Your Worships and ever, or mine owne never, till death,

William Hawkridge.

[JOSIAS HUBART'S REPLIES.]

My answere to the first demaund[1] under your favour: I think it not amisse to search this River, if God give strength to our Men, before our departure from it, to have the knowledge how farre it doth extend; and that we may meet with some Inhabitants, which may further our expectations, but I cannot thinke of any profit to be made by it.

My answer to the 2 Demaund[1] is to search to the Northward about this Westerne land, untill, if it be possible, that

[1] From these replies by Josias Hubart, we may infer that two queries or "demands", to somewhat the following effect, had been made by Captain Button *before* that given above:—(1) "What is the best thing to do whilst we are imprisoned for the winter?" and (2) "What is the best thing to do when we get free in the spring?" Perhaps these two queries were only put to Hubárt: at any rate, he was the only one whose replies to them have been preserved. His reply to the "second demand" was in every way a shrewd reply, and quite the best he could have given. Foxe, in one of his quaint side-notes, exclaims: "Well guest, Hubart!" while Barrow remarks (*Arctic Voyages before 1818*, p. 199) that: "The answer given by one James Hubert [*sic*], the pilot of the *Resolution*, to the question: 'How the discovery might be best prosecuted when they should be able to go to sea?' shows the sound notions entertained by this man respecting the true mode of searching for the passage." It appears from Purchas (*his Pilgrimes*, vol. iii, lib. IIII, p. 827) that Hubart, who was, like Foxe, a Hull man, had sailed with Hall on his second and third Greenland voyages in 1606 and 1607 (see p. 95), and also that he afterwards sailed with Bylot and Baffin on their second north-west voyage in 1616, and that, in all probability, he was also with them on their first voyage in the previous year. We are not told what position Hubart held when sailing under Button. Forster (*Voyages and Discoveries in the North*, p. 345) says he was "mate", though on what grounds I know not.—C.

we may finde the Flood comming from the Westward, and to bend our courses against that flood, following the ebbe, searching that way for the passage.[1] For this flood which we have had from the Eastward, I cannot be perswaded but that they are the veynes of some head-land to the Northwards of the *Checks*, and by the Inlets of Rivers which let the floods tides into them; which Hedlands being found, I do assure myselfe that the tide will be found to come from the Westward.

Herein I have shewed my opinion so farre [as] my Iudgement will afford, untill further reasons induceth me to the contrary.

Per me Iosias Hubart.

From the *Durses*[2] in Ireland, being in 52 Lat. to *Cape Farewell* in *Groenland*, Lat. 58.56, the course is W.N.W. W-terly, and the distance is	460 Leagues. Variation at Cape *Farewell*, 11 deg.

The Southermost part of the Iland of Resolution is in Lat. 60 *d.* 34 *m.*

From *Cape Farewell* to the Iland of *Resolution*, the course is W. and by N. and the distance is	208 Leagues. Variation 29 d.[3]
Sir *Dudley Diggs* his Iland is in Lat. 62 D. 40 M. and is in distance from the Ile of *Resolution* upon a W. and by N. ½ Northerly	180 Leagues. Variation 3 points.
The Checkes lye in 61 D. 17 M. Lat.[4] from Sir *Dudley Diggs* his Iland, thereto the course is W. and by S. and the distance is	190 Leagues. Variation 3 d.

[1] Well guest Hubart!—F.

[2] Dursey Island, off the coast of Cork (see p. 157).—C.

[3] A great mistake.—F.

[4] This is probably wrong, for both Prickett (p. 165) and Glanville

Our wintering being in the Lat. of 56 *d.* 58 *m.*[1]

From the Cheekes to our wintering place, the course is S. and by W. ½ Westerly,[2] and the distance is	87 Leagues Variation 2 d. different.

The 27 of November, I made an observation of the Moone and the planet Mars, and for that I stand in doubt, for the houre to be exactly founde out by any Diall, Clock, or other Instrument, to hang a plummet to find when the foremost Guard was right under the Pole starre, at which instant I found ♂ and ☾ to be one degree and 41 minutes asunder, by which working I suppose or deeme it to bee as followeth: this our wintring place 111 degrees and 15 [minutes] of longitude from our Meridian of the Citie of *London.*[3]

Per me Iosias Hubart.

In the name of God, Amen.

[EDWARD GLANVILLE'S REPLY.[4]]

Of the courses from the Misson Head[5] in *Ireland*, being bound towards the Northwest passage, Captain *Thomas*

(p. 175) give the latitude of "the Checks" or "Hopes Checked" as 60° 40′. If Hubart is right, the spot lies a little to the north of Cape Esquimaux.—C.

[1] The exact spot has already been indicated (see p. 166).—C.

[2] He should have said S. by *East*. Hawkridge fell into the same error (see p. 170).—C.

[3] Hubart's calculation was very much out. The wintering-place was about Long. 92° 40′ W. This error probably accounts for the fact that both Hubart and Hawkridge describe Port Nelson as lying south by west (instead south by east) from the Checks.—C.

[4] We are not told what position this man held. Probably he was a superior officer of some kind.—C.

[5] Mizen Head is another of the prominent headlands on the ragged coast of Cork.—C.

Button, Gentleman, being our Generall, in the good ship called the *Resolution: John Ingram*, Captaine and Master of the Pinace called the *Discoverie*, 1612.

Imprimis: from the Misson head in *Ireland* to Cape *Discord* in *Groanland*, the course is N.W. by N. Northerly, and the distance is	360 Leagues.[1]
From the *Misson-head* to Cape *Discord* in *Groynland*, the course is N.W. by W. Northerly by the compasse, the Lat. 59 d. 20 m. and the distance is	380 Leagues.[2]
From the *Misson-head* to Cape *Desolation* the course lyeth W.N.W., & the distance is	490 Leagues.
From the foreside of Cape *Discord* to Cape *Farwell*, the course lyeth S.W. Southerly by compasse, distance	58 Leagues.
From Cape *Farwell* to the Westerne part of this Head Land, by Cape *Desolation*, the course is W.N.W. halfe Northerly, 100 Leagues distant, and from this Headland to *Desolation* is 10 Leagues distant: in all, from *Farwell* to *Desolation*, the distance is	100 Leagues.[3] Variation 23 degrees, as hee judged.
N.N.E. by compasse; betweene Cape *Farwell*, and the foresaid Head Land, there set a very great current to the Westward.	

[1] Latitude 60 d. 30 m.—F.

[2] The information here given seems to be repeated, apparently through a printer's error.—C.

[3] There is some confusion here, probably due to Foxe's printer. —C.

From Cape *Desolation* to the Ile of *Resolution*, the course lyeth W.N.W. Westerly, altitude 62 d. 30 m., and the distance	120 Leagues. Variation 29.
From *Resolution* to *Salisbury* Ile, W. by N., Altitude 63 d. 15, and from the Iland to *Wostenholme's* Cape, the course lyeth W.S.W. Southerly,	140 Leagues. 12 Leagues.
And from this Cape to *Diggs* his Iland [the distance is]	3 Leagues.
From *Resolution* to *Wostenholme's* Cape, the course lyeth W. by N. Westerly, and the distance is	153 Leagues. The Variation of this No. 34.
From *Resolution* to *Diggs* his Ilands, the course is W. by N. Northerly, and the distance	156 Leagues.
Altitude, 63 d.	
From Sir *Dudley Diggs* his Ilands to *Nottingham's* Iland, [the course is] N. by the compasse, and the distance is	7 or 8 Leagues.
From Sir *Dudley Diggs* Iland to *Swann's* Iland,[1] [the course is] W. by S.	40 Leagues.
From *Diggs* his Iland to *Hopes Checkt*, the course is W.S.W. a little Westerly, and the distance is	200 Leagues.
The Altitude is 60 d. 40 m.	
From *Hopes Checkt* to the Broken Land,[2] where our Admirall received a great storme: the course lyeth S.W. 49 Leagues; Altitude 59.	49 Leagues.

[1] This apparently means Cary's Swan's Nest.

[2] Clearly this was near the Checks (see p. 166).—C.

From this broken Land to the head *Northerland*,[1] the course lyeth W., & the distance is	8 Leagues.

The Headland is the entring into this Bay called New Wales.[2]

From this Head-land unto the Roade of the harbour, the course lyeth South 42 Leagues, and from *Hopes checkt* to this Roade, the course lyeth N.E. and by N.[3]	86 Leagues.

Hitherto the Lord of his mercy hath blessed, preserved, and kept us from all dangers whatsoever,[4] which wee beseech him to bless us of his mercy, and to send us well forth againe, *Amen.*

Per me Edward Glanvile.[5]

---

[1] Evidently by this name Cape Churchill is meant.—C.

[2] This name has, I believe, never since been applied to the *Bay*. The *land* was so named by Button.—C.

[3] Apparently he meant S.E. by S.—C.

[4] As the questions to which these answers were given are dated December 22nd, it appears from the above statement that, at any rate up to that date, the hardships which are said to have caused the death of many of Button's men (see p. 167) had not commenced.—C.

[5] Here the information Foxe obtained from Captain Hawkridge comes to an end. What follows was supplied in manuscript by Sir Thomas Roe, who had abstracted it (as Foxe says) from Button's own journal, which is now lost. It does not commence until July 15th, 1613, when Button was back at Hopes Checked. After getting free from the ice in his wintering-place in Port Nelson, he had wisely taken Hubart's advice and sailed to the northward with the idea of exploring along the shore. How Sir Thomas Roe came to obtain access to Button's journal is not clear, as he was not one of the adventurers who sent out Button. However, he was one of the leading promoters of Foxe's voyage in 1631 (see Introduction), and probably his high position and great influence enabled him to obtain access to the journal (which had been withheld from others) and to make extracts therefrom for Foxe's use.—C.

What I received from Sir *Thomas Roe* was an abstract copy taken out of Sir *Thomas Button's* owne Iournall, be[ginn]ing at his return to this *Hopes Checkt*, and not before. As from the beginning of his Voyage, or his wintring, I have nothing but by report, and thus he proceeded from thence.

July 15 [1613]. This day, being there,[1] he plyed up with contrary winds and fogg, having diversity of depth; altogether standing Eastwards off, sometimes to 30 Faddome and more, and standing in to 7 or 6, thus traversing and anckoring until the 23 day, finding the Tyde to come from N.E. by N., which is an especiall argument that the land doth lye so, and not [observing] any thing else of note.

23. Having thus plyed up to the N. E.-wards untill this day, and standing into the shore, hee anckored for to try the tyde, thinking to send his Boate on Land, to see what Land it should be. It bare on him from the N.E. by N. to the N.W. by N., and the depth 42 Fadome not above 4½ League from the Land. This was the highest land hee had seene since hee came from Sir *Dudley Diggs* his Iland the last yeare. This Land he named *Hopes Advance*.[2]

---

1 That is, at Hopes Checked.—C.

2 It is, at this point, difficult to follow the narrative with precision. It appears not unlikely that the land Button thus named was Cape Esquimaux. However, on Foxe's map, and also on Briggs's map in *Purchas* (part iii, p. 852), Hopes Advance is shown to the north of Brook Cobham (now Marble) Island, which seems to identify it with Baker Foreland, or perhaps with Cape Fullerton. The former locality of these two is to some extent indicated by Foxe's statement (p. 258, original edition) that he stood northward "into the latitude of 64° 10′ which is to the north of that high land called Hopes Advanc'd". In any case, the name (which does not appear on modern charts) must not be confused with *Cape* Hopes Advance on the south shore of Hudson's Strait. Rundall has clearly misread the above paragraph, for he says (*Voyages to the North-West*, p. 89): "The 23rd of July, Hopes Advance, *which was seen and named on the former part of the Voyage*, was *again* fallen in with." Button had certainly not visited this spot before. Rundall was thinking of Hopes *Checked*, which

24. This morning proved thick; easte winde; and it fell to lighten and thunder, so as hee held it not fit to adventure the Boat from the Ship.

25. He stood off from midnight until this day 10 leagues N.E. by N. to anchor; but he saith the winde was N.E. by N. The ground was more even in standing off and on then before. He stood off 4 leagues more into 87 Fathom. This evening, at the Sunnes setting, he took him with two several instruments, and found him to goe downe 33 d. to the W. of North by one, and by the other 30 d. They tooke the middest betwixt both [these observations].

26. After midnight, he stood in againe N.N.W. 5 leagues. The wind came to the S.W. The weather faire and cleare, and the best he had [had] since he came from his wintering place. About 9 this morning, he sees land; it bore on him from the N.E. by North to the N.W. by North, about 8 or 9 leagues off. At noone, hee steered to the Easterne point, to get hold thereof, and had a good observation in 62 d. 42 m., and his depth 74 fathoms.[1] (*Who will protract this Voyage, to try whether this* ut ultra *proved a Bay or no, must*

Button had named in the previous year (see p. 165), and which was also in this part of the Bay. Rundall also (*loc. cit.*) confuses with *Hopes Advanced* a certain other locality which (although not mentioned in Foxe's account of Button's voyage) is referred to as *Hubbart's Hope* in *Purchas his Pilgrimes*, part iii, lib. IIII, p. 848 (by Purchas himself: *not* by Briggs, as Rundall evidently thinks), for he says, arrived "in 60°, [they found] a strong race of a tyde running sometimes Eastwarde, sometymes Westwarde, whereupon Josias Hubbarde, in his platt [which is, unfortunately, lost], called y^t^ place Hubbart's Hope" (cf. *North-West Foxe*, original edition, p. 161). Hubart's Hope was undoubtedly Port Churchill, which is about in Lat. 60° N., whereas Hopes Advance of Button was several degrees further north.—C.

[1] Here, again, the narrative leaves some uncertainty as to the exact point reached; but it seems as if Button had now reached the eastern point of Marble Island.—C.

begin here, and proceed to Cary's Swan's Nest.[1]) From this noone untill midnight he stood N.E. by N., 18 leagues, with much winde and growne Sea. From midnight, being in 15 fathoms, the weather hazy, the winde S.W., he stood off N.E. by North 12 leagues, till noone. Seeing the weather like to be bad, and at 3 in the morning he found his depth not answerable to his expectation, for he had but 25 fathom, and at ½ past 3, 21 fathom. Hee stood S.E. till four that morning, and had 30 fathom.

27. About 6 this morning, hee gave order to take in his Skiffe; the weather clearing, he saw land N. and by E. about 5 leagues off, and [it] was as the other land[2] that hee last sawe, something higher; and, having edged in with the land from this morning 6 untill noone, the land was N.E. and by E. from him. From noone, untill 6 in the evening, he steered E. one point without the land, 4 leagues; at 6 he anchored, 32 fathom. It was then slake water[3]; for at 8 came a strong set from N. by E.; at 11 the winde [came]

---

[1] This is obviously a comment by Foxe, who evidently means that, if a north-west passage through Hudson's Bay existed at all, it must be through the great bay between Hopes Advance and Cary's Swan's Nest, which Button was now about to explore for the first time. This bay was known for some time after as *Button's Ne Ultra*. Luke Foxe, not satisfied (as we shall see hereafter) that it was really a bay, explored it more carefully in 1631, since which time it has been known as Sir Thomas Roe's Welcome, the name he gave to it. The name *Ut Ultra* (as beyond), which Foxe uses in several places, seems meaningless. Probably he meant *Ne Ultra* (no further), for he was evidently an uneducated man; or it may be a printer's error. At all events, on Briggs' map in Purchas (1625, see p. 177), and on some others, even of nearly a century later date, Button's furthest point at the north of Sir Thomas Roe's Welcome is marked as *Ne Ultra*.—C.

[2] Possibly he was now off Baker Foreland; or perhaps it was Cape Fullerton.—C.

[3] *Slack-water:* the interval between the flux and the reflux of the tide.—C.

S.S.W. and blew much; hee weighed and stood away N.E. with his Fore and Mision[1] Sayles.

28. At midnight he put out all sayles and stood N.E. away, 4 leagues, untill morning 4, having depth then 73 fathom; he edged in N. and N. by E. upon that tyde hee found the night before, and runne from this morning 4 untill noone, 6 leagues N. somewhat Easterly. This noone, with bould winde S., he loost up[2] for the shore to get an opportunity, if possibly, to send his Boat on land to have found which have beene the flood, and which the ebbe; but when he came within 2 leagues of shore, hee found the Bay all broken ground, and the Sea full of breaches a good way off the shore, and some within one mile under his Lee. The weather growing suddenly very thick, he came to an anchor at one a clock in the afternoone and sent the Boat on Land; but they had not beene from him halfe a Glasse, but it grew so thick that he caused Muskets to bee shot to call the Boat back againe. At the end of 5 Glasses they returned, saying that, having rode halfe an houre at anchor, they could discerne no tyde: no more could he in the Ship. In this place, it is a Bay, full of small Iles neere the shore under the high land; and, as he rode, [it] bore on him N.N.E. Easterly, and S.W.[3] At clock 4, the winde began to blow very much; hee wayed and stands E.N.E., with a S.S.W. winde, until 6 that night. The winde encreasing to a great storme, he takes in his maine and top sayles and stands away E.N.E. with a foresayle.[4] At 8 a clock, he had 63 fathom, and so, standing

---

[1] No doubt a misprint for mizzen.—C.

[2] This obsolete expression, which means that he *set sail* or *weighed anchor*, occurs several times in The Acts, ch. xiii, v. 13, and ch. xxvii, vv. 13 and 21.—C.

[3] The description seems to indicate one of the bays under the high land of Cape Fullerton.—C.

[4] Begin here to protract and find whether it be a Bay or no.—F.

one houre longer, it began to shoald contrary to expectation, he came into 50 fathom, for that to fore running Eastward, hee alwaies depened his water, which now shoalding, he edged off E. till midnight, [when] they had but 44 fathom; which, perceiving, he found that he was imbayed. Therefore, to prevent further danger, he strook sayle and lay to Hulle.[1]

29. From midnight, until 3 this morning, he lay on Hull, having the water to shoald; and at 3, the weather clearing a little, with the increase of the daylight, hee saw land from the E. by S. to the N.N.E. Easterly; which, perceiving, hee set his Sayles, and flats[2] about his Ship head, and stood W. off till noone, 4 Leagues.

Thus, at noone, it falling calme and hazie, he anckors with his kedge Anckor.[3] Hee feared all the coast along to bee rockie ground and, being thus at anckor at 23 Fadome, he had at 2 a Clock a little cleare and saw the Land not 2 Leagues from him, from the N. to the W.S.W., which, together with the Land which he saw in the morning, from S. to the E. by S., N.E. Easterly, gave him assurance that he was imbayed, for that he could not be off that Easterne Land above 7 Leagues.[4]

Being thus at a *Nonplus*, himselfe by observation the day before, & by his reckoning since, to be as far Northerly fully as 65 d., it troubled him something, and especially to see the Easterne Land to wend away Southerly. Well, the wind comming about to N.N.E., hee came out of this hee called a Bay, and saith hee was enforced therein by

---

[1] To lay to hull is to strike sail and drift without oar or rudder, generally during a storm. Foxe always spells it with a capital H.—C.

[2] This means that he put his ship's head round. Smythe (*Sailors' Word-Book*, p. 304) explains the meaning of the term *to flatten in* sails with the object of turning the ship's head.—C.

[3] See p. 71.—C.

[4] 64 deg. 30 m.—F.

extremity of a S.W. wind, and yet hee saith that by his not far standing from the West side, this Bay cannot be above 16 or 17 Leagues over, from Land to Land, and in his running, hee found the depth of it to be not above 4 or 5 and 30 Leagues.

But, as a wise Gentleman,[1] and one well understood in the Rules of Navigation, who having exactly surveyed these journall writs in the Margent, [says]: I cannot find that it is proved a Bay, nor is it one by any thing herein written, and, for other things knowne, is none.[2]

About 5 in the afternoone, the wind encreased to a great storme. He tooke in his Sayles and went away with a fore and sprit sayle, having 46 Fadome, and betweene 7 and 8 hee came after to 62 Fadome, and in a cleere [interval] hee saw the Land E. and by S., 3 Leagues off. From the time that he set sayle this afternoone untill 8 at night, he runne 7 Leagues S.S.W., the wind at N.N.E. At Clock 10, hee stood S.E., hoping to have found the Land to wind away, and, running but $\frac{1}{3}$ of an houre, he came to smooth water upon a suddaine, and had but 26 Fadome. And the

---

[1] Foxe gives us no clue as to this "wise gentleman's" identity, but probably it was either Mr. Henry Briggs or Sir Thomas Roe.—C.

[2] *Vt ultra* not proved a bay.—F. The reader must here take note that the foregoing remarks are comments by Foxe on Button's narrative. It is clear enough that, up to this point, Button had been sailing up the western side of the Strait which was afterwards more fully explored by Foxe in 1631, by whom it was named Sir Thomas Roe's Welcome. Here, on the 30th of July 1613, Button reached his most northerly point in about 65° N., commonly spoken of as his *Ne Ultra*, after which, fancying himself in a bay, he turned south-east and followed downwards the western shore of Southampton Island, being now fairly started on his homeward journey. Notwithstanding Foxe's observations, more than a century elapsed before the Welcome was anything like thoroughly explored. In 1741-42, Captains Christopher Middleton and William Moor attempted the delineation of its northern extremity, but their proceedings did not meet with approval, and in 1746-47 Captains Moor and Smith were sent out, but without any great result.—C.

weather, being yet hazie, hee saw the Land faire by him, bearing E.S.E., very high Land[1]; then he edg'd off untill midnight S.S.E. by Easterly, having come since 8 a Clock, 5 Leagues. From *Midnight,* [with] Land still in view, he edg'd off W.N.W., and at 2 in the morning, hee came into 65 Fadome; then hee edg'd off W. and S.W. and S.S.W. untill 4 in the morning, when the storme was so violent that he was driven to steere away before it, with his fore course [set]. At 5 this morning, by the extremity of the Sea, he sunk his long Boate.

[30.] All this morning, he steered untill 8 S.S.W., 5 Leagues along the shoare, and every flache[2] hee saw the Land not above 3 Leagues from him; from 8 untill Noone, hee steered S. 5 Leagues; then the storme began to lessen, and he put out more Sayle. And this Easterne Land hee saw to beare S.S.E. Easterly from him, and [it] lyeth, as neere as he could Iudge, S. and Westerly. From Noone, untill Midnight, hee stood away S.S.E. 13 Leagues; and from Midnight untill this noone S.S.E., 6 Leagues, and 4 Leagues S.S.W.

31. This day at Noone it was cleere, and by his observation this was in 62 d. 57 m. Latitude, the wind comming to the East S.E. and E. by S.; he stands to the N. wards, desirous to keepe that Land still in sight, or at least to get the shoalding thereof. At Clock 2 the wind encreased, to the taking in of both top Sayles; about 6 it turned about

---

[1] Probably Cape Kendall. Apparently he named this Cape Philips, for that name appears about in this position on Briggs's map in Purchas (see p. 177). As Button and Munk were the only persons who had visited this part of the Bay at the time Briggs's map was printed, and as the latter does not speak of having named it, while his voyage was apparently unknown to Purchas, it is a fair conclusion that Button gave the name, although Foxe does not mention the fact, and it is not known whom it was named after.—C.

[2] Presumably this means that, with every *flash of lightning*, he saw the land.—C.

to E.N.E. and to N.E. by E.; he tacks about to the South-wards, having run N. wards 4 Leagues, N. by E.; about 8 it blew to both Bonnets[1] off; and hee stood with two courses,[1] making way untill Midnight S. and by E. 2 Leagues, and other 2 Leagues, untill 4, S. and by E.

August 1. This morning he lay to *Hull*, the weather extreame [—][2] and hazie, and so thick that he could not see a pistal-shot from him; in this time, hee drives 2 Leagues S.S.E., and this is to be noted, that he had 3 stormes in 4 dayes, and for 17 or 18 dayes last past he had not had past 6 houres cleere at any time, nor beene 20 houres without a stiffe blowne cold and English storme. After 3 this afternoone, it fell flat calme. Hee sets both his Sayles, and stood to the N. untill Evening 6, the wind about N.E.[3] At 6, seeing the wind would not permit him to seize[4] in that N. shoare, he stood to the Southward, the weather thick and exceeding uncomfortable. His way was 1 League N.N., W.S.W. by W. 2 Leagues, and one League S.S.E.

2. From Midnight, a League S.S.E., the weather thick and foggie, the Wind Easterly. From this day noone untill Midnight, E. by S. 4 Leagues, and 2 Leagues N. and by E., the wind hanging still to the Northwards, and hazie weather, his depth betweene 70 and 75 Fadome. He coveted still to hold sight of the N. shoare, but it pleased not God it should be so. Therefore, with patience,

---

[1] See p. 55.—C.

[2] A word seems to have been omitted here: probably either *hot* or *cold*.—C.

[3] He was now probably off the entrance to Fisher Strait, between Southampton Island and Coats Island, but he does not seem to have observed it.—C.

[4] This probably means that he was unable to reach the shore, but the word is certainly now obsolete in this sense. Webster, however, gives "to fall or rush upon suddenly, to lay hold of, to gripe or grasp suddenly, to reach and grasp." He quotes Spenser—"For by no means the high bank could he seize."—C.

he stood away as neere as he could; but, by his standing Northwards, his water still shoalded, so as he guest the Land still winded to the Eastward.

3. From Midnight untill Noone, 7 Leagues N.E., from thence 4 Leagues E. by S. till Midnight.

4. From Midnight untill 5 this morning, N. and by E. 3 Leagues; from thence he stood E.S.E.; foggie weather with some cleeres.[1] From noone until 10 at night, 6 Leagues E.N.E.; his depth shoalding from 65 to 40 fathom; the weather thick and bad; hee stood 2 leagues West by North.

[5.][2] At 2 this morning, hee stands 2 leagues N.E., and untill noone 7 leagues S.S.W., and past 4 this morning he sees land[3] about 2 leagues off, bearing from E. to S. He writeth that the sight of it grieved him much, so that now he made himselfe assured of that which he did but doubt before; which was, that they joyne to the Easterne part of the Bay from whence he came: but I doe otherwise beleeve.[4]

---

1 No doubt *clear intervals* are meant.—C.

2 From the 5th of August to the 17th, all dates have been omitted, apparently through a printer's error. I have restored them as well as the information given allows, but this cannot be done with any certainty.—C.

3 And seene from the 31 [of July] untill this day.—F.

4 The reader must again note that the opinion here expressed is Foxe's. The land which Button had now encountered was, with very little doubt, the south-eastern part of Coats Island, though it was not then known by that name, as Button believed it to be a part of Southampton Island. He evidently believed he had (as Foxe says in his side-note) been skirting an unbroken coast since he first turned southward, about a week before. Button had now explored from his wintering-place at Port Nelson up the entire western coast of Hudson's Bay; he had ascended what is now known as Sir Thomas Roe's Welcome to about 65° N.; he had turned back south-east, and had explored down the western side of Southampton Island almost to Cary's Swan's Nest, a locality he had named when sailing into the bay the year before; yet he had seen no signs of a navigable passage leading in the desired direction. As a matter of fact, he had

[This 5th], all the afternoone, he stood along the shore, edging into 7 fathom, and crossed a Race,[1] which set N.E. and S.W. and continued about halfe a Glasse. At 4 a clock, the N.W. point of the land did beare from him N.W. by N., about a mile off; then, steering within lesse then one mile of this Cape-land, for so it was, and a faire one of a low one as ever hee saw, you shall have 9 and 10 fathom, and shall open a very safe Bay, the Easterne land whereof will beare from you E. by N., 4 leagues off. In this Bay, he handed all his sayles,[2] thinking to have ridde and watered; but, his anchor being downe and his Boat almost out, one of his Master's Mates said he descried land from the Top-Mast-head, S. and by E., the newes whereof made him wonder, for that they all knew this was the land they had seene the last yeere,[3] which they took

missed Frozen Strait, which leads from Sir Thomas Roe's Welcome, to the north of Southampton Island, into Fox's Channel (then undiscovered), and also Fisher Strait, which also leads, on the south of Southampton Island, into Fox's Channel; but, had he detected these, he would not have been very much nearer discovering what he sought. Thus it was that the sight of this land "grieved him much". It showed him that the bright hopes offered by Hudson's discovery were groundless, and that it was useless to search further for the passage—at least, through Hudson's Bay; for that Bay he had now for the first time proved to be a bay, or, rather, a vast land-locked sea. It is true that there still remained to be explored the southern shore of the Bay, between his wintering-place at Port Nelson, on the west, and Cape Henrietta Maria on the east (which we may regard as the limit of Hudson's discoveries westward; see p. 131); but there was little or no prospect of a navigable passage leading from that part of the Bay westward through the mainland of the American Continent. This undiscovered portion of the southern shore of the Bay is left blank on Briggs's map in *Purchas his Pilgrimes* (see p. 177), published in 1625, and was only explored in the year 1631 by Captains Foxe and James.—C.

[1] This, doubtless, was off Cape Southampton, where the Admiralty Chart says the "Tides [are] supposed to meet".—C.

[2] Apparently this means that he took in all his sails.—C.

[3] Undoubtedly Cary's Swan's Nest (see p. 165). He was now in the small Bay between it and Cape Southampton.—C.

to be an Iland. He sent up Captaine *Ingram*,[1] who assured him it was land; whereupon he hoysed up his anchor and set sayle, and stood S. and by E. and S. and came into 30 fathom: so, missing this land,[2] hee anchored (night at hand) in 30 fathom.

[6.] At 2 this morning, he weighed to better his deepe and to goe neere the land he thought he had seen; in 2 Glasses, hee lessed his deepe 3 fathom; so, steering S.S.E. he came to 46 and 56 fathom in two Glasses. About 5 in the morning, the wind came about to the N. by E.; he stands E. close upon a wind, and came to 125 at noone, without sight of land; whereby hee was assured that it was the imagined last yeeres Iland.[3]

[7?] From yesterday noone untill this noone, 14 leagues S.E. by E. and 6 leagues E.N.E.; the weather somewhat cleare, but not fit to make observation; after noone the Gale hards on[4]; he strooke his lofty sayles; the Sea somewhat growne; the day cleerest and fairest of 16 or 17 dayes before. From noone this day until midnight, E.N.E. 8 leag.; this night, 11, the weather grew bad, and hee saith the Sea beat off his Beaks[5] head, at which time he took in sayles and came to *Hull*.

[8?] From midnight before untill this day noone, hee drive on *Hull* 3 leagues S.S.E., the weather foggie, raine and wind, but somewhat lesse.

---

[1] Ingram was master of the *Discovery* (see p. 174).—C.

[2] A land of Fogge.—F. Foxe evidently means that a fog-bank deceived them, as there is no land in the direction indicated.—C.

[3] This passage is unintelligible. Probably it means that Button was assured the land he had just left was Cary's Swan's Nest, which he had named the year before. The narrative contains no mention of any other island sighted in the previous year.—C.

[4] Apparently he means that the gale increased and it began to blow harder.—C.

[5] A small platform at the foremost part of the upper deck, over the figurehead.—C.

[9?] At 2 in the morning, the storme being broke up and Sea downe, he set sayles, and had driven since noone before S.S.E. 4 leagues, the winde betwixt N.N.E. and N.E. At 8 this morning, he had 50 fathom; the thick was cleered, and he see land to beare on him East.[1] From 2 this morning, the time of his setting of sayle, hee made N.E. by N. Northerly, 7 leagues; the morning was foggie, but the day proved faire. At noone, they had a good observation, and all that did observe agreed in one, and to be in 61 d. 38 m. Latitude.[2] Hee made way to the shore, E. by N. Northerly, 2 leagues, and came to an anchor at clock 2 in 17 fathom, the land bearing from N.E. by N. to E.S.E. After 3 this afternoone, the winde comming to the E.N.E., he set sayle and stood to the Northward; and, about 5 this evening, coasting along the shore, the North-

---

[1] He named this Mancel Island.—F.

[2] This clearly proves that the land he now encountered was the island which (as Foxe says in a side-note) he named "Mancel's Island", the southernmost point of which is almost exactly in the position indicated. This island Button undoubtedly named after his near kinsman, neighbour, and contemporary, Sir Robert Mansel, Knight, Vice-Admiral of England, Treasurer of the Navy, and Member of Parliament for Glamorganshire, who was born at Margam about 1573 and died in 1653. Button had married a daughter of Sir Walter Rice, of Dynevor, whose wife was sister to Sir Robert Mansel, and this had naturally brought them into close connection, especially as both were members of the naval service. The name of Mansel Island has appeared as *Mansfield* on almost all later maps and charts. It even appeared in its corrupted form only twelve years after Button's voyage, namely, on Briggs's map published in *Purchas his Pilgrimes* (see p. 177), published in 1625, and the island is similarly named on Foxe's map in 1635. It also appears as Mansfield Island on the latest Admiralty Charts; and thus a public servant, who rendered very distinguished services to his country in his day, is deprived of an honour that was intended for him. The Admiral's name, however, seems to have been sometimes spelled *Mansfield*, even in his own day. The best account of him is to be found in Mr. G. T. Clark's privately-printed pamphlet, entitled *Some Account of Sir R. Mansel, Kt., . . . and of Sir T. Button, Kt.* (Dowlais, 1883).—C.

land seemed like small Ilands and broken lands; the sight whereof made him desirous to put in amongst them, but the winde would not permit. At clock 5, the weather thickned, and the land bore from N.E. Easterly to S.E., from him 4 leagues. This land was a very low and a smooth land; from hence he stood off to the Westward till midnight; his depth from 14 to 95 fathom, his course W.N.W Northerly 7 leagues. At Midnight, having the same depth, the wind was at noone N. northerly; he stood about to eastward 10 Leagues E.N.E.

[10?] Somewhat before noone this day, it cleered and in 12 Fadome hee saw the Land 2 Miles off, and upon the same hee saw 5 white Beares; so, standing off, the coller of his main-stay broke, which caused him to beare up with the S.-most part of the Land he saw, to the intent to send his Boate on Land for water, and to try from whence the flood came. At 1 in the afternoone, he anchored in a Bay in 4 Fadome, smooth ground but rockie, for he could see it under him. He sent his boates well man[ne]d and armed, who found water instantly. They found, not farre from them, 2 old houses,[1] broken and fallen downe to the ground, wherein were the skulls and bones of dead men, Images, and toyes, which as[2] they found by digging with their hands under the ruines of the decayed houses, with some dozen of small Mors teeth.[3] His opinion is that this

---

[1] From what is said, it is clear that these were the remains of houses of some kind built by the *natives*, not by white men; indeed, no Europeans are known to have landed on Mansel Island before this date, though the island is clearly marked on Hudson's chart. There is not the slightest reason to suppose that the ruined houses had been built by those who had been abandoned with Hudson, two years before.—C.

[2] The word *as* seems to have got in by accident, and to be injurious to the sense.

[3] The "tusks", or canine teeth, of the Morse or Walrus (*Trichecus rosmarus*), which are usually from eight to fourteen inches in length,

were the ruines of some who, by mischance, had miscarried there by wrack of their Boate, or being inforced to winter there, made the best provision they could to endure it; but, the extremity being so strong[1] for them, and the place neither affording meanes for them to repaire their Canooes (the ruines whereof he found some) nor fuell for fire to comfort them in Winter, hee guesses this killed them; for, had they beene any other then such as were thus enforced by such extremity, they would not have left such things behind them as they found; for there nature is, wheresoever they come, to leave few things of worth or value behind them; much lesse doth hee thinke they would have left their Images, which hee perceiveth they account their gods. And there was Mors' teeth, for which and the treyne[2] thereof they make all their Sommers travells and labours. Thus, his Boate having brought him water, he rides all that day with wind off Land, and observed that the flood came from N. and the ebbe from S., making a full sea that day betweene 3 and 4 a Clocke.

[11?] Having watered, he weighed and steered untill 4 this morning N.W. by W. 7 Leagues, the wind N.N.E. It grew thick with fogg; at 4 in the Evening, hee tackt about to the Eastward, till [Clocke] 8, 3 Leagues N.E. by E. The thick caused him to take to the W. ward, and till Mid-night he made way N.W. and by W. 3 Leagues.

[12?] From midnight untill morning, 6, he stood in E.N. ½ northerly 4 Leagues. He tackt about untill 10 a Clock 3 Leagues N.W. and by N. At which time, hee came to an anchor in 85 Fadome, and found the tyde of

---

and are valuable because they yield a very dense white kind of ivory.—C.

[1] Probably *too strong* is intended.—C.

[2] Train-oil (which has no necessary connection with railway-trains) is oil obtained from the fat of animals, usually from the blubber of whales. Hakluyt (1589) speaks of "trane-oyle".—C.

Flood to come from the N. and ebbe from the S. and ride within 6 or 7 Leagues of shore, it bearing N.E. and by E. This Evening, 8, he stood N. and by E., winde N.W. and by W., 4 Leagues untill midnight. At noone before, hee had an observation 62 d. 19 m.[1] This day hee saw great store of geese fly to the Southwards, which he tooke to be a token that the winter did approach; the weather was very faire and cleere.

[13?] From midnight until 7 this morning, his way W. 5 Leagues N.W. and by W.; the wind as before; the weather faire and cleere, and most likely to prove so to continue of any he had so seene since he came from his wintering. From 7 this morning untill 2 in the afternoone, hee stood to the E.-ward, and made way 4 Leagues N.E. and by E. From 2, the wind wearing, he stood to the E.-ward, and made way untill midnight 7 Leagues N.W.

[14.] From midnight until morning, hee held the same course, 4 Leagues N.W. northerly; it grew calme, and he was within 5 Leagues of a faire Head-land; hee came to Anchor in 65 Fadome; the Land bare from the N.N.W. to the W.S.W. Both lands[2] he deemed to bee distant 10 Leagues. It being calme, with windward Tyde, hee wayed anchor, hoping to get to the Northward of this faire Headland, and that the height of that Land would bring him into deeper water. After he was loos'd, he was sagged[3]

---

[1] He appears to have been on the west side of Mansel Island in this latitude.—C.

[2] The "faire headland", of the height of which he speaks, was undoubtedly Cape Pembroke, at the north end of Coats Island. The other land of which he speaks was probably Leyson Point, on the south of Bell Island; or it may have been the north end of Mansell Island, which he had just left. The course pursued since leaving Mansel Island seems to have been very irregular, and cannot readily be followed.— C.

[3] This means that he was drawn or that he drifted in to leeward. On shore, the verb *to sag* means to bend, sink, be drawn or pressed downwards in the middle.—C.

into the Bay from 62 to 50 fathome; a small gale comming on, hee stood off into 60 fathome; when it fell calme, he anchored againe. This day was faire and cleere; he obserued in 62 d. 38 m.[1] This Cape was a very faire Headland, and the northerne part is much higher then the westerne, but it is as all the other Land is of this straight, except 15 leagues on this side his wintering place, which was woody, else on this side it is all barred and rockie, but a bold shoare to saile along.[2] This forenoone calme, but faire and cleere weather, and the onely Somers day hee had since his comming from his wintering. Being at anchor, betwixt the Cape and the W. point, he saw the Land to make with two Bayes,[3] the further point of the southerne Bay being from the northerne Land of that Bay W. and by S. southerly, and the North point E. and by N. northerly: And the other Bay from this point to the Cape it selfe lieth E. northerly, and S. westerly. At noone he observed in 62 d. 42 m. At 4 [of] the afternoone he set sayle, with small winde. At 5 he stood with the Cape. The weather being cleere and faire, hee sent to the Cape to try the Tyde, being thwart of the Cape; the point is low with an underland, and some 2 Cables length from shore

---

[1] He appears now to have been between Mansel Island and Coats Island.—C.

[2] Some error, probably of the printer, here obscures the sense, although I have made one correction, and the word *barred* should probably be *barren*. Foxe appears to mean that the shore had been wooded for fifteen leagues to the north of Button's wintering-place in Port Nelson, but that everywhere else he had been it had been bare of trees. This, at any rate, agrees with known facts. Dr. Robert Bell's *Map showing the General Northern Limits of the Principal Forest Trees of Canada* (Report of the Geological Survey of Canada, 1879-80) shows that several trees (namely, the Balsam Poplar, the Larch, and the Spruces) have their northern limit on the western shore of Hudson's Bay at Port Churchill, which is not far from the spot above indicated.—C.

[3] It is by no means clear what two bays are meant.—C.

is a low flat little Iland; the land you shall open to the Northward, lieth N.W. Westerly; the boat coming on board, he edged off and anchored within one mile of the land, having opened the land to the westward of the Cape; he anchored [in] 15 fathom, and rid there all night to try the tyde, for that his Boat had brought word that it was ebbe tyde and that it set to the Northward, which did agree with the tyde he found on the Eastland where hee watered[1]; but, from the time he now anchored, being betweene 10 and 11 at night and 3 the next morning [when] he set sayle, hee could not finde any certainty but what followed.[2]

[15.] This morning was calme, but the night before was full of strange Harbours,[3] as they call them, which is a streame in the Element, like the flame that commeth forth [out of] the mouth of a hot oven; which, upon this Coast, how faire soever the weather bee, when you see them, yet it is an infallible signe of a storme to follow within 24 houres after, as it proved by this and divers times before. At 3 this morning, without certenty of the Tyde, a small gale S., he wayed and advised with Captaine *Gibbins*, and Captaine *Ingram*, and with the rest, what course was best to bee taken. They resolved (this Land falling away N.W. and by W. westerly, and having 113 Fadome within a Mile of the shore) to stand away N.N.W. alongst the shore, resolving not to leave this Land untill he were fully satisfied. Standing thus untill 8 at night, being some 7 Leagues off the Cape, he saw an Iland of the westermost Land,[4] that bore from him W.N.W., 7 or 8 Leagues off; hee had then 100 Fathome, and stood N.N.W. as before.

---

[1] On Mansel Island (see p. 189).—C.

[2] The last few words seem meaningless.—C.

[3] Undoubtedly Auroras are alluded to, but I cannot learn the derivation of the word. Perhaps it is a misprint for *Harbingers*.—C.

[4] It is by no means clear what island this was.—C.

At noone the weather was close and began to blow. Hee was then 15 Leagues N.N.W.[1] from the Cape (he doth not write whether true course or no, but in all the former from *Hopes Advance* unto this Cape, I have writ by him according to the true course) and had 95 Fadome, which made him assure himselfe that the land winded away more northerly, and thereupon at noone hee stood away N.E. and by N., hoping that course would have bettered his depth, but on the contrary, for in 2 glasses he shoalded to 60 Fathoms, then he hayled away E.S.E., assuring what experience had often shown him that, as the water shoalded, so the Land winded. At 3 in the afternoone, the weather thick, the wind increasing, and hee in 60 Fathome, knowing there was no better depth to the N.-ward, hee stood E.S.E. till 8 at night, having 50 Fathome. The weather bad, and night at hand, hee stood about and kept it up with short sayle all that night betweene S.W. and by W. and W.S.W.

[16.] From midnight till 7 this morning, as from 8 last night till midnight, thus standing brought him into 80 fathom. He cast about to the Eastward, with much winde at S.S.W.; hazy and thick weather; he heeled it up in courses and Bonnets till clock 2, they being [then] starke calmed, as it is a special note[2] every Blower ends with a starke[3] calme in those parts. Being then in 65 fathom, he anchored and rid untill 4 in the evening, when a small gale rising at S.S.W., he waighed and stood S.E., guiding himselfe by his depth, for the winde would not give him leave to better his hopes.

17. From midnight to 8 this morning, as the night

---

[1] Sir *Thomas Button* doth allow 2 points variation.—F.

[2] Presumably this means "as it is especially to be noted".—C.

[3] This word is still current in some connections, meaning completely or absolutely, as "stark mad" or "stark naked", but few modern writers would use it as above, though Smythe mentions it.—C.

before, & after that time, there is small worth the note to write upon traverse; and his greatest depth 140 fathom, untill the next day in the afternoone, and then he had sight of the same Cape he sent his Boat unto the 14 day before.[1] It bore from him N.N.W. by compasse about 7 leagues, so he steered S.S.W. untill midnight, having runne, since he set sayle at 2 in the afternoone, 10 leagues S.E. and by E.

Concluding, he writeth that he came to 43 fathom, which shoalding was upon the North part of the Iland he watered upon[2]; and that this Iland, and the said Cape where his Boat was at the 14 day, lyeth S.S.E ½ Easterly, and N.N.W. ½ Northerly, about 10 leagues betweene both.[3]

[18.] This morning [at] day light, he see the land bearing from S.W. by S., at which time he had 65 fathom.

19. From 2 this morning untill noone, he made away E. & by N. 10 leagues. At 7, he saw land on head, at least 12 leagues off him; he judged it to bee the high land of the Maine within Sir *Dudley Diggs* his Iland, and yet [for it] to beare off him as it did, it was strange unto him; but he saith that Captaine *Gibbons*' confidence was such as that hee must not let passe unspoken of; for, albeit that hee is so neere in blood, as that modestie will not allow of his speaking too much of his merit,[4] yet hee will boldly say

[1] This evidently means the 14th day of the month (August), and thus helps in supplying the missing dates (see p. 185). The cape was apparently Cape Pembroke.—C.

[2] Mansel Island (see p. 189).—C.

[3] *Mansells* Ile & *Ca. Pembrok* 10 Leagues distant.—F. See p. 191.—C.

[4] He was Button's cousin. According to Barrow and Rundall (see *ante*, p. 167), he was, like Captain Hawkridge, a volunteer on the expedition, but the source of this piece of information is not obvious. In the year following Button's return, he made an altogether fruitless voyage to the north-west, particulars of which follow hereafter. The conclusion of the paragraph is evidently Foxe's wording, and is, as usual, very badly expressed.—C.

thus much of his sufficiency, as hee assures himselfe he shall make it appeare at his returne, if God please to let him live so long, as that [in ability] he is not short of any man that ever yet he carried to Sea. All that he can say of him further is that, for his Countries good, and for the advancement of this businesse we have in hand, he could wish his body were answerable to his other abilities; which, if it were, not himselfe [only], but many, and his Country most, would bee the better for it.[1] But that God that made us all of dust will not faile to raise up some good spirits, he hopes, for the further prosecution of this businesse; as that, by their honest endeavours and religious Resolutions, they will effect that which as yet is not ripe for his Sickle; but that God, which best knowes what the truth of his endeavours have beene in this businesse, he hopes will not faile to give a blessing to some that shall follow; and, for his part, he desires to be blest no otherwise than as he hath sincerely laboured in it; and, therefore, he must conclude and ever beleeve, according to the word, that *Paul* plants, *Apollo* waters, and God gives the increase. So that, untill his good will and pleasure is, all that we doe cannot, in this [or in] ought else, prevaile.

At noone, the weather faire and cleare, he had a good observation, by which Captaine *Gibbons* was well assured; but himselfe and some others discenting from him in the bearing of the land from them, were deceived in the set of the tyde, which, in his going out last yeere, mightily carried him to the Northward, as now as much to the Southward; which then, nor till now, was not found by any of them all.[2] He steered open of the Land N.N.E., with tyde 5 leagues,

---

[1] This very confused sentence appears to mean that Captain Gibbons had exceptionally good ability, but poor bodily health.—C.

[2] Apparently this means that this tide had not been observed by any of those who had previously sailed over the spot.—C.

untill evening 4, and at 2 he had 307 fathom. At 6 in the evening, small winde, *C. Wolstenholme* bore S.E., 7 leagues off; the Westerne point of the westerne Iland, S.E. by S. ½ S., 5 leagues off. He kept it up that he might get about the Westermost point of *Nottingham's* Iland (it lying 8 leagues from *Diggs* his Iland), there thinking to trie the tyde; and from thence he purposed to goe to the N. maine, betweene *Salisbury* and *Nottingham's* Ile, to try [the tide in] the distance betweene them, it being all he could doe for this yeere.[1]

[20.][2] About 8 this morning, the weather thick, and he not above 4 leagues from Ile *Nottingham* in 64 fathom, he anchored, and, as he remembreth to his comfort, being ebbe, hee found it a very strong one from S.S.E. to S.E. and by S.[3] At noone slack, it was a fresh gale at S.W.; hee waighed to get about to the Westward of the Iland; about 2, it blew hard and was thick weather, with raine and thunder, and [he was] within a mile of the shore. It fell starke calme upon a sudden; then he was in a great ripling in 20

---

[1] Button's instructions informed him that, between Salisbury Island and the North Main he was "like to meet a great and hollow billow", by which, of course, was meant that reported by Hudson's survivors (see p. 150). Button, no doubt, now thought that the only thing left for him to do that year was to further observe this billow.—C.

[2] From this point, to the end of the narrative, the dates are again missing. I have supplied them as well as I have been able to do from the narrative.—C.

[3] But was deceived.—F. Button was fully persuaded that this was (as he afterwards says) "the true channel tyde"—that is to say, a tide or current flowing through an opening into the main western ocean: hence the comforting influence it had upon his feelings. Button himself, after his return home, and others, made much of the discovery of this reputed tide, and there can be no doubt that it was the main incentive for the dispatch, in the following year (1614), of the fruitless expedition under Captain Gibbons, as well as of that of Bylot and Baffin in 1615, to be noticed hereafter. It was, of course, only the ordinary tide flowing into and out of the then-undiscovered strait now known as Foxe's Channel.—C.

fathom water, the Sea all breaches round about him like a Race, which, descrying, he stood off with little winde along the Iland, and found his drift more in lesse then in one Glasse.[1] Then hee could run ahead before [it] with both top sayles on trip, and a stiffe gale; in 2 houres, hee came to anchor to be fully satisfied of the tyde. But, when the tyde came, it came with such force & strength out of the N.W. and by N. as hee had much adoe to ride at [anchor], and could not have ridde it had not he steered the ship all the tyde time.[2] The sight, hee writes, did comfort him for what was to be done hereafter; for by the course of this tyde, and his owne knowledge of the land, being to the Westward of that place 200 leagues & 66 on the Sea, and might very well see one degree more on land to the Northward; hee then saw good reason for it. At 8 this night, the weather being a little cleere with ebbe, he wayed and plied to windward to get about the N.W. end of the Iland; and, being about the West point, the ebbe being done, he saw another point open upon him that bore N.; the winde at N.W., in 33 fathoms, the weather thick and bad, he anchored, where, in lesse than one houre, the tyde of flood came most strong as before, from N.W. and by N., whereby he concluded, having brought the Northerne point N. from him, that it was the true Channell tyde. For, had it beene otherwise, it would have come as the land lay, which was N., but now, being open of the land, and finding it to come from the N.W. and by N., he saith in his judgment that course and N.N.W. must direct whomsoever shall seeke this passage hereafter. And, the rather to continue himselfe in this opinion, he now, too late, found that those that were this way first, & himselfe the last yeere, were all of them

---

[1] Half an hower.—F. The wording hereabouts, which is evidently Foxe's, is almost incomprehensible.—C.

[2] A strong tide. Not so.—F.

deceived of the set of the tyde within Sir *Dudley Diggs* his Iland, for there they found it come more Westerly, which was caused by many broken Ilands that lye to the Westward of it, which he never sawe untill his returne homewards. And, upon this tyde (if I can judge, saith he), we cannot be deceived, for this caveat he doth give to whomsoever shall succeed him in this discovery:—That, whensoever he loseth his strong tyde, or finds ground in 200 fathoms, let himselfe know he is out of his direct course for finding of this Voyage.[1] So this his experience upon his unhappy counter-course taught him that, whensoever it is to be found, it must bee in deepe water and in a strong tyde; and, in this course that he took, he hopes it will not be imputed an errour of his for what he did in the directing of it; for it was to follow the letter of his instructions.[2] For, albeit he was precisely tyed to stand with *Hudson's* Westerland in 58 d.,[3] yet he never came much to Leeward of 61 d. till he was encountered with land 200 leagues Westward from Sir *Dudley Diggs* his Iland.

Thus much in effect I received in a Manuscript *from Sir* Thomas Roe, *besides divers others, towards the furtherance of my Voyage.*[4]

But further from *Abacuk Pricket*, who saith they came

---

[1] This is clearly a misprint for *Passage*.—C.

[2] Button meant that he hoped he should not be blamed for not having first sought for a passage in this direction; for, in searching elsewhere, he was only carrying out his very precise written instructions, which are printed herein.—C.

[3] The question whether Hudson did, or did not, explore any part of the western shore of Hudson's Bay proper has been already discussed (p. 131, *note*), and the conclusion was that he went no further north in this direction than Cape Henrietta Maria, in lat. 55°, although his map would lead us to believe that he had reached lat. 60°.—C.

[4] See note, p. 163.—C.

not through the maine Channell of *Fretum Hudson*, nor thorow *Lumley's* Inlet, but that he came through into the *Mare Hyperborum* betwixt those Ilands first discovered and named *Chidley's* Cape by Captaine *Davis*, and the North part of *America*, called by the Spaniards, who never saw the same, Cape *Labradore;* but it is meet by the N.E. point of *America*, where there was contention amongst them, some maintaining (against others) that them Ilands were the [Island of] Resolution, which *Josias Hubbart* withstood untill he stood himselfe into the danger of displeasure; but at length it proved a new streight, and very straight indeed to come through, which resolved all doubts[1]; but hereupon all their plots and Iournalls [are silent[2]]. This part which came into my hands I have writ, thinking there may be some that will protract the same. He met no Ice in his home comming untill he came into *Fretum Hudson*, and but little there.

[Their] iournals were taken from them, and therefore who doth desire any further satisfaction from this Voyage must seeke it from Sir *Thomas Button*, onely *Pricket* saith that they were at home in 16 dayes.[3]

---

[1] Apparently this means that he left Hudson's Strait and emerged on to the open sea, not by the main channel to the south of Resolution Island, but by the channel between Cape Chidley and the islands which have since been known as Button's Isles.—C.

[2] These words seem to be required here in order to make sense.—C.

[3] The exact date of Button's return to England has not been recorded; but the following passage, which occurs (folio 76) in the MS. *Autobiography of Phineas Pett* (see Introduction), places it about the third week in September:—"The 27 day of this month [September 1613], my second son Henry departed this life at Chatham, and at ye very instant my noble worthy friend Sir Thomas Button (then Capt. Button) alighted at my house, newly being returned from the dangerous Voyage of the North-west Passage, where he had winte ed."—C.

Concerning the Voyage of Captaine Gibbons, *with a Ship called the* Discovery, *vitled for* 12 *Monethes, in the yeare* 1614.[1]

LIttle is to be writ to any purpose, for that hee was put by the mouth of *Fretum Hudson* &, with the Ice, driven into a Bay, called by his Company *Gibbons* his hole,

[1] This is another voyage of which we have no other record than that given here by Foxe (meagre as is the information he supplies), except a passing mention of it in *Purchas his Pilgrimage* (2nd ed., 1614, p. 746, and 3rd ed., 1617, p. 926). The voyage, however, failed as ignominiously as any voyage could, and added absolutely nothing to the geographical knowledge of the northern seas and coasts. This was the more disappointing, bearing in mind the very high opinion expressed about its leader by his cousin Sir Thomas Button (see *ante*, p. 196), whom he had accompanied on his voyage in the previous year. Well may Foxe speak of Gibbons' voyage (in a side-note which has not been reproduced) as "a caveat for others". We have no information as to the course Gibbons intended to follow in his exploration ; but there can be little doubt that, had he not been unfortunate at the outset, he would have searched to the north of Nottingham and Salisbury Islands, up what is now known as Fox's Channel, in which direction Sir Thomas Button had just before strongly stated that, in his opinion, anyone should search who, after him, sought for a North-West Passage (see *ante*, p. 198). We are able to gather from outside evidence (see Markham's *Voyages of Baffin*, p. 111) that Gibbons had with him, in some capacity, on this voyage, Robert Bylot, who had brought home Hudson's ship after the mutiny in 1611, and who had also sailed under Button in 1612-13. Although Foxe gives us no information as to the individuals at whose cost this ill-fated voyage was undertaken, we know, from Purchas (*loc. cit.*), that it was promoted by "The Company of Merchants Discoverers of the North-West Passage". Moreover, it was made in the *Discovery*, which belonged to that Company, and which was now making the fourth of her six recorded trips to the Arctic Seas, Weymouth having previously had her in 1602, Hudson in 1610-11, Button in 1612-13, and now Gibbons in 1614. Foxe does not record the dates of her departure or return, but Purchas (*loc. cit.*) says she sailed in the month of March. Forster (*Voyages and Discoveries in the North*, p. 348) and Barrow (*Arctic Voyages before 1818*, p. 205), both identify "Gibbons Hole" with the Bay of Nain, on the coast of Labrador ; but it seems

in Latitude 58 & ½, upon the N.E. part of *America*, where hee laid 10 weekes fast amongst the Ice, in danger to have beene spoyled, or never to have got away, so as the time being lost, he was inforced to returne.

The Voyage of Robert Bilot, *set forth by Sir* Dudley Diggs, *Mr.* Iohn Wolstenholme, *Alderman* Iones, 1615, *in the* Discovery, *of* 55 *tunnes burthen. Written by* William Baffine.[1]

THis *Robert Bylot* had beene in this ship all the 3 voyages before, viz. [those of] *Hudson* (as you finde

---

more likely to have been one of the many inlets on the coast lying further to the north, nearer to the entrance to Hudson's Strait, for which the navigator was undoubtedly making. The error is, doubtless, due to a misprint in Foxe's work, wherein the latitude of Gibbons's Hole is given as "about 77°". This being obviously incorrect, most writers have taken it to mean 57° N., which is about the latitude of Nain. They have, however, overlooked the fact that, in his *Errata*, Foxe says it should be read as lat. 58½°, which makes it probable that Gibbons's Hole is identical with Saglek Bay. Nor is this the only misprint in the few lines Foxe gives about Gibbons' voyage; for, in this errata, he also pointed out that the word he prints as *Stinenia* should be read as *America*, and that *20 weeks* should be *10 weeks*.—C.

[1] This William Baffin occupies (as Mr. C. R. Markham says) "a deservedly high place in the list of our early navigators. Although he is only known to us during the last twelve years of his life, and his previous history is an absolute blank, yet the record of those later achievements secures for him an honourable niche in England's Temple of Fame. He was a daring seaman, a scientific observer, and a great discoverer." The only collected account we have of his achievements is contained in Mr. C. R. Markham's *Voyages of William Baffin, 1612-1622* (London, Hakluyt Society, 1881). Therein reasons are given for surmising that Baffin was a native of London or Westminster; that he had relations living in the parish of St. Margaret; and that he had a home and family in the city—probably in a street near Queenhithe, in the parish of St. Thomas Apostle, for a certain "Susan Baffen, daughter of William Baffen", was baptised in the church of that parish on October 15th, 1609. We first meet with Baffin as pilot of the *Patience*, one of the three ships which sailed

by *Pricket*), Sir *Thomas Button*, and *Gibbons*, and therefore was a man well experienced that way.[1] His company

---

under Captain James Hall on his fourth and last voyage for the discovery of Greenland in 1612. The partial account of this voyage, as printed by Purchas (see Foxe's abstract, p. 96), was written by Baffin. On his return, Baffin entered the service of the Muscovy Company, and in 1613 was pilot of the *Tiger*, one of seven ships sent out by that Company to prosecute the Spitzbergen whale-fishery. In 1614 he was again engaged in the same service ; but, in the following year (1615), we find him (as already stated) sailing with Bylot in the employ of the Company of Merchant Adventurers for the Discovery of the North-West Passage. There can be no doubt that the object of the voyage was to test the correctness of Button's opinion that the tide coming down Foxe's Channel showed that a passage existed in that direction, upon which Gibbons's futile voyage in the previous year had thrown no light whatever. The whole history of this voyage was written by Baffin in considerable detail, and his account of it came into the hands of Purchas, who printed considerable extracts (*his Pilgrimes*, 1625, part iii, pp. 836-842), but omitted the accompanying map and certain portions, which he says "were somewhat troublesome and too costly to insert". The portions omitted, and a copy of the map, were first printed by Rundall (*Voyages towards the North-West*, pp. 97-135), and still more fully by Mr. C. R. Markham, in his work above quoted, from the originals which are still preserved in the British Museum (Additional MSS. No. 12,206). The omission by Purchas of these particulars was exceedingly regretable, and afterwards led to some most unfortunate geographical misconceptions ; but the fault cannot be said to lie wholly with Purchas, as is implied by the remarks of Messrs. Thos. Rundall and C. R. Markham. Knowing, as we do, that Purchas lived and died in very straitened circumstances, on account of his devotion to geographical research, we should rather praise him for what he did, than blame him for not doing what he was clearly unable to do. Those desiring the fullest information about the two joint voyages of Bylot and Baffin, must consult the works of Rundall and Markham. The following narrative has been a good deal condensed by Foxe from the original, and it appears to have been rather carelessly abstracted, the misprints being very numerous. I have corrected, without comment, a good many of those which were obvious. By some oversight, too, many of the dates have been omitted. Foxe has made a few trifling additions of his own to Baffin's narrative as it appears in Purchas, but nothing of any consequence.—C.

[1] This is the same man of whom we have already heard as having

consisted of 16 men and 2 boyes; he anchored in Lee road[1] the 18 Aprill.

May 6. Vpon this day he had sight of *Groenland*, on the East side of Cape *Farewell*. That night he had a great storme, but hee kept southerly, to get cleare of the Ice that lay on shore. Hee kept his course untill the 17 day, seeing many great Ilands of Ice. Some doth affirme that there is not above one-7 part of the Ice above water; saith *Baffine*, hee observed one peece to be 140 Fathome above water. This day he came to the firme Land of Ice as hee supposed, being in 61$^{d.}$ 16$^{m.}$, the Latitude of the S. part [of] the Ile *Resolution;* then hee asked opinion concerning putting in amongst the Ice, saying the Sea was on the N. side of the South channell and much Ice hee must passe, and if he could get but 2 or 3 Leagues within the Ice, it would open every Tyde, and so hee should get

---

sailed under Hudson, Button, and Gibbons. He was now sailing as Master, and was commanding the *Discovery*, in which he had sailed with all three of the above-named explorers, and which was now making her fifth recorded voyage in search of a North-West Passage. As to his personal antecedents, nothing seems to be known; but his experience must have well fitted him to take charge of the ship on this occasion. The expedition was set forth at the charge of "the worthy Aduancers and Aduenturers for the finding of a Passage by the North-West", of which Sir Thomas Smith, Sir Dudley Digges, and Mr. (afterwards the elder Sir) John Wolstenholme were the most prominent members. These gentlemen had previously organised the expeditions commanded by the three navigators named above, and this was their fourth attempt to discover a passage. The voyage was distinctly Bylot's, and Foxe ascribes it to him; but it is often spoken of as Bylot and Baffin's (as by Rundall, *Voyages to the North-West*, pp. 97-135), and sometimes even as Baffin's alone. Mr. Clements R. Markham includes it as one of Baffin's voyages in his admirable work on those voyages (Hakluyt Society, 1881). Baffin, however, was only the pilot, or (as he himself says) the captain's "mate and assotiate".—C.

[1] See p. 114.—C.

something on his way, having all the channell to the S. of him ; and, with this resolution, he put in with E.N.E. wind. This first entrance *Baffine* liked not well, finding scarce a place to put the Ships-head into, being 30 Leagues from any Land. Towards evening, they were fast amongst the Ice. Sometimes ere day[1] the Ice would something open, and so made what way hee could to the N.W. in for the shore untill this day, the wind all South ; yet hee could see plainely so that he set to the Southward, doe what he could.

22. This day the wind came up at N.N.W., and hee determined to stand forth againe; for, if the wind had come'd at N.E., it had beene impossible for him to have fecht any part of the channell againe,[2] for he thought he drave fast to the southward with South wind, yet he had not seene the Land.

23. Hee was also determined to spend 20 or 24 dayes in *Fretum Davis*, to see what hopes would bee that wayes, supposing there would be little good done in *Hudson's* straights.[3] For the time limited, hee plyed to get to Seaward, and at Clock 8 in the night hee was cleared from the Ice. Hee then changed his opinion and stood to the N. all hee could, as the Ice would give him leave, comming, 13 Leagues to N.E. by N. in Latitude 61 d. 50 m. At Clocke 6, the wind came N.N.E.

26. This day was faire and coole, but the after noone was close and hazie ; hee tooke in his sayles and huld[4] untill

---

[1] *Ere day* should be read as *each tide*.—C.

[2] This means that Baffin feared they would be driven so far to the southward by the ice that they would be unable to reach the mouth of Hudson's Strait.—C.

[3] This is wrong. The original has it that he hoped, by the end of the time named, the ice would allow him to enter Hudson's Strait.—C.

[4] That is, he *lay to hull* (see p. 181).—C.

morning 4. All this day he past by many beds of Ice, having great quantity to the N. of him, and having run about 21 Leagues upon a true W. course.

27. This 27 was close foggie weather, with much snow, freezing his shroudes and tackling; but at Clock 4 it cleared and he saw Land, it being the Ile *Resolution*, bearing W., about 13 or 14 Leagues off. He stands to and fro, as Ice would suffer him; when night came with W. wind, he made fast to a peece of Ice.

28. Faire weather all this day,[1] hee being fast to a peece of Ice, with W. wind, and hee could well perceive that hee set faster into the straights with the flood, then the ebbe could take him back againe.

29. This day the weather was faire and wind variable; hee sets sayle and tacks to and fro along the Iland; the next morning 2 the wind came to S.S.E., but he was so pestured with Ice that, with faire wind, he could doe little good; the wind continued a stiffe gaile all day and night, for it was not darke, and so [he] was set within the point of the Iland, so as now hee was within the straights.

[31.] This day was faire weather, the wind N.W.; hee saw *Button's* Iles beare S. by compasse, but S.S.E. with variation allowed, which was 24 degrees.

June 1. Some snow in the morning, but very faire in the afternoone; the wind at W.N.W. Hee, perceiving the Ice to open close to the shore, made way to get into anchor, and by Clock 7 he was in good harbour on the W. side of *Resolution*, where an E.S.E. Moone makes a full Sea or halfe an hower past 7 on the change day; the water doth rise and fall neere 4 Fathomes; the compasse doth vary 24 d. 6 m., and his Longitude from *London* 66 d. 35 m. The breadth of the S. channell is 16 Leagues, and the breadth

[1] Whitsunday.—F.

of the N., or *Lumley's Jnlet*, is 8 miles wide in the narrowest place.[1]

He found here no signe of inhabitants, but the track of Beares and Foxes; Rocks and stony ground, hardly any thing growing thereon; it is indifferent high Land to the N., having one hill or hummoke to the N.E., but to the South it falleth away very low.

[2.] This morning the wind came to the E.S.E., with much snow and foule weather. At noone he wayed anchor and stood about by the Iland side, as well as the ice would give him leave, to get to the N. shore, with much variable wind and weather, but stood fast in continuance amongst Ice untill the 8 day the wind fell contrary and, being somewhat neere a point of a Land, or rather a company of Ilands, which hee called Savage Ilands, having a great Sound or Indraft betweene the N. shore and them. At Clock 6, hee came to Anchor neere one of them, being the E.-most save one; but, whiles he was forling his saile, hee heard and saw a great company of Dogs, howling and barking, that it seeming very strange; after he had mored his Ship, hee sent his boate neare shore to see if they could discern any people, who, returned, said there were Tents, and Canowes, and Doggs, but for people they saw none. This writer[2] (being fitted), after Prayers and supper, went on Land to their Tents (with 7 others), where, finding no people, they marched up to the top of a hill (being about a flight shot),[3] where they saw a great Canow, which had

[1] By the South Channel is meant that between Resolution Island and Button's Isles. By the North Channel is meant that to the north of Resolution Island, called Gabriel Strait on the Charts. The words "or Lumley's Inlet" have been added by Foxe.—C.

[2] Baffin, that is.—C.

[3] Nares (*Glossary*) says that "a *flight* or *flight-shot* was frequently spoken of as a measure of distance", and quotes Leland, who, in his *Itinerary*, defines it as "about equal to the breadth of the Thames above London Bridge".—C.

about 14 men therein, being on the N.W. part of the Iland, and about a Musket shot from them; so called to them in Groenlandish speech, making signes of friendship; they did the like to them but, being fearefull, and he not trusting them also, made signes of a knife and other trifles, which he left upon the top of the hill, and returned to there tents againe, where he found to the number of 30 or 40 Whale finnes,[1] with a few Seale skins, which hee tooke with him, leaving for them knifes, beades, and counters. Hee found a little Bag where[in] were the Images of men, and one the Image of a woman, with a child at her back, which he brought with him.

Amongst these Tents, being 5 in number, all covered with Seales skins, were running 35 or 40 Dogs, the most of them muzled; there were of a Mungril Mastiffe,[2] being of a brinded black colour, looking almost like Wolves: those Dogs they use instead of Horses, or as the *Laplanders* doe their Deere, to draw their Sleedes,[3] which are shod or lyned with bones of great fishes to keepe them from wearing; their Dogs have collers and furniture very fitting.

Their apparell, Boates, and Tents, with other necessaries, are much like to those of *Groneland*, but not so neate and artificiall; they seeme to bee more rude and uncivill,

---

[1] In the original it is 14 or 15, but Purchas has it 40 or 50. The "finnes" of the whale (which are often mentioned in the narratives of Baffin's voyages) are the numerous horny plates of baleen or whalebone which depend from the upper jaw. Fotherby gives a lucid account of the way whales were killed and their oil and baleen obtained from them in his day (see Markham's *Voyages of Baffin*, p. 72), and another appears in Purchas (*his Pilgrimes*, part iii, p. 470).—C.

[2] The original has it, "They are most of them about the bigness of our mungrell mastives."—C.

[3] *Sled* is the older and correct form of the word now commonly written sledge. The form *sleede*, which Foxe here uses (though, perhaps, only a misprint), still survives in Essex, where a plough-runner is often called a *plough-sleed*.—C.

travelling up and downe as their fishing is in season; for, in most places where they were on Land, they see where people had beene; but, where their habitation or winter abode is, they know not, nor cannot conjecture.

This Iland lyeth in 62 degrees 32 minutes, and in longitude West from *London* 72 degrees,[1] or neere there about, being 60 Leagues from the entrance of the straights. The compasse doth vary 27 degrees 30 minutes, and South East Moone 4 degrees East maketh full Sea. It floweth almost as much water as at *Resolution*; the Tyde commeth from the Eastwards.

[10.] This day, morning 6, he set sayle with North winde, which continued not, but was variable, till noone, [when] it came to North West; hee, having sayled along the shore some 7½ leagues North North West, the Ice lying so thick in the offing that he could not well get out of it, he perceived a good Harbor betweene two small Ilands and the maine,[2] and went in, wherre he moord and stayd untill the twelfe day in the evening.

In this place, a South East Moone make a full Sea; Latitude 62 degrees 40 m. The tyde doth come from South East; every point hath his set and eddy. In this place hee could perceive of no people.[3]

[16.] Lying still in the Ice, the weather close and hazy, as it had beene for 6 dayes, being neere a great company of Ilands (the winde West North West), he stood in amongst them and at evening, moored to one of them in a small Cove, the better to defend her from the Ice. Here hee stood[4] all the 17 day; the 18, being almost calme, he set

[1] These observations are not quite correct. See Markham's *Voyages of Baffin*, p. 119.—C.

[2] Apparently it was Big Island, not the mainland.—C.

[3] The narrative from the 11th to the 15th, which contains nothing of importance, is omitted by Foxe.—C.

[4] In the original it is *stayed*.—C.

sayle, the better to get forth. Here was a great company of Ilands, each whereof hath his severall sets and eddyes, which drive the Ice to and againe with such violence that hee was in greater danger here then if he had beene further off. The Latitude of this Ile he lay at was 63 d. 26 m., longitude neere 72 d. 15 m.[1] from *London*, Variation 27 d. 46 m., ½ past 9 the change day maketh full sea. This evening and morning he had a fair gale at South East, and he stood along by the land, it being all small broken Ilands, to a point about 12 leagues distance from the Ile he set from, it being all broken land, so calls it Broken Point.

[19.] This day, 12, he was about 4 miles from the foresaid point, fast amongst Ice; and he saith he might well have called this *Fairenes*[2] or *Faire Point*, for from this day to the 30, the weather was so faire, and almost altogether so calme, that in few places elsewhere fairer weather could not be; and, untill the 27, hee was so fast inclosed amongst the Ice that one could not dip water by the Ship's sides. Vpon the 21 day, he see the Sunne and Moone both at one time, as indeed in faire weather is usuall in those parts.

[22.] Being faire and calme, the Sea almost as steady as on shore, with his Instruments for Variation, hee went to worke to take the time of the Moones comming to the Meridian, and had a Quadrant of 6 foote *Semidiamiter* ready to take *Sols Almicanter*, having taken the Variation of his Needle as properly as he could, which was 28 *d.* 10 *m.*[3]

The Sunnes *Almicanter*, at the instant when the Moone

---

[1] The MS. has it 74° 25′, which is obviously wrong.—C.

[2] By which name it is still known.—C.

[3] For critical remarks upon the following very complicated observations, see Markham's *Voyages of Baffin*, pp. 124-126.—C.

was upon the Meridian, was 26 *d.* 40 *m.*, the Sunnes declination 23 *d.* 6 *m.*, by which 3 things given, he found the houre to be 5 a clock 4 *m.* 54 *secon.* $\frac{14}{34}$, or 67 *d.* 13 *m.* 16 *s.* of the Equinoctiall afternoone; and, according to *Searls Ephemerides*[1] the Moone came to the Meridian at *London* at 4 a clock 54 *m.* 30 *s.*; and, after *Origanus*, the Moone came to the Meridian at 4 a clock 52 *m.* 5 *s.* at *Wittenberg* the same day. Now, having this knowne, it is no hard matter to know the Longitude of this place sought; for, according to the Moones meane motion, which is 12 *d.* a day, it is in time 48 *m.*, and to this account, if shee bee on the Meridian at 12 of clocke this day, to-morrow it will be 48 *m.* past 12, so hee having the time found by observation at this place, *viz.* 5 houres 4 *m.* 52 *s.* $\frac{14}{34}$, but in this he needeth not come so precise, and at *London* at 4 houres 54 *m.* 30 *s.*, which, subtracted from the former, leaveth 10 *m.* 22 *s.* $\frac{14}{34}$. Now the Moones motion that 24 houres was 12 *d.* 38 *m.*, which, converted into time, is 50 *m.* 25 *s.* 20 *th.*; then the pro-[por]tion standeth thus: if 50 *m.* 25 *s.* 20 *th.* give 360 *d.*, what shall 10 *m.* 22 *s.* $\frac{14}{34}$ give? The proportional will be 74 *d.* South, which is West of *London*, because the Moone came later by 10 *m.* 22 *s.*; and, by the working of *Origanus* his *Ephemerides*, the distance is 91 *d.* 35 *m.* West of West[2]; but whether be the truer, hee leaves it to others to judge; for, if those workings bee not carefully looked unto, there may be great errour committed, as in the observation, and in the Moones comming to the Meridian to the place for which the Ephemerides was calculated for, and it may be in the *Ephemerides* themselves; in all which, the best and most judicious may erre.

[He mentions another observation which he made on] the 26 of Aprill, outward bound at Sea, by the Moones

---

[1] See p. 97.—C.

[2] Apparently this should read *West of Wittenberg*.—C.

comming in a right line with two fixed Starres ; the one was the Lions heart, a Starre of the first magnitude; the other in the Lions Rumps, of second magnitude, as followeth.

The Circumference or outward edge of the Moone being in a right or straight line with those two Starres before named, at the instante he tooke the Altitude of the South balance, 2 *d.* 38 *m.*, because he would have the time, but in this it is good to waite a fit time, as to have her in a right line with 2 Starres not farre distant, and those not to be much difference in longitude, because the Moone will soone alter the angle or position, and such a time would bee taken, when the moon is in the 90 of the Ecliptique above the Horizon, for then there is no parallel of Longitude, but onely in Latitude. But who is painefull in these businesses shall soone see what is needfull, and what is not. His observations were as followeth.

		degr. min. secon.
Lion's Heart,	*Right Ascention* .	146.—28.—30.
	Declination . .	13.—57.—30.
	Longitude . .	24.—29.—45.
	Latitude . .	00.—26.—30.
	Almicanter . .	33.—40.—00.
Lion's Rumpe,	*Right Ascention* .	163.—23.—00.
	Declination . .	22.—38.—00.
	Longitude . .	05.—53.—45.
	Latitude . .	14.—20.—00.
the Moon's Paralax,	*Paralax* . .	00.—47.—46.
	Latitude . .	03.—20.—00.
	Almicanter . .	37.—00.—00.
	Latitude of the [*place*]	56.—43.—00.

This note, hee saith, is set downe for any that can, and are disposed to, spend their time therein themselves, having spent some, and would have spent more if leasure had served : but, finding it not to his minde, he hath set downe the particular worke as he received it from Mr. *Rudstone.*

27. Lying here, enclosed now among the Ice, with faire and calme weather (as before is said), untill the 27 day, at evening, he set sayle, the winde South East, an easie gale. All the 28 and 29, he made way through the Ice; but the 29, it was more open then [it had been] before in 10 dayes; at noone, *Salisbury* Ile bare West from him.

July [1]. This day was close foggie weather, with much raine; the winde S.S.E.; at noone he was 3 leagues from the land, but had much Ice by the shore. He stood to the N. and the next morning hee was faire by another small Iland, or rather a company of Ilands, which he afterwards called *Mill* Iland, by reason of grinding the Ice, as he had proofe. The Lat. is 64. Driving here to and fro untill clock 7, the Ice began to open and separate; hee had not past along the Ile by the East side thereof, but the Ice came driving with the flood-tyde from S.E. with such swiftness that it overwent his Ship, having all sayles abroad with a reasonable gale of winde, and put him out of the streame into the eddy of the Ile.

This Iland or Iles, lying in the middle of the Channell, having many sounds running through them, with many points or Head-lands encountering the force of the Tyde, causeth such a rebound of the Ice and water, which ran one way and the Ship another, the Ship having met with Ice, with the first of the flood put him neere the shore, that hee was in the partition betweene the Ice, which the eddy caused to runne one way, and the streame another, where shee endured great distresse. Thus hee continued untill towards high water, which [was] about one a Clock; then, with no small trouble, hee got into the Channell and stood to the North West-ward. After hee had past some distance from this Ile, hee found the Sea more open then it was since he put into the straights, and sayled all the next day with a South wind thorow an indifferent cleare Sea. At Clock 8 in the morning, hee was come againe into much Ice, and this Ice was thicker and bigger then any he had before, where he began to be enclos'd, 26 Leagues distant from *Mill* Ile North West by West, true course. Being fast amongst the Ice, hee perceived a great Tyde to set to and fro, and had 120 Fathome ozie ground.

[5. Standing northwards, in the forenoon he had ground] at 80 Fathomes; the wind comming to the North and setting him somewhat Southward, [he] had 110 Fathomes. Thus, seeing great abundance of Ice in this place, and the more he got to the North Westward the shoalder it was, the Ice being foule and durty, as not bred far from shores, hee determined to stand to the Eastward, to be better informed of the Tyde.

July 6. This morning, standing to the Eastward, hee brooke a planck and two timbers in his ship's bow, which after hee had mended, he proceeded to the East along the North shore, which Land stretcheth along from *Resolution* within the Streights, and is the West[1] side thereof.

7. This day he saw the Land, it being but low, and the Sea shoald in respect of other places, having 10 or 12 Fathome about a League from shore, and some 30 or 35 Fathomes 5 or 6 Leagues from shore, having very good channell ground, some 18 or 20 Leagues off, as small stones and shells, but the farther of the more ozie.[2] Also here runneth a very great Tyde to the Northward, which this evening hee found to be the Tyde of ebbe. For, comming neere the shore about Clock 7, hee went on

---

1 Query *North*. It is so in Purchas.—C.

2 These depths are not given in the original. They seem to have been added by Foxe from his own experience, gained when he sailed over the spot in question in 1631.—C.

Land with his Boate and found it so; he stayed on Land about an houre and a halfe, in which time the water fell about 3 foote and a halfe, and a South South East Moone maketh a full Sea. They saw no signe of people to have beene here this yeare, but other yeares before they could well see by divers places where their Tents had stood, and perhaps their time of fishing was not yet come, there being such great abundance of Ice as yet.

8. This day the wind was West.

9. This day was almost calme, and he reckoned to be neere the shore.

10. This day hee determined to stand to Ile *Nottingham*, to trye the Tyde there; the wind South West, so as hee turned it up untill night. The wind came to the North North West; 150 Fathome deepe,[1] so as hee stood away to the Westward and left the search of *Nottingham's* Ile, having a great swelling Sea out of the West, with the wind that had blowne, which put him in some hopes.

11. This morning hee saw Land West from him,[2] and had no ground at 130; standing along by the Land, which then lay North West, and by West, the next morning hee was thwart of a Bay[3]; then, standing over to a faire Cape or Headland hee saw in the afternoone, it was almost calme; [and,] being about a League from Land, hee sent his Boate to try the Tyde, and they stayed about 3 houres, going at 5, and returning at 8, and brought word that it was falling water, and that whilst they were on shore it had ebbe two foote: also they affirmed that the flood came from the N. in this place, which he perceived by the Ship,

---

[1] Here, again, Foxe seems to have added the depth from his own experience.—C.

[2] This appears to have been Terror Point.—C.

[3] It was, in reality, no bay, but the northern entrance to the channel separating Bell Island from Southampton Island.—C.

shee setting a pace to the N., although it was no wind: also they might see by the Rocks that the water was faln. This made him doubtfull[1] of a passage that way.

[13.] Master *Bylot* named this Cape *Comfort*, for the reasons before [stated][2]; and not a league from Land, is 140 Fathome water; here a S.E. Moone makes a full Sea; the Latitude is 65, and 86 d. 10 *m.* W. from *London.*[3] But this suddaine comfort was soone quailed, as hee saith; for, the next day, having doubled the Cape, and proceeded not above 10 or 13 Leagues, but hee saw the Land trent from the Cape to the Westward, untill it bare from him N.E. and by E. and very thick[ly] pestered with Ice.[4] And the further he proceeded N.-ward, he found shoalder water, and more Ice, and small show of any Tyde. At 6, he had 130 Fathom, soft ozie, and at noone [he had] had 150 Fad.

This was the furthest of this voyage, being in Latitude 65. 25., and Longitude from *London* 86. 10.;[5] for, seeing the Land so farre to the E. of him, 9 or 10 Leagues off, and the Ice, hee was fully perswaded this was but a Bay, and so turned the ship homeward without any further search.[6]

---

[1] This should be *hopeful*, as in the original.—C.

[2] That is to say, because he was comforted by the hope of finding a passage.—C.

[3] Cape Comfort is still so called. Baffin gives its position as 65° N., 85° 20′ W. Purchas, for some reason, alters this to 65° 26′ N., 86° W. Foxe again alters the figures, apparently to make them accord with his own experience. According to Parry (*Voyage of the Fury and Hecla*, 1821-23, p. 33) the exact position of Cape Comfort is lat. 64° 54′ N., and long. 82° 57′ W.—C.

[4] He was now in the mouth of Frozen Strait.—C.

[5] The latitude and longitude have been added by Foxe from his own experience. The latter should be about 84° W.—C.

[6] Baffin's reasons for turning back at this point seem inadequate, and the omission by Purchas (whom, of course, Foxe follows) of a passage in Baffin's manuscript makes them appear still more so. Baffin says (see Markham's *Voyages of Baffin*, p. 132):—"The land which we sawe beare north and north-east was about 9 or 10 leagues

14. The wind at S.E., that he could make but small way back againe. The next morning it was foule weather, and hee Anchored in a small *Inlet* neere Cape *Comfort* on the N. [side thereof]. When here, he found a S. and by E. Moone to make a full Sea, but could not discerne from whence the flood came, for it was bad weather at Sea.

15. The afternoon, the wind came N.W., stiffe gale, and hazie. He wayed and stood along the shore.

By the 16, at noon, he met with a great quantity of Ice, lying within the point of Land; amongst this Ice, hee saw a great number of Sea *Mors*,[1] not seeing any in all the streights but in this place, and those very fearfull, not suffering any Ship or Boate to come neere them. By clock 8, he was come to this S. shore point, which he called Sea-horse *Pointe*,[2] where he came to Anchor in open Sea, the better to try the Tyde, where he and all his company apparently found that in this place the Tyde came from S.E. and the Ebbe from N.W. He wayed after he had found this, and stood over with a stiffe gale of wind, which continued all day, and at night it was very foule weather and sowre stormes. By Clock 10, he was come to Anchor on the N.W. side of *Nottingham's* Iland, where 2 or 3 small Iles lye off from the greater, which make very good sounds and harbour.[3] About this Ile, he had store of Ice, but nothing as in other places. He staid here[4] untill the 27

from us; and shurely, without any question, this is the bottom of the baye on the west side; but howe far it runneth more eastward is yet uncertain."—C.

[1] The Morse, or Walrus (see p. 189), is often called the sea-horse. —C.

[2] It is still so called.—C.

[3] These islands are not shown on the latest Admiralty Chart, though clearly marked on Baffin's Chart, as reproduced in Markham's *Voyages of Baffin*, p. 103.—C.

[4] In the original it is not *here*, but *about this island*, which is obviously correct.—C.

day, with much foule weather, many storms, often fogg and uncertaine winds; many times he wayed Anchor to goe to that side of the Iland where the Ship rod,[1] when Cap. *Button* was in her,[2] finding in other places of this Ile the Tyde of flood came from the S.E.-ward, and the time of high water upon the change of day to bee at 10 and halfe an houre past, and not after, as they supposed before. In [the] 10 dayes he stayed about this Ile, he fitted his Ship with ballast and other necessaries.

26. This day, being indifferent faire weather, hee passed betweene *Salisbury* and *Nottingham's* Iles, at the S. point thereof, where lie many small low Iles, without the which had beene a fit place to have Anchored to have found out the true set of the Tyde[3] : but the Mr., being desirous to have come to the same place where he had rod before,[4] stood along by this Ile to the W.-ward and came to an Anchor in the Eddie of this broken grounds, where the Ship rod at no certainety of the Tyde.

27. This morning was foule weather, with much raine and wind, that the Reager[5] Anchor would not hold at 80

---

[1] This word, of course, should be *rode*. It occurs several times in the following paragraphs.—C.

[2] Bylot was with Thomas Button.—F.

[3] These islands form Port de Boucherville, so named by Lieut. A. R. Gordon, R.N., in 1884, after an observer whom he stationed there for the winter. A description, plan, and view of the harbour will be found in Lieut. Gordon's *Reports on the Canadian Hudson's Bay Expeditions of 1884, 1885, and 1886.* It is also shown in detail on the Admiralty Chart No. 1221.—C.

[4] That is, when he was with Button (see above).—C.

[5] *Kedger* (see p. 71) is meant. Foxe follows here a misprint in Purchas's text. This and other circumstances tend to show that Foxe did not make these abstracts with his own hand. As a seaman, familiar with nautical terms, he would hardly have copied from Purchas so obvious a misprint. It is a likely surmise that the abstracts are those he tells us in his "Preparations" were made by his friend, Mr. Thomas Sterne.—C.

Fathomes scope, but was driven into deeper water, and [he] enforced to set saile, the wind at E. and came about to N.E., with fowle weather. He stood away towards Sea Horse Point; he was perswaded that there might be a passage betweene that Land and the Land they called *Swan* Iland[1]; so this afternoone hee saw both Sea Horse Point and Ile *Notingham*; the distance betwixt both is not above 15 or 16 Leagues; they lye one from another S.E. and N.W.

28. In the morning he saw Sea Horse Point, and the Land to stretch away W.S.W., so far as he saw, and with Ice, wherefore he tackt about and stood away S.E. and by S.[2]

29. This day, 11, he came to anchor at *Diggs* his Iland, having very foule weather; at this place where he rod, it lyeth open to the W., having 2 of the greatest Iles to breake off the force of the flood Tyde; for, after the water was risen an houre and a halfe by the shore, then would the Ships ride truly on the tide of flood all the Tyde after; now the time of high water on the change is at ½ past 10, or thereabouts.

[30.] This day was faire weather; he wayed & stood close by *Diggs* his Iland, where presently he perswaded[3] the Salvages to bee close [hid] upon the Rocks; but, when they saw he had espied them, divers of them came running down to the water side, calling to him to come to anchor, which he would have done if hee could; but in this place the water was so deep as it is hard to find a place to ride in; which seeing, he lay to and againe with the ship, whilst some of his men with the Boat killed about 70 Fowles; for, in this place is the greatest store of those fowles, which

---

[1] Cary's Swan's Nest (see p. 165).—C.

[2] He sought no passage then.—F.

[3] A misprint for *perceived*.—C.

we call Willicks,[1] that in few places else is not to be seen; for, if need were, he might have killed many thousand, almost incredible to those that have not seene it. Here he had sufficient proofe of the tyde; but, when his Boate returned, he set sayle homewards.

[*August* 3.] This day he was forced to anchor 30 leagues within *Resolution*, upon the N. shore; the next day he weighed, and the 5 day he passed by *Resolution*, but see it not.

[*September* 6.] He had a sight of Cape *Cleere* in *Ireland*. He came into *Plimouth*, all his men alive, but 3 sick, which presently recovered.[2]

---

[1] See pp. 107 and 168. This is the place where Hudson in 1610, his survivors in 1611, and Button in 1612, had killed so many of these birds.—C.

[2] These last two paragraphs are, in Foxe's work, placed after the "Observation", probably through a printer's error. I have moved them to what seems their right place. After them, in Baffin's original MS., there is a passage which is worth reprinting here. Purchas (and of course, Foxe after him) omitted it, but Rundall (*Voyages North-Westward*, p. 130) and Markham (*Voyages of Baffin*, p. 137) both print it. It is as follows:—

"And now it may be that som expect I should give my opynion conserninge the passadge. To those my answere must be that doubtles theare is a passadge. But within this strayte which is called Hudson's Straytes, I am doubtfull, supposinge the contrarye. But, whether there be or no, I will not affirme. But this I will affirme: that we haue not been in any tyde then that from Resolutyon Iland, and [that] the greatest indraft of that commeth from Dauis Straytes; and my judgment is [that], if [there is] any passadge within Resolution Iland, it is but som creeke or inlett; but the main [passage] will be upp fretum Davis. But, if any be desirous to knowe my opynion in pertyculler, I will at any tyme be redy to showe the best reasons I cann, eyther by word of mouth or otherwise."

The general correctness of Baffin's opinion has since been conclusively shown. It was in consequence of it that, on his voyage in the following year (1616), he searched up Davis Strait.—C.

The Observation.[1]

He set forth the 18 of April, and he saw the land of *Groynland* the 6 of May; he made *Resolution* the 27, where, on the N. side, he found a good Harbor, where it flowes on E.S.E. moone and neere 4 fatho. He found people at Salvage Iles. He was much troubled [with ice], but especially at Mill Isle. He made Cape *Comfort*, and found a tyde, but knew not from whence it come. The land to the N. trents about him to N.E. by E. The water, the farther Northward, was but more shallow and dirty. He returns homewards the 10 of Iuly, his greatest deep at 180 fathoms. In his return, at *Sea-horse* Point, he and all his people saw that plainly the tyde came from S.E., as also at Isle *Nottingham*. He broke in a planck and timber of his ship amongst Ice. He might have killed thousands

---

[1] This is a sort of summary by Foxe of the main results of the voyage. The whole of the track sailed over is shown accurately on Baffin's original chart, which, though suppressed by Purchas (see p. 204), is preserved in the British Museum, and is reproduced roughly by Mr. Rundall (*Voyages towards the North-West*, p. 97) and in exact *facsimile* by Mr. Markham (*Voyages of Baffin*, facing p. 103); but the route could easily be followed on a chart from Baffin's admirably-clear narrative. As Mr. Markham observes (*Voyages of Baffin*, p. lii): "Baffin's work in Hudson's Strait does not amount to discovery, but it was a painstaking and valuable survey, and was recognised by Sir Edward Parry as praiseworthy and highly creditable." The only adverse criticism that can be passed upon the conduct of this voyage is that no adequate reasons are given to explain why Bylot and Baffin did not proceed further to the northward than they did up the Strait now known as Foxe's Channel; for the little way they did proceed in this direction was over fresh ground. The excuse (p. 216) that they found the water becoming shallower and the ice dirtier was inadequate. The northward extension of this Channel was not further explored until nearly twenty years later, when Foxe sailed up it.—C.

of Fowle at *Diggs* his Island. His greatest Variation was 27 *d.* 46 *m.*; his greatest Lat. was 65 *d.* 25 *m.* He saw many Sea Mors at Cape *Comfort.* His Longitude from *London* was 86 *d.* 10 *m.*

The next yeere, being againe imployed in discovery, amongst other instructions,[1] he received this: For your course, you must make all possible hast to Cape *Desolation*, & from thence you, *William Baffyn,* as Pilot, keep along the Coast of *Groenland* and *Fretum Davis* untill you come towards the height of 80, if the land will give you leave; then, for feare of imbaying by keeping off too Northerly a course, shape your course W. and Southerly, so far as you shall thinke it convenient, untill you come to the Lat. of 60; then direct your course to fall with the land of *Yedzo*, about that height, leaving your further sayling Southward to your owne discretion, according as the time of the yeere and the winds will give you leave; although our desire be, if the Voyage be so prosperous, that you may have the yeere before you, that you goe so far Southerly as that you may touch the N. part of *Iapon*, from whence, or from *Yedzo*, if you can compasse it without danger, wee would have you to bring home one of the men of the Country; and so, God blessing you, with all expedition, to make your returne home againe.[2]

---

[1] These are copied verbatim from Purchas (*his Pilgrimes*, vol. iii, p. 842), with the exception of some obvious misprints, which I have corrected.—C.

[2] Baffin's interesting letter to Sir John Wolstenholme, which Foxe introduces here, I have removed to its proper position after the narrative of Bylot and Baffin's second voyage (see p. 233).—C.

The Journalls of his Voyage set forth at the charge of the right worshipfull Sir Thomas Smith, *Knight, Sir* Dudley Diggs, *Knight, Mr.* Iohn Wolstenholme, *Esquier, and Mr. Alderman* Ioanes, *with others, in the* Discovery; Robert Bylot, *Mr., himselfe Pilot.* 1616.[1]

HE set [forth] from *Gravesend* [on March 26, 1616], and went about by the W., by *Ireland*, and the first Land hee saw was within *Fretum Davies*, in 65 de-

---

[1] The account of this voyage is abstracted by Foxe from *Purchas his Pilgrimes* (part iii, pp. 844-848). The expedition was sent out by the same enterprising gentlemen who had sent out Bylot and Baffin in the previous year, and Hudson, Button, and Gibbons before them, namely the Company of Merchant Discoverers of the North-west Passage, this being, therefore, their fifth expedition. It was undertaken by Bylot, as master, and Baffin, as pilot, in the Company's old ship *Discovery*, which was thus making her sixth and last recorded voyage in search of a North-west Passage. Inasmuch as this voyage did not add anything to the knowledge of Hudson's Bay or Strait (with which I am herein chiefly concerned), I have not paid much attention to its annotation. It was, however, a remarkably well-carried-out and successful voyage. Baffin sailed up the channel which had been known as Fretum Davis or Davis Strait since Davis had discovered it in 1585-86. He discovered, and largely completed the geographical exploration of, the vast bay which has ever since borne his name, Baffin's Bay. The voyage was a very important one, for on it Baffin first discovered (though he did not explore) both Smith's Sound and Lancaster Sound, through which not a little of our modern Arctic exploration has been conducted. These discoveries, and others, says Baffin, "the map doth truly describe"; but, unfortunately (as was the case with that of the voyage in the previous year), Purchas was obliged through want of funds to omit this map, and (although the map of the former voyage has been preserved, see p. 221) this map seems to have been irrecoverably lost. The serious consequences of this omission on the part of Purchas are discussed by Mr. Markham in his *Voyages of Baffin* (pp. liv-lvii). It is not very clear why Baffin's Bay should not have been known as *Bylot's Bay*, seeing that Bylot, not Baffin, was the chief of the expedition;

grees 20 minutes. On the 14 May, the forenoone, 6 of the people, being a fishing, came to him, to whom hee gave small peeces of Iron, and they, keeping him company with great love, suppose he had intended to come to anchor, but, when they see him stand off from shore, they followed a while and then went away discontented, as he thought.

Hee prosecuted his voyage, being loth to Anchor as yet, although the wind was contrary, but plyed to the N.-ward, untill he came into 70 d. 20 m. There he entred a faire sound neere Capt. *Davies* his *London* Coast[1]; the people, espying [them] with great wonder and gazing, fled away in their Boates, and after this night hee saw them no more, yet they left their Dogs running to and fro upon the Iland.

At this place hee stayed two dayes, taking in water and doing things needfull; the Tydes not rising above eight or nine foote made him dislike of the passage; it floweth ¼ past 9 upon the change day; the Tyde commeth from the South.

[*May* 22.] At N. Sun, hee set saile and plied to the Northward with flood Tyde.

[26.] Hee found a dead Whale[2] about 26 Leagues from the shoare; hee made fast and got some finne out of her; the next day he was enforced by storme to leave her; having stood from her not above 3 Leagues N.W., hee came to Ice; [he] then tackt into the shoare againe, and a great storme ensued.

---

but Baffin, as chronicler of the voyage, seems to have usurped the honour of the commander, and perhaps not without much right to do so. Even Foxe, who speaks of the voyage in the previous year as *Bylot's*, calls this "*Master Baffyn his Voyage*".—C.

[1] See p. 76. Probably the sound was one of those to the north of Disco Island.—C.

[2] Baffin again refers to this whale in his letter to Sir John Wolstenholme (see p. 236).—C.

[30.] This day he came faire by *Hope Saunderson*,[1] Captaine *Davies* his farthest, betweene 72 d. and 73; that evening, at N. Sunne, hee came to Ice, which hee put into, plying all the next day to get through it.

June 1. He was cleere off the Ice, and not farre from shoare; the wind North East, hee put in amongst divers Ilands; the people, seeing him, fled away, leaving their tents behind, and under a small Rock hidden two young Maids or women. The Ship riding not farre off, the Master with some others went on Land, they making signes to bee carried to the Iland, where their Tents were there adjoyning. When they came there, they found two old women more, the one to estimation 80 yeares, the other younger. There was also one woman with a child at her back, who had hid her selfe amongst the Rocks untill the others had told her how kindly they had used them in giving them peeces of Iron and other trifles, in change whereof they gave them Seale skins; other things they had none, save dead Seales and fat blubber, which the poore women were very diligent to carry to the Boate and put into their Casks, making shew that the men were over at the Maine and at another small Iland something more Eastward. Then they made signes to them that hee should shew them his shippe, and set them where the men were; the foure youngest came into their Boat; and when they were in the Shippe they much wondred; and they gave them of his meate, which they tasting would not eat; two of them hee set to the Iland where they found the men to be; the other two he set to their Tents againe; those that went to seeke the men could not find them, but came backe to the Ship againe, and were set over to the other side.

This place they called Women's Ilands[2]; it lyeth in 72

[1] Discovered and so named by Davis in 1587 (see p. 76).—C.

[2] The islands surrounding the Danish settlement of Upernivik.—C.

degrees 45 minutes; the Flood commeth from the S.-ward at neepe Tydes; the water riseth not above 6 or 7 foote: S.S.E. Moone makes full Sea. The Inhabitants being[1] very poore, living chiefly upon Seales flesh dried, which they eate raw, and cloath themselves with the skinnes, as also they doe cover their Tents and Boates therewith, which they can dresse very well; the women differ from the men in apparell, and are marked in the face with divers black streakes or lines, the skinne having beene raised with some sharpe Instrument when they were young and black colour put therein, so growne in that by no meanes it can be got forth.

Concerning their Religion he saith little, onely they have a kind of worship or adoration to the Sunne, to which they will point, striking their hand upon their breasts and crying *Elyoute*[2]; their dead they bury on the sides of the hills where they live, making a pile of stones over them, yet not so thick but that he could see the dead body, the aire being so piercing that it keepeth them from stincking savour; so likewise hee hath seene there doggs buried in the same manner.

[*June* 4.] This day hee set saile from thence with faire weather; the wind contrary, yet hee plyed it up betweene the Ice and shoare, as it hath beene in a channell of 7 or 8 Leagues broade; on the 9, hee was in 74. 4., much pestered with Ice, neere 3 small Ilands lying 8 Myles from shoare, where hee anchored.

Those Ilands hee taketh to be frequented by people at the latter end of the yeare, as it seemed by the houses and places where their Tents had stood, but as yet they were not come. The flood Tyde was very small, not rising above 5 or 6 foote, yet the ebbe runneth with indifferent

---

[1] *Are*, instead of *being*, would make better sense.—C.

[2] See p. 67.—C.

stream, caused by the melting snow from off the mountaines.

[10.] Seeing that as yet hee could not proceed, hee determined to stand in for the shore, there to abide untill the Ice were more consumed, which he plainely saw to wast very fast. Hee came to anchor [in] 73 d. 45 m.[1] Here hee continued 3 dayes, without any shew or signe of people.

[*June* 15.] This day there came 42 of the Inhabitants in their Canoes; they gave him Seale skinns and many peeces of the horne of Vnicorne, and shewed him divers peeces of Mors teeth, making signes that to the N.-ward were many of them; hee gave them in exchange thereof peeces of Iron, glasses, and Beads; 4 severall times they repayed to him, bringing alwayes of those commodities aforesaid, by reason whereof he called this place *Horne Sound*.[2]

[*June* 18.] Here hee staid 6 dayes, and on this day at night sets sayle with little wind; hee stood to the N.-ward, the winde having been contrary the most part of the moneth; but it was strange to see the Ice so much consumed in so little space, for now hee could come to the 3 Ilands formerly named, and stand off to the Westward almost 20 Leagues without let of Ice, untill hee came to 74. 30 m. Latitude; then hee put amongst scattered Ice, plying all this moneth, every day gaining somewhat, nothing worthy of Note, hoping that hee might see many of those fishes with long hornes which they call *Sea Vnicornes*.[3] The weather variable, few dayes without snow and

---

[1] Near Cape Shackleton.—C.

[2] This name has not survived.—C.

[3] Foxe, or whoever it was that made these abstracts for him, though careful and correct as a rule, has here made an error in transcription. In Purchas, the passage says "every day we got something on our way, nothing worthy of note happening, but that at divers times we saw sea unicorns" (see p. 45).—C.

freezing: but Midsummer day his sayles and tackling were frozen that they could not handle them, yet the cold is not so extreame but that it may bee well endured, hee being still within sight of Land.[1]

July 1. Hee came into an open Sea in 75. 40. Latitude, which newly revived his hope of a passage; and, because the wind was contrary, he stood 20 Leagues off the shore, and then he stood in againe, and anchored to try the Tyde, but found small comfort. Shortly after, the wind came to South East, and blew very hard, with thicke and foggy weather; he set sayle and runne along the shoare the second day; and the next day hee came by a fayre Cape, hee named Sir *Dudley Diggs* his Cape,[2] in Latitude 76. 35. It hath a small Island close adjoyning to it. The wind still increasing, he past by a goodly sound, 12 leagues distant from the former Cape, having a small Island in the middle, making 2 currents; under this Island he Anchored 2 houres. The Ship drove, although he had two Anchors on ground; whereupon he was forced to weigh and stand forth. He called this sound *Wolstenholme's Sound*.[3] It hath many Inlets or smaller sounds, and is a fit place for killing of Whales.

[4.] This morning a storme began at West that blew away his fore-course, continuing so that he was not able to beare any sayle, but lay to *Hull*[4]; and, when it cleared up, he found himself imbayed in a great Sound: then he set sayle and stood over to the S.W. side, and Anchored in a Bay, where he lost both Cable and Anchor, the winde blowing so extremely from the hill tops that he could get no place to Anchor in; but was forced to stand too and againe in

---

[1] The last few words are added by Foxe.—C.

[2] This name is still used.—C.

[3] This name is still retained.—C.

[4] See p. 181.—C.

the sound, the bottome being all froze over. In the afternoone, it was lesse wind and hee set forth. In this sound was great number of Whales; hee called it *Whale-sound*, in latitude 77. 30.[1]

[5.] This day was faire weather and he kept along by the land untill he came unto a great banke of yce which was backt with land, which hee, seeing, determined to stand backe againe some 8 leagues to an Island he called *Hakluits Ile*.[2] It lyeth betweene two great sounds, the one *Whale-sound*, and the other Sir *Tho. Smith's Sound*[3]; this last runneth to the N. of 78, and is admirable in one respect, because in it is the greatest of variation of the Compasse of any part of the knowne World. For by divers good observations hee found it to bee above 5 points, or 56 d. varied to the N.-ward.[4] This Sound seemeth to bee good for killing of *Whales*, it being the greatest and largest in all the Bay; the cause why he minded to stand to this Island was to seeke for *Whale*-Finnes; this night he Anchored with foule weather that his Boate could not land; the next day, the Wind [was] more outward, [and] the Sea growne so that he wayed, spending two dayes before he could get a good place to Anchor in.

[8.] This day it cleared up and he spyed a company of Islands lying 12 or 13 leagues off from shoare: he minded to go to them, but the Wind tooke him short; and hee, being loath to spend more time, tooke the opportunity

---

1 This name also is still retained.—C.

2 Baffin had now attained, and had commenced to return from, his most northerly point, which we may judge to have been nearly 79° N. —a remarkably high latitude for the time and place.—C.

3 This name also survives.—C.

4 This is a misprint for *West*ward. See Markham's *Voyages of Baffin*, pp. 145 and 154.—C.

thereof, and left the search of those Isles, which he called *Caryes Isles.*[1]

[10.] Then he stood to the W.-ward with open Sea and a stiffe gale of wind untill this day, when it fell calme and foggy, he being neere the entrance of a faire Sound, which hee called Alderman *Iones* his Sound.[2] This afternoone, it being cleare and faire weather, he sent his Boate on land, the ship being under sayle, but the wind beginning to blow, they returned, saying they saw many *Sea Mors* by the shoare amongst the yce, but no signe of people, so farre as they were, nor any good place to Anchor in. Then, having the wind E.N.E., hee run along the shore, it beginning now to trent to the South and shewing it selfe to bee a Baye.

[12.] This day he was open of another great Sound, he called Sir *Iames Lancaster's Sound*[3]; here his hope of passage began to lessen every day more then other; for, from this Sound to the Southward, he had a ledge of Ice betweene the shoare and him, but the Sea was cleere to

---

[1] No doubt named after Mr. Alwyn Cary (see p. 165, *note*), ship's-husband for Baffin's previous voyage, and also, says Mr. Markham (*Voyages of Baffin*, p. 146), for this. The islands are still so called. —C.

[2] This name is still retained.—C.

[3] Baffin had now dutifully named, after the five leading members of the Company which sent him out, prominent geographical features in the great Bay he had discovered, all the names he gave being still retained on our modern charts. Thus we have Sir Dudley Diggs' Cape, Wolstenholme Sound, Sir Thomas Smith's Sound, Alderman Jones Sound, and Lancaster Sound, all of the gentlemen commemorated being at that time munificent patrons of Arctic research. Of Sir Dudley Digges (see *ante*, p. 118, *note*), Sir Thomas Smith, and Alderman (afterwards Sir Francis) Jones, brief notices may be found in the Introduction to Markham's *Voyages of William Baffin* (pp. vi-xvi): of Sir James Lancaster, a notice appears in *The Voyages of Sir James Lancaster to the East Indies*, edited by Mr. Clements R. Markham (Hakluyt Society, 1877); and of Sir John Wolstenholme (the elder) I have spoken in my Introduction.—C.

the S. of it ; hee kept close to this Ice vntill this day,[1] and then hee was in 71 deg. 16 m. and plainely perceiued the land to 70 deg. 30 min. ; then, hauing much Ice about him, hee stood to the E.-ward, supposing to haue beene soone cleere, and to haue kept on that side the Ice vntill he had come into 70 deg., and then to haue stood in againe. But it prooued quite contrary to his expectation ; for hee was forced to runne aboue 60 leagues through Ice, and many times fast that hee could goe no way, although so hee kept this course due E.; and, when hee had gotten into the open Sea, hee kept so neere the Ice that many times hee had much to doe to get cleere, yet could not get neere the land vntill hee came to 68, when hee see the shoare, but could not come to it by 7 or 8 leagues for the great aboundance of Ice. This was on the 24 day of *Iuly;* then hee spent 3 dayes to see if hee could anchor to trie the Tide, but the Ice led him into 65. 40. min. lat., where hee left the west shore, because that then hee was in the Indraft of *Cumberland* Inlet. He knew no certainties nor hope of passage could bee there.

Now, seeing that hee had made an end of his discouery, and the yeere being too farre spent to goe for the bottome of the Bay to search for drift Finnes,[2] hee determined to goe for the Coast of *Groenland* to seek for refreshing for his men, Mr. *Hubart*[3] and two more hauing kept their Cabbins 8 dayes, besides his Cooke, which dyed the day before, and diuers of his company so weake that they could not labour.

---

[1] July 14th.—C.

[2] In the original, it is *drest finnes.* Foxe (apparently taking the word for a misprint) has altered it to *drift*, which seems to have more meaning. Presumably "drift finnes" are the finnes of whales (see p. 208) which have been drifted ashore.—C.

[3] In Purchas, it is *Master Herbert.* No doubt this was the Josias Hubart who had sailed with Hall in 1606, and with Button in 1612-13 (see p. 171).—C.

So, the winde fauouring, hee came to anchor in 65 deg. 45 min., [on July 28th,] in a place called *Cockin Sound.*

[29.] The next day, vpon an Island, wee found great store of *Scuruie-grasse*, with *Sorrill*[1] and *Orpen.*[2] The *Scuruie-grasse* hee boyled in Beere, by meanes whereof, with God's blessing, his men were in perfect health in 8 dayes and so continued vntill his Arivall in *England.*

[*August* 1.] Heere hee rode 3 dayes before any of the people came to him; this day came 6 in Canoes; they brought *Salmon-Peale* and such like, which was good refreshing for his men; the next day following, the same 6 came againe, but they saw them no more vntill the 6 day, when hee had weighed anchor and was almost cleere of the harbour; the same 6 and noe more came and brought of the like commoditie: for which they gaue them Glasses, Beades, Counters, and small pieces of iron; which they doe esteeme as wee Christians doe Gold or Siluer.

In this *Sound* was such Skulls of *Salmon*[3] swimming too and fro that it was much to bee admired: it floweth aboue 18 foote water. It floweth on the change [day] till seauen a clock. It is a very good Harbour, and easie to be knowne, having three high round hills like Piramidies close adjoyning to the mouth thereof; and that in the midst is the lowest. All this coast along is full of good Harbours, by reason of so manie Ilands that lie from the Mayne.

[*Aug.* 6.] By 3 this day hee was cleere of this place, hauing a N.N.W. winde, faire weather. So God sent him

---

[1] See p. 127.—C.

[2] This was, no doubt, some member of the genus *Sedum* deriving its name from the yellow colour of its flowers. In England, the common *Sedum telephium* is generally known as "orpin", but the plant Baffin found was probably some other member of the genus inhabiting the region in question.

[3] See p. 71.—C.

a speedy passeage, for, in 19 dayes after, hee saw the coast of *England*.

The 30 he anchored in *Douer* Roade.

Master Baffyne *his Letter to the right Worshipfull Sir* Iohn Wolstenholme, *one of the chiefe Adventurers for the discovery of a passage to the North-west.*[1]

WOrthy Sir, there needs no filling a Iournall or short Discourse with Preamble, circumstance, or complement; and therefore I will onely tell [you] I am proud of my remembrance, when I expresse your worth to my conceit; and glad of any good fortune when I can avoyde the imputation of ingratitude, by acknowledging your many favours; and, seeing it is not unknowne to your Worship in what estate the businesse concerning the North West hath beene heretofore, and how the onely hope was in searching *Fretum Davis;* which, if your selfe had not beene the more forward, the Action had wel-nigh beene left of; now it remaineth for your Worship to know what hath beene performed this yeare[2]; wherefore I intreat you to admit of my custome and pardon me if I take the plaine highway in relating the particulars, without using any refined Phrases or eloquent speeches.

Therefore, briefly thus, and as it were in the Fore-front, I

---

[1] This letter of Baffin's is reproduced almost verbatim from the pages of Purchas (*his Pilgrimes*, part iii, p. 843), the few alterations being merely verbal and no doubt unintentional. I have corrected, without comment, a few obvious misprints. Though placed by both Purchas and Foxe before the account of Baffin's second voyage north-westward, it was obviously written after it, and forms a sort of report on it. I have followed Mr. Markham in removing it to its correct position (see p. 222).—C.

[2] That is, 1616.—C.

entend to shew the whole proceeding of the Voyage in a word as, namely, there is no Passage nor hope of Passage in the North of *Davis* Straight; wee having coasted all or neere all the Circumference thereof, and finde it to be no other then a great Bay, as the Voyage doth truly show.[1] Therefore I cannot but much admire the worke of the Almighty, when I consider how vaine the best and chiefest hopes of men are in things uncertaine; and (to speake of no other matter than the hopeful passage to the North West) how many of the best sort of men have set their whole endeavours to proove a passage that way? not onely in their Conference, but also in writing and publishing to the worlde: Yea, what great summes of money have beene spent about the Action, as your Worship hath costly experience of? Neither would the vaine-glorious Spaniard have scattered abroad so many false Mappes and Iournals, if they had not beene confident of a passage this way; that, if it had pleased GOD a passage had beene found, they might have eclipsed the worthy prayse of the Adventurers and true discoverers. And, for mine owne part, I would hardly have beleeved the contrary, untill my eyes became witnesses of that I desired not to have found, still taking occasion of hope on every likelihood, till such time as wee had coasted almost all the Circumference of this great Bay. Neither was Master *Davis* to bee blamed in his report and great hopes, if hee had anchored about Hope *Saunderson*,

---

[1] This very positive statement by Baffin that there was no hope of finding a passage by way of Davis Strait is interesting, though erroneous. It will be remembered that, as a result of his voyage in the previous year, he had expressed a very strong opinion that no passage was to be found through Hudson's Strait, and that, if one existed at all, it lay up Davis Strait. Now we find him stating positively that there was no possibility of finding a passage in that direction either. Later discoveries have shown that Baffin was wrong, but his opinion sufficed to check the search for nearly twenty years after.—C.

to have taken notice of the Tydes. For, to that place, which is 72 deg. 12., the Sea is all open, and of an unsearchable depth, and of a good colour; onely the Tydes keepe a certaine course, nor rise but a small height, as eight or nine foote; and the Flood commeth from the Southward; and in all the Bay beyond that place the Tyde is so small, and not much to bee regarded: yet, by reason of Snow melting on the Land, the Ebbe is stronger then the Flood, by meanes whereof, and the windes holding Northerly the fore part of the yeere, the great Iles of Ice are set to the Southward, some into *Fretum Hudson*, and other into *New-found-land;* for, in all [places] where the Channell is open, are great quantities of them driving up and downe; and till this yeere [it was] not well knowne where they were bred.

Now that the worst is knowne concerning this Passage, it is necessary and requisite your Worship should understand what probability and hope of profit might here bee made hereafter, if the Voyage might bee attempted by fitting men. And, first, for the killing of Whales: certaine it is that in this Bay are great numbers of them, which the Biscayer calls the Grand Bay Whales, of the same kinde which are killed at *Greeneland*[1]; and, as it seemeth to mee, easie to bee strooke, because they are not used to bee chased or beaten. For wee being but one day in Whale-sound (so called for the number of Whales wee saw there, sleeping and lying aloft on the water, not fearing our Shippe or ought else) that, if wee had beene fitted with men and things necessarie, it had beene no hard matter to

[1] The species of whale Baffin saw was, no doubt, the Greenland Right Whale (*Balæna mysticetus*). It is distinct from, though nearly allied to, the Atlantic Right Whale (*Balæna biscayensis*), which, until the end of the sixteenth century, was a very important source of wealth to the Basque whale-fishers inhabiting the ports in the Bay of Biscay, but which is now very rare.—C.

have strooke more then would have made three Ships a saving Voyage; and, that it is of that sort of Whale, there is no feare. I, being twice at *Greeneland*,[1] took sufficient notice to know them againe; besides a dead Whale wee found at Sea, having all her Finnes, or rather all the Roughs of her mouth, of which with [no] small labour wee got a hundred and fifty[2] the same evening wee found her: and, if fowle weather the next day had not followed, no doubt wee had got all, or the most part of them; but, the winde and Sea rising, shee broke from us, and wee were forced to leave her.[3] Neither are they onely to be looked for in Whale Sound, but also in *Smith's* Sound, *Wolstenholme's* Sound, and others, &c.

For the killing of Sea-mors I can give no certainty, but onely this: that our Boate being but once a shoare in all the North part of this Bay, which was in the entrance of Alderman JONES his Sound, at their returne our Men told us they saw many Mors along by the shoare on the Ice; but, our Ship being under sayle, and the winde comming faire, they presently came on board without further search. Besides, the people inhabiting about 74 degrees tould us by divers signes that towards the North there were many of those Beasts having two long teeth; and [they] shewed us divers pieces of the same.

As for the [Sea] Vnicorne, it being a great Fish having a long horne or bone growing forth of his forehead or nostrils (such as Sir MARTIN FROBISHER in his second Voyage found one), in divers places wee see of them: which, if the horne be of any good value, no doubt but many of them may be killed.

As concerning what the shoare will yeeld, as Beares

---

1 He evidently means Spitzbergen.—C.

2 In the original, it is one hundred and *sixty*.—C.

3 Baffin has previously mentioned this whale (see p. 224).—C.

skins,[1] Mors-teeth, and such like, I can say little, because we came not on Land in any of those places where hope was of finding them.

But here some may obiect why wee sought that Coast no better? To this I answere that, whilest we were thereabouts, the weather was so exceeding fowle wee could not; for first we anchored in *Wolstenholme's* Sound, where presently we drove with two anchors on head; then we were forced to stand forth with low sayle; the next day, in Whale Sound, we lost Anchor and Cable and could fetch the place no more; then wee came to Anchor neere a small Iland, lying betweene Sir *Thomas Smith's* Sound and Whales Sound; but the winde came more outward, [so] that wee were forced to weigh againe. Neverthelesse, if wee had beene in a good Harbor, having but our Ships Boat, we durst not send her farre from the Ship, having so few men (as 17 in all), and some of them very weake: but the chiefe cause why wee spent so little time to seeke a Harbor was our great desire to performe the discovery; having the Sea open in all that part, and still likelihood of a passage. But, when we had coasted the land so farre to the S.-ward that hope of a passage was none, then the yeere was too far spent, and many of our men very weake, and withall we having some beliefe that Ships the nexe yeare would be set forth about the killing of whales, which might doe better then we.

And, seeing I have briefly set downe what hopes there is of making a profitable Voyage, it is not unfit your Worship should know what let or hindrance may be to the same. The chiefest cause is that some yeares it may happen, by reason of Ice lying between 72½ and 76, that the Ships cannot come into those parts untill the middle of Iuly, so

---

[1] This, in the original, is *beach-finnes*, which presumably are the same as the "drift finnes" already spoken of (see p. 231).—C.

that want of time to stay in the Country may be some let; yet they may well tarry untill the last of August; in which space much businesse may be done, and great store of oyle made. Neverthelesse, if store of Whales come in, as no feare to the contrary, what cannot be made in Oyle may bee brought home in Blubber, and the Finnes may arise to good profit. Another hindrance will be because the bottome of the Sound will not be so soone cleare as would bee wished; by meanes whereof now and then, a Whale may be lost. The same case sometime hapneth in *Greeneland*. Yet I am perswaded the Sounds before named will be cleare before the 20 of Iuly. Wee this yeere were in Whales Sound the 4 day [of July], amongst many Whales, and might have strooke them without let of Ice.

Furthermore, there is little wood to bee expected, either for fire or other necessaries; therefore Coales and other such things must be provided at home; they will bee much the more ready there.

Thus much I thought good to certifie to your Worship, whom I hope will conceive that much time hath not beene spent in vaine on the like businesse, nor carelessly neglected[1]: and, although wee have not performed what wee desired (that is, to have found the passage), yet what wee have promised (as to bring certaine and a true description) truth will make manifest that I have not much erred.

And I doe boldly say (without boasting) that more good discovery hath not in shorter time (to my remembrance) beene done since the action was attempted, considering how much Ice we have passed and the difficulty of Sayling so neere the Pole upon a traverse.

And, aboue all, the variation of the compasse, whose

---

[1] This is nonsense. In the original the words are: "I trust you will conceive that much time hath not been spent in vain, or the business over carelessly neglected".—C.

wonderfull opperation is such in this Bay, encreasing and decreasing so suddainely and swift, being in some part, as in *Wolstenholme's* and *Smith's* Sounds, varied above 5 points, or 56 d., a thing almost incredible, and almost matchlesse in all the world besides; so that, without great care and good observations, true description would not have beene had.[1]

In fine, whatsoever my labours are or shall be, I esteeme [them] too little to expresse my thankfull mind for your many favours, wherein I shall be ever studious to supply my other wants by my best endeavours, and ever rest at your worships command.

William Baffyn.

¶ *A briefe Discourse of the Probabilitie of a Passage to the Westerne or South Sea, with Testimonies; [and a brief Treatise and a Map] by Mr.* Henry Briggs.[2]

I Thought good to adde somewhat to this Relation of Mr. *Baffyn*, that learned vnlearned Marriner and Mathematitian, who, wanting Art of words, so really

---

[1] Mr. Markham inserts an interesting note concerning Baffin's observations on variation.—C.

[2] These documents are copied from Purchas (*his Pilgrimes*, part iii, pp. 848-852) with very slight alteration. I have corrected a few obvious misprints. The whole is *not* by Henry Briggs, as the reader might assume after a hurried examination of either Purchas or Foxe. It is a comment by *Purchas himself* on the results of the voyages of Button, Baffin, and others, backed up by "testimonies" (which we now know to be worthless) from Thomas Cowles and Michael Lok, and a brief treatise and a map by Henry Briggs. This is shown by the title, which, in Purchas, reads: "*A briefe Discourse of the probabilitie of a Passage to the Westerne or South Sea, illustrated with Testimonies: and a briefe Treatise and Mappe by Master Briggs.*" Foxe, by omitting Briggs' treatise and map (because, as he says, "later truths have proved them to be but the imagination of men"), without materially altering Purchas' heading, makes it more than

jmployed himselfe in that jndustrious worke, whereof heere you see so euident proofe. His Mappe and Table would much haue jllustrated this Voyage, if trouble, cost, and his owne dispaire of passage that way, had not made vs willing to content ourselues with what followed[1] of that learned (and in this Argument, three times thrice jndustrious) Mathematitian, Master *Henry Briggs*, famous for his reading in both Vniuersities, and this honourable Citie,[2] that I make no further Voyage of Discovery to find and follow the remote Passage and extent of his name. Mr. *Baffyn* told mee that the Tide from the N.W. about *Diggs* his Island was misreported, by mistaking the houre 8 for 11, and that hee would, if hee might get imployment, search the Passage from *Iapon* by the coast of *Asia*, or (*qua data porta*) which way hee could. But in the *Indies* hee died, in the late Ormuz businesse, slaine in fight with a shot as hee was trying his Mathematicall conclusions.

For the discouerie of Sr. *Thomas Button*, I haue sollicited for his notes, and receiued of him gentle entertainement and kind promises; but, being then forced to stay in the City about necessarie and vrgent Affaires, he would at his returne seek & impart them. Since [then], I heare that weighty occasions haue deteined him out of *England;* and I cannot deliuer that I could not receiue; which, if I

---

ever appear as if these observations were by Briggs. This omitted treatise by Briggs seems to have been reprinted *verbatim* by Purchas from Waterhouse's *Declaration of the State of Virginia, etc.* (1622), which has already been alluded to (see Introduction). It will be noticed that Foxe has retained the first person singular in which Purchas wrote, which he has not done in the case of any of the preceding narratives. Mr. Markham, who has reprinted this "Discourse" and the "testimonies" at the end of his *Voyages of Baffin* (p. 155), adds several interesting comments.—C.

[1] In Purchas the words are: "with that mappe following," alluding to the map by Briggs (see p. 177), which Foxe, of course, omits.—C.

[2] London.—F. For a notice of Briggs, see the Introduction.—C.

doe, I purpose to giue them out of due place, rather then not at all. Once hee was very confident, in conference with me, of a Passage that way, and said that he had therein satisfied his Maiestie, who from his discourse in priuate inferred the necessitie thereof. And the mayne Argument was the course of the Tyde; for, wintering in Port *Nelson*, hee found the Tide rising euery 12 houres 15 Foote. Whereas in the bottome *Hudson's Bay*, it was but 2 Foote; and in the bottome of *Fretum Davis*, discouered by *Baffyn*, but one; yea, and a West winde equalled the neepe Tydes to the Spring [tides], arguing the Neighbourhood of the Sea which is on the West-side of *America*. The Summer following, hee found, about the lat. of 60., a strong Race of a Tyde running sometimes Eastward, sometimes W.-ward. Whereupon *Iosias Hubbart*, in his Plot,[1] called that place *Hubbart's Hope*. Now, if any make scruple why this discouery was not persued by Sr. *Thomas Button*, let him consider that, being Prince *Henries* Seruant, and partly by him imployed ([after] whom I thinke [he] named the Countrey *New Wales*[2]), the vntimely death of that Prince put all out of ioynt; nor was hee so open, that others should haue the glory of his discouery.[3]

If any man thinke that the Passage is so farre as the Mappes use to expresse *America* running into the West, it is easily answered that, either of negligence or ouerbusie dilligence, Mappes by *Portugalls* in the East, and *Spaniards* in the West, haue beene falsely proiected. Hence that fabulous streight of *Anian*, as before by *Frances Gaule's* testimony and Nauigation is euident; and hence

---

[1] This chart, or whatever it was, does not seem to have been preserved in any form (see p. 178).—C.

[2] More likely Button named it after his native land (see p. 170).—C.

[3] This is, of course, an allusion to the fact that Button's journals were kept secret after his return (see p. 163, and also Introduction). —C.

the *Portugalls*, to bring in the *Moluccas* to the Moietie of the World agreed vpon betweene the *Spaniards* and them, are thought to haue much curtailed *Asia* and the longitude of those Islands, giuing fewer degrees to them then in just longitude is due. So the older Mappes of *America* make the land from *Magelane Streights* to the South Sea, running North-west,[1] when they are rather contracted somewhat Easterlie from the North. And the like is iustly reported of their placing *Quinera*, and I know not (nor they neither) what Countries they make in *America* to run so farre to the N.W.-ward, which Sir *Francis Drake* his Voyage in that Sea (his *Nova Albion* being little further North-ward, then *Aquatulco*) plainely conuinceth to bee otherwise. Yea, the late Mappe of *California*, found to be an Island, the *Saluage* discourses of all the Countries North-ward and West-ward from *Virginia;* the Fame whereof filled my friend Mr. *Dormer*[2] with so much confidence that, hearing of strange Shipps that came thither for a kind of Vre[3] or Earth, the men vsing Forks in their Diet, with Caldrons to dresse their Meate, and things nothing suitable to any parts of *America*, hee supposed them to come from the East, neere *China* and *Iapon*, and therefore hee made a Voyage purposely to discouer; but, crossed with diuers disasters, hee returned to *Virginia*, frustrate of accomplishment that yeere, but fuller of confidence, as in a Letter from *Virginia* hee signified to me, where death ended his designe soone after. But how often are the usuall Charts reiected by experience in those Nauigations in this worke[4] recorded? Painters and Poets are not alwayes the best Oracles.

---

[1] In the original these words read "runne much west".—C.

[2] In Purchas the name is *Dermer*.—C.

[3] In Purchas this word is *Yre*.—C.

[4] That is, of course, *Purchas his Pilgrimes*.—C.

For further proofe of a passage about these parts into the W. or S. Sea (as it is called from the first discouerie thereof to the South, from the parts of new *Spaine*, whence it was first described by the *Spaniards*), there is mention of a *Portugall* (taken in a Carracke in *Queene Elizabeths* dayes, of Famous Memorie) confirming this opinion. Sir *Martin Frobisher* also, from a *Portugall* in *Guinie*, receiued Intelligence of such a Passage: hee saying hee had past it. The Pilots of *Lisbone* are said generally to acknowledge such a thing, and the Admirall of *D. Garsia Jeaffery Loaisa*, of *Citie Royall*, in the time of *Charles* the *Fifth*, is reported, by the Coast of *Baccalaos* and *Labradore*, to haue gone to the *Mollucas*. *Vasco De Coronado* writ to the Emperour that, at *Sibola*, hee was 150 leagues from the South Sea and a little more from the North. *Anthonio de Horera*, the King's Coronista Maior, maketh with vs also in the distance described. But, to produce some Authority more full, I haue heere presented *Thomas Cowles*, a Marriner, and Master *Michaell Lock*, Merchant, and, after them, a little Treatis ascribed to Master *Briggs*. And, if any thinke that the *Spaniard* or *Portugall* would soone haue discouered such a Passage, this will answere, that it was not for their profit to expose their East or West *Indies* to *English*, *Dutch*, or others, whom they would not haue sharers in those remote Treasures by so neere a Passage.

First, *Thomas Cowles* auerreth thus much[1]:—

I, *Thomas Cowles* of *Bedmester*, in the County of *Somerset*, Marriner, doe acknowledge that, Six yeares past, being at *Lisborne* in *Portugall*, I did heare one *Martin Chacke*, a *Portugall*, reade a Booke of his owne making, which hee

[1] These are the "testimonies" mentioned above. They are reprinted by Foxe with but slight verbal alterations, except at the end of Lok's discourse, where various unimportant particulars are omitted.—C.

had set out 6 yeeres before that time, in Print, in the *Portugall* tongue, declaring that the said *Martin Chacke* had found, twelue yeeres now past, a way from the *Portugall Indies* through a Gulfe of the *New-found-land*, which hee thought to bee in 59 deg. of the N. Pole. By meanes that hee, being in the said *Indies*, with 4 shippes of great Burthen, and hee himselfe being in a small ship of 80 Tunne, [was] farre driuen from the companie of the other 4 shipps with a West winde. After which hee had past along by a great number of Islands, which were in the Gulfe of the said *New-found-land;* and, after hee ouer-shott the Gulfe, he set no more sight of any other land vntill hee fell with the N.-west part of *Ireland;* and from thence hee tooke his course home-ward; and by that meanes hee came to *Lisbon*, 4 or 5 weekes before the other shipps that were separated from his Company. And, since the same time, hee[1] could neuer see any of those Bookes, because the King commanded them to bee called in and no more of them to bee Printed, least in time it would bee [to] their hinderance.

In witnesse whereof I set to my hand and marke, the 9 day of *Aprill*, 1579.

¶ *A Noate of* Michaell Locke, *touching* Fretum Anian, *through the North-west Passage of* Meta Incognita.

WHen I was at *Venice*, in *Aprill*, 1596, happily arriued there an old man, aged about 60, called commonly *Juan de Fuca*, but named properly *Apostollos Valerianos*, of Nation a *Greeke*, borne in the *Island* of *Sepholonica*, of Profession a Marriner, and an ancient Pylot of shippes. This man came lately out of *Spaine*, arriued first at *Legorne*, and went thence to *Florence*, where hee found out *Iohn*

---

[1] In the original it is *I*, meaning Thomas Cowles.—C.

Dowlas, an *English*-man, a famous Marriner, ready comming for *Venice*, to be Pylot of a *Venetian* ship for *England*. They came both to *Venice* together; and *Iohn Dowlas*, being well acquainted with mee, gaue me notice of this Greeke Pylot, and brought him to my speech; and, in conference, this Pylot declared, in the *Italian* and *Spanish* tongue, these words following:—

First, hee said that hee had beene in the West *India* of *Spaine* by the space of Forty yeeres, and sayled too and fro, as Marriner and Pylot, to many places thereof in the seruice of the *Spaniard*.

Also he said that hee was in the *Spanish* shipp which, in returning from the Islands of *Philipinas* and *China*, was robb'd neere Cape *Calliformia* by Captaine *Cauendish*, an *English*-man, whereby hee lost 60 Thousand Duckets of his owne goods.

Also hee said that hee was Pylot of 3 small Shipps which the Vizeroy of *Mexico* sent from thence, armed with a 100 Souldiers, vnder a Captaine *Spaniard*, to discouer the Streights of *Anian*, along the Coast of the South Sea, and to fortefie in that streight, to resist the Passage of the *English* Nation, which were about to passe through the streights into the South Sea, and that by reason of a Mutinie which happened amongst the Souldiers, for the Sodomie of their Captaine, that Voyage was ouerthrowne, and the Ship turned backe from *California* to *Noua Spania* without any effect of things done in that Voyage. And at their returne the Captaine was punished at *Mexicoe*.

Also, hee saith that, after the said Voyage was so ill ended, the Vice-roy set him out againe in 1592, with a small Caravell and a Pinnace, armed with Marriners onely, for discovery of the said Streight. And hee, following his course W. and N.W. in the South-sea, along the coast of *Nova Hispania* and *California* and *India*, now called North *America* (all which voyage hee signified vnto me in a great

Mappe and Carde of my owne, which I laide before him), vntill hee came to the Latitude of 47 degrees, and that there, finding the land to trent N. and N.E., with a broad Inlett betweene 47 and 48, hee being entred thereinto, sayling therein more then twenty dayes, and found the land trenting still sometimes N.W. and sometimes N.E. and also S.E.-ward, a farre broader Sea then at the said entrance; and that hee passed by divers Ilands in that entrance.[1] And that, at the entrance of this said Streight, there is, on the North-west coast thereof, a great Head-land or Iland, with an exceeding high Pinnacle or spired Rocke, like a piller, there-vpon.

Also, he said that hee went on land in divers places; and that hee saw some people on land clad in Beast-skinns; and that the land was very fruitfull, and rich of gold and silver and Pearles and other things, like *Nova Hispania.*

Also, hee said that hee, being entred thus farre into the said Streight, and being come into the North-Sea allready, and finding the Sea wide enough every where, and to bee about 30 or 40 leagues wyde in the Streight where hee entred, hee thought he had now well discharged his office & done the thing which he was sent to doe: and that he, not being armed to resist the force of the *Saluage* people that might happen to assult him, therefore hee set sayle and returned towards *Noua Hispania,* where he arrives at *Aquapulco, Anno* 1592, hoping to be well rewarded of the Viceroy for his voyage so performed.

Also, he said that he was greatly welcomed to *Mexico* by the Viceroy, and had promise of great reward; but, staying there 2 yeares to his small content, the Viceroy told him he should be rewarded in *Spaine* of the King, and therefore willed him to repayre thither, which he did performe.

[1] This word in Purchas is "sayling".—C.

At his comming thither, he was greatly welcomed at the King's Court, in words, but after long suite he could not get any reward there to his content. And therefore at length he stole away and came into *Jtaly*, to get home to live amongst his kindred in his owne country, he being now very old.

Also, he said that he thought that the cause of his ill reward had of the *Spaniards* to be, for that they did understand very well that the *English* Nation had now given over all their Voyages for the discovery of the N.W. passage, wherefore they feared not them any more to come that way into the S. Sea; and therefore they needed not his service therein any more.

Also, he said that, in regard of his ill reward had of the *Spaniard*, and understanding of the Noble mind of the Queene of *England*, and of her warres maintained so valiantly against the *Spaniard*, hoping her Matie. would do him Iustice for his goods lost by Capt. *Cavendish*, he would be content to goe into *England*, and serve her Matie. in that Voyage to discover the N.W. passage into the S. Sea, and would put his life in her Maiesties hands to performe the same, if shee would furnish him with one ship of 40 tonnes and one Pinnace; and that hee would performe the same [in thirty days] from the one end of the Straits to the other. And he willed me so to write into *England*.

Whereupon, after this twice conference, I did write to the old Lord Treasurer *Cecil*, and to Sir *Walter Rawleigh*, and to Mr. *Richard Hackluit*, that famous Cosmographer, praying that 100 pounds might be sent for the charge of sending this Pylot into *England*. I received answer from some of my friends that the action was well liked of, if the money could be procured. After one fortnight, he went from mee into his owne countrey, where he dyed.

There are divers other things written in the Originall, as enticing perswasions to those undertakings, and is to be read as in Sir Humfrey Gilbert *and others in Mr.* Hackluit ; *and also other wch were his collections (after his death) inserted into the latter end of the 4h Booke of the 3 part of* Purkas *his works ; but, because these latter truths have proved them to be but the imagination of men, I omit them as things needlesse to this ornament ; for, although I have beene carefull to be as compendious as I could, yet I feare me my readers will thinke me to tedious.*[1]

Concerning Capt. William Hawkridge, *of whom I find nothing written by himselfe, but what hath come to my hands by manuscript or relation, as followeth here.*[2]

HEe went bent by the West,[3] and the 29 of *June* [1619] he found himselfe betwixt land and land,

---

[1] This passage is clearly by Foxe. The "original" that he speaks of is, of course, *Purchas his Pilgrimes.*—C.

[2] This is another of those voyages of which we have no other account than that given by Foxe, whose meagre narrative here follows. It affords no certain information as to the year in which the voyage was made ; the name of the ship or ships employed ; or the person or persons at whose expense the expedition was made. Rundall, however, produces evidence which goes to prove (*Voyages towards the North-West*, p. 150) that it was made in the year 1619 ; that Sir John Wolstenholme was the prime mover in the matter ; that there were, associated with him, some of his friends ; and that the East India Company made a grant of £200 towards the expenses. It is a little difficult to see what can have been the exact incentive to the voyage, after Baffin's very strongly-expressed opinion that there was no passage to be found through Hudson's Strait ; but it seems from Rundall (*loc. cit.*) that Sir John Wolstenholme based his hope of

[3] That is, he went by the west coast of England, and not up the east coast, like Hudson, Foxe, and some of the other north-west explorers. The word *bent* (which is certainly now disused in this sense) may, I think, be read as *directed* or *inclined* towards the west. —C.

and thought he had bin in the great channell or *Lumleyes Jnlet,* where it pleased God, by the clearing up of the

---

finding a passage, mainly on the previously-observed rise and fall of the tide in "Botton's Bay". It is quite probable that the voyage was undertaken in the *Discovery*, and that it was chiefly promoted by the Company of Adventurers for Discovering the North-West Passage; but the last voyage which we know with certainty to have been made under the auspices of that body, or in the vessel named, was Baffin's in 1616. Of Hawkridge, we only know that he had accompanied Button in 1612-13 (see p. 167); and that, like Button's other companion, Gibbons (who failed so ignominiously in 1614), he accomplished nothing of importance on his voyage. The account of the expedition given by Foxe is so full of dates, distances, directions, soundings, and other particulars, that there can be no doubt that either Hawkridge himself supplied a full abstract of his log, or allowed Foxe to make such an abstract for himself. But, when one comes to examine the narrative critically and in detail, one finds it unsatisfactory in the extreme. I confess that I am totally unable to follow on a chart the course sailed over. The difficulty of following out the narrative is due, not so much to the absence of any determinations of latitude and longitude, as to the fact that not a few of the observations which are given appear from the context to be erroneous. On various occasions he narrates that he sailed, from a certain point, so many leagues in a certain direction, and that he arrived at a certain other point which he states to be, perhaps, as much as a degree to the north or to the south, as the case may be, of the point we might suppose him to have reached. His surmises as to the lands which he sighted from time to time are of the vaguest description and often, apparently, quite wrong. The only thing the narrative does make clear is that, wherever he went, he cruised about in an apparently aimless way, first in one direction then in another, and that he frequently crossed over his own track. Altogether, we are driven to the conclusion that he was a most incompetent navigator, in spite of the experience he might be presumed to have gained whilst sailing with Sir Thomas Button. His entire voyage is quite valueless as a piece of discovery (for Baffin, in 1615, and others before him, had discovered all the coasts he can have visited); and his whole narrative scarcely contains a single contribution to a precise geographical knowledge of the region visited or an observation of value to any later explorer. At the end of the narrative, the confusion is made worse by the fact that the printer's "copy" has evidently got mixed (see p. 256).—C.

weather, to deliver him from a rocke he might have indangered himselfe upon to the E.-ward.

30. This day he plyed it out againe to the E.-ward, finding that he was in the N. or wrong Channell,[1] where he saw 3 Rocks; he wondred he had escaped, for he had runne in amongst them.

July 1. This day was fayre and cleare weather, the wind at N.W. He stood out againe S.E. all the forenoone amongst yce, loosing for one and bearing up for another, the current setting to the W.-ward. He tooke marks upon the land and by the logge the Ship run after[2] 5 leag. a watch, and for all that got nothing; he observed by his Astrolob[3] and was in 62 d. 25 m., having had Fogs and Mists for 6 dayes before, so as he could not observe; and this day he had 29 d. variation Westward; the magneticall Amplitude 83 d.; true Amplitude 54.

2-3. These dayes were fayre cleare weather.

The 4 was foggie.

5. This day the winde was E.

6. This day he plyed to windward to weather *Resolution;* the wind at E.N.E.

7. This day was foggie; wind N.E. He lost sight of his Pinnace.

The 27 of *Iune* he made the *Resolution*, and the 8 of *Iuly* he was come backe againe out of the N. channell betwixt *Resolution* and Cape *Elizabeth.*[4]

9. This day he met againe with his Pinnace and thought to have borne up; but, the fogge taking him, he plyed to

---

[1] Meaning evidently Lumley's Inlet or Frobisher's Bay.—C.

[2] It seems as if this word were a misprint for ***about*** or ***above.***—C.

[3] The sea-astrolabe was a graduated brass-ring with a movable index, for taking the altitude of stars and planets. It derived its name from the armillary sphere of Hipparchus, at Alexandria. A brass *armilla tolomæi* was one of the instruments supplied to Sir Martin Frobisher in 1576, price £4 6*s.* 8*d.*—C.

[4] Hee was the first that entred *Lumleys Inlet* so farre.—F.

the Eastward, to the intent he might get into the great channell.[1]

10. This day, standing to the N. shore, with very foggie weather, he was taken with an indraft of a strong tide and drawne in amongst divers Ilands about Cape *Elizabeth*, and was in more danger then he saw; but, having a swelling Sea from the E., hee followed and so escaped all dangers.

11. This day he had a strong ripling of a tide; his Latit. 61 d. 30 m., and the body of *Resolution* bore N.W. by N. from him.

The 12, 13, and 14 dayes he made account that he had kept that latit. and rather to the N.ward, but he was horst[2] with a current (he could not tell how) above 1 d. & 30 m., wch all men know is 30 leag., so that he was to the S.-ward of *Button's* Iland.

16. This day, when he came to observe, thinking he had bin in the mouth of the strait, it proved otherwise, the wind coming contrary, as at W. and by S. Some hopes were taken away that he should not insist any further for that yeare, but that himselfe had framed a sound resolution to continue and persevere.

22. From the 16 untill this day, he plyed to the West and was faire by land, not 3 leag. of the sounding, he had no ground in latit. 61 50.[3]

23. The wind N.N.E. and N.E.; cleare weather.

24. This day he espied land on the S. shore, nere Cape *Charles*,[4] but to the E.-ward he espyed a little Iland, where

---

[1] That is, Hudson's Strait.—C.

[2] This is probably a misprint for either *forst* (forced) or *hoist*. Either word would make fairly good sense.—C.

[3] Apparently this should read: "not 3 leag. off. Sounding, he had no ground, etc." He seems now to have entered Hudson's Strait and to have been in Gabriel Strait.—C.

[4] This cannot have been Charles Island of Hudson. From the latitude given, it seems to have been one of the Savage Islands.—C.

he stood into a Bay to water, and anchored in 25 fath.; fine fishing ground (but catcht none). He had land bore round from the N.N.W. to the E. by N.; the Ilands Lat. 62 d. 19 m., variat. 3 d. 9 m.; and here he had a tyde which, minding the setting and flowing, may cause some Argument of strong consequence to prove a passage that way. On this Iland, he caught Ducks; here he found it to flow 21 foote water, the tyde setting S.E. and the flood from N.W., and in this place a S.E. Moone makes a full Sea.

He sent the Mrs. Mate and Carpenter, with others, in the boat to rowe about the Iland, and, when it bore S.E. of them, they had 74 fath., halfe a mile from land, & a strong set tide from E.

27. This day he set sail from this Iland; the wind E.S.E.; much wind all day.

30. This day, in the morning, he met with much yce.

31. This day he run 35 leag. W. by N., but the last day at night he reckned himselfe to be at the westermost Cape on the N. side.[1]

August 1. This day he run 25 leag. W. by N., but the last day he sawe land, and thought it had bin *Salisbury* and *Nottingham's* Ilands, but it proved the N. Mayne; he sailed along W. by N. and W. by S., with a stiffe gale at E.N.E.; he found this land to be thicke with yce and very low,[2] and run 30 leag. along by it.

The first of *August* he espyed this land and sailed along in 67, 89, 10, 11, 12, 13, 14, 15, 16, fath., rocky ground; as he stood to the S.-ward, it was white sand; his lat. was 63 d. 30 m., variat. 27; he found an Iland sayling along the shore, the depth betweene which and the Mayne was 70 fath.

---

[1] That is, at King Charles's Cape, but it seems from what follows that the Cape was more likely Fair Ness.—C.

[2] I know none such.—F. It is difficult to imagine what land it can have been.—C.

2. This day he sailed along the N.W., with fayre weather and easie wind.

3. This day he stood away S.W. and anchored in 45 fath., clay ground.

4. This morning he weighed and stood N.W. 3 leag. and, being faire weather, the water shoalding to 30 fath., he anchored againe, still finding a pretty soaking current setting most an end N.W. and S.E.

5. This morning, clock 8, he anchored againe, having but new weighed because of the Fog.

6. This morning, clocke 4, he weighed and stood away W.N.W., with S.E. wind, true course 5 leag., and was in 63 d. 50 m., having a swelling Sea out of the Westerboard[1]; the water waxed deepe from 30 to 50 fath., and the ground was hard channell ground. From 12 to 4, he sailed N.W. by N. 6 leagues, but found no ground and supposed he saw the W. land bore N.N.W. from him. From 4 untill 8, he stood away N. and by E., 5 leag.; at 8 he sets, tacks aboard, and stood N.E. & by S. in a deep gut; this day he had 45 fath. at clocke 4, at 5 he had 8 fath., and at 8 he had 65, the land bearing N.N.W. and N.W.

7. This day, at the dawning, the land bore N.W. and he stood along in N.E. true course; he had sounding 25 and 30 fath., and anchored at clocke 8, and weighed againe presently; the land bore E. and had sounding all day; the further N.-ward the deeper water. This writer saith he iudged it to be *Salibury* Ile; he sailed N.E. the forenoone some 7 leag. & in the afternoone N.N.E., for so the land did lye. Towards the bottome of the Bay, the latit. was

---

[1] I have submitted this word to the Rev. Professor Skeat, who has kindly criticised it as follows:—"*Star-board* = steer-board, the side on which the steersman sat (see the Bayeux Tapestry). *Larboard* = leer-board, *i.e.*, the empty side (on which he *didn't* sit. *Wester-board* is new to me, and rare; but is made up of *wester*, as seen in *wester-ly*, and *board;* and therefore means the western side or quarter."—C.

64 d. 30 m.,[1] variat. 23 d. 10 m.; the part of this land bore from him N.N.E.; fine low plaine land.

The 8, in the morning, he was perswaded it was a bay, but that he will not say. He was this day calming and did thinke that there was no tide here, but, sending his boate on shoare, found 20 foot ebbing and flowing, and sport enough for them all, for in 3 houres space he saw in conscience as good as 300 Deare, as fat as butter, but caught none, for his intent was to travell as good as 2 miles, hoping to have seene the Sea on the other side, but could not.

9. This day, clocke 8 to 12, he run 6 leag. W.N.W. From 12, he steered away as the land would give him leave, W.S.W. and W., a fine shoalding coast and dainty sounding, shelly ground, from 10 to 16 fathomes; here he had a little current, set W.N.W. This, as he judged, is all broken land; latit. 63 deg. 40 min., variat. 23 d. 30 m.

10. The wind was at S.E. and by E.; the farther to the westward, the shoalder water; they judge themselves to be shot to[2] farre to the W.-ward as *Sea Horse* Poynt, because of the coast trenching[3] to Southerly; his boate rid with her Grapnel[4] and found a pretty streame.

11. This day he was in 63 d. 40 min. latit., the land bearing from him S.W. and trenching along to the S. Being in this lat., they thought themselves farre shot to the westwards, within Sea Horse poynt,[5] and so returned backe

[1] If this is correct, he was now near King Charles's Cape.—C.

[2] Apparently this is a misprint for *so*.—C.

[3] No doubt a misprint for *trending*, meaning the direction in which a coast bends or inclines.—C.

[4] A grapnel is a small anchor, having a ring at one end and four palmed claws at the other. It is used for boats.—C.

[5] There is some duplication here, the facts being stated twice over. It seems he was now near Sea Horse Point, but it is by no means clear how he got there. The "Bay" he speaks of seems to have been the southern end of Fox's Channel.—C.

againe for the Bay, where they were in almost 65 deg. to the N. wards; but he altered his mind and stood for *Diggs* his Iland, to try the tyde N.E. by N. ½ Northerly; from thence, where he turned out of the bay of Sea Horse 24 leag.

12. This day the wind was E. and by S.; thick weather.

13. From the last day to this day noone, he was becalmed in thicke weather.

14. From the last day noone till this, he made way 9 leag. E.S.E. and 2 leag. N.W.

15. He tryed the tyde, and found as strong a streame at this time as you have here in the *Thames*; it set S.E. and N.W. He followed it to see whether it would carry him; at clocke 8 at night he anchord, and wayed againe at 8 next morning, and to 12 he run 4 leag. N.W. by N.; from 12 to night 10 leag., and he had sounding 60 and 70 fath., but anchored in 30.

[III.[1]] 16. This morning, 4, he wayed and stood to the N.-ward, but thought he was stopt by land and therefore bore up the helme for *England*, not for that he was out of hope of a passage, for that he will never say.

17. From the last day untill this, S.W. 8 leag. He tackt to the N.-ward; this morning he was in sounding 70, 60, 59 fath.

18. From the last noone to this, he drove N.E. 6 leag.; both these last dayes were thick weather.

19. To this day noone, 20 knots S.E. and 10 knots S.W.

20. To this day noone, 20 leag. S.E.; the wind W., and foggy.

21. To this day noone, he run 20 leag. S.E.; thicke weather, and he was in 62 d. 40 m. latit., and he saw land to the N. off him, and had sounding from 45, 40, 36 fath.

---

[1] See note on p. 257.—C.

23. This day he was in latit. 62 d. 00 m. In the morning the land bore S. off him, and they judge it to be the N. shore or Cape.[1] He had sounding 9 and 19 fath., and had run from last day 22 leag. E. by N. and 6 leag. S.; he had sounding along the land 17 or 18 fath.; fine beach land, and a stiffe gale at N.W.

24. From the last day to this, 23 leag. S.E. & 9 leag. E. by S. and this morning he fell with land, which he tooke to be the King's *Forland*[2]; it bore S.E. 9 leag. off, and latit. 61 d. 30 m. This day the Pinnace stole from them, as they thinke upon purpose.

25. From this day at noone, N.N.W. 8 leag. & N.E. by N. 9 leag. Sounding was 40 & 45 fath.

[I.] 26. This day at noone, they saw the same breach that they parted from,[3] & was by observation in latit. 62 deg. 10 min., variat. 26 deg.; the wind was at S.E., and they thought themselves on the W. side of *Mansfield* Ile,[4] 2 leagues off, & had deepe 16 or 18 fath.

---

[1] He appears to have been in Hudson's Strait in this latitude, but the confusion is hopeless. It is impossible to understand how he can have supposed the land to the south of him to be the north shore!—C.

[2] Hudson named one of the capes on the south side of Hudson's Strait the King's Foreland (see p. 125). Dr. Asher (*Henry Hudson*, p. 105) identifies it with Cape Weggs, but it seems to me to have been one of the capes further to the east.—C.

[3] I know of this breach.—F.

[4] If the narrative may be relied on, this was a preposterous delusion. But it seems that some of the dates are entirely wrong, and that several portions of the narrative have been altogether misplaced. We read that, on August 16th, Hawkridge was near Sir Dudley Diggs' Isle, and that he then turned his ship's head homeward. After this, we can follow him with fair ease, sailing south-eastward on his homeward course, until the 25th, when he says he was in Lat. 61° 30′ N., and apparently about Long. 69° or 70° W. Then, on the following day, he says he was in Lat. 62° 10′, and supposed himself on the *west side of Mansell Island*. This would make his longitude about 81° W., which

27. From last noone to this, he run 27 leag., true course N. & by E., & were in latit. of 60 d.[1]; the wind at E.S.E.

[IV.] This night at clock 10 the fogge came; the next morning it cleered, but he had no ground at 100 fath.; he tackt about to the S.-ward till next morning, & then to the N.-ward, but at noone could have no observation.

---

is impossible. Yet subsequent statements undoubtedly relate to his movements in the neighbourhood of Mansell Island; for, on the following day (the 27th), he had somehow got into Lat. 60°, which would hardly have been the case had he been passing through Hudson's Strait. Proceeding, we find that, from the 28th to the 31st, he seems again to be sailing in a general easterly direction, apparently through Hudson's Strait. Under the last-named day, however, commences a fresh paragraph which suddenly takes us back again to Digges Island, and on the following day (Sept. 1st) we hear of Salisbury and Nottingham Islands being sighted. To make matters worse, we find that from the 2nd of September to the 10th (when the narrative ends), Hawkridge appears to be again continuing on his homeward voyage past Resolution Island and across Davis Strait. In fact, it is quite obvious that some inexplicable muddle has arisen. Apparently Hawkridge made a perfectly futile voyage, and Foxe further marred Hawkridge's reputation by publishing an inexplicably confused narrative of it. I have caused a white space to be left between those portions of the narrative which appear to have been misplaced, and have prefixed to them Roman numbers [within square brackets], indicating the order in which it appears to me the various fragments make the best sense. I can only suggest that the later sheets of the MS. "copy" became disarranged and were so set up in type, the dates being afterwards "cooked" to make them consecutive. That Foxe himself was conscious of the confusion is clear, for in one of his side-notes he makes the comment: "A confused accompt." The pages on which these transpositions occur are those to which Sabin (*Dictionary of Books Relating to America*) refers as follows: "After page 168, are two leaves, paged 170, 172, 170, and blank, which are said to be cancelled leaves." In one copy in the British Museum they are paged 172, 170, 171 and blank. This fact may have some connection with the muddle.—C.

[1] If so, he was now well on towards the centre of Hudson's Bay, but it is impossible to understand how he could have got into this latitude by sailing N. by E. from Mansell Island.—C.

28. This morning, 8, he tackt to the S., for he saw a firme land of Ice; from last day to this, N.E. · 12 leag. and 7 leag. S.E., because of the Ice; at clock 10 he had 80 fath.

29. From last day to this, 10 leag. S.E. & 3 leagues N.E., and sounding had no ground.

30. From last day to this was fog; they got but little to the E.-ward, and sounding had no ground, and latit. 62 d. 40 m.

31. From the last to this 10 lea. E. by S. & S.S.E. by E., the wind at N.; at 8 in the evening, calme but cleare weather; they were close about the N. shore in 80 fath.; he had a rippling of a Tyde to the E.-ward; the land was something low towards the water, but double height within land; it lyeth W.N.W. and E.S.E.

[II.] This day he past by an Iland they tooke to be the W.-most end of the Straight, & see Sir *Dudley Diggs* his Iland, being high land, & see the S. shore & a gut when it beareth S.W. seemeth to be 4 miles over. They also see Nottingham's & Salisburie's Ile, & a channel between them of 8 leag., & this strait is over about 15 leag. They stood away this day E. by N.

September 1. To this noone from the last 21 leag. E. by S., latit. 63. Salisbury bore N.N.W., and the W. Cape on the N. shore bore N.E., variat. 28 d. The wind N.E. by N.

[V.] 2. To this noone 21 leag. E. by S. & S.S.W. 5 leagues; this morning he was close aboard the N. coast; it seemeth high ragged land & full of guts; he was becalmed & befogged, & stood S.-wards into the channell, having 110, oazie ground.

There is nothing else of note untill the 7 day, when *Resolution* bore N.W. from him.

The 9 day he was open in Davis his Straite, in 59 d. 25 m., & had variat. 26 deg.

10. This day, by storme, his Pinnace threw over their boate, & he lost sight of her.

[*The Observation.*]

This is all that is to be observed, that he entered Fretum Hudson *the* 22 *of July, & returned from his Search the* 16 *of* August, *having been no further to the N. then almost* 65^{d}, *& upon the S. side of* Fretum Hudson, *neare Cape* Charles ; *the Tide came from S.E., as it doth on the N. side; the rest is, he was set at his entring the mouth of the straite* 30 *leag. to S. of Cape* Chidley.

FINIS.

COurteous Readers, the Printer, but especially the corrector, craves your patience for this long Errata following, which is also my request, although to my unknowledge, and in my absence, they (not being acquainted with the methode of our Sea tearmes) have committed all these mistakes in 6 sheets (*beginning at N. & ending with S.*[1]), being sent to another Presse for expedition ; &, for others, I desire thy good construction, promising they shall be amended at the next edition.

[1] This refers to the six sheets or "signatures" from N to S, which Foxe says were printed at another press. In most copies there is a marked difference in what printers know as "colour" between these sheets and those that precede and follow them. The errata in these sheets (pp. 149-233 in this reprint) are numberless. All those which Foxe himself indicated, as well as a great many he overlooked, have been corrected herein.—C.

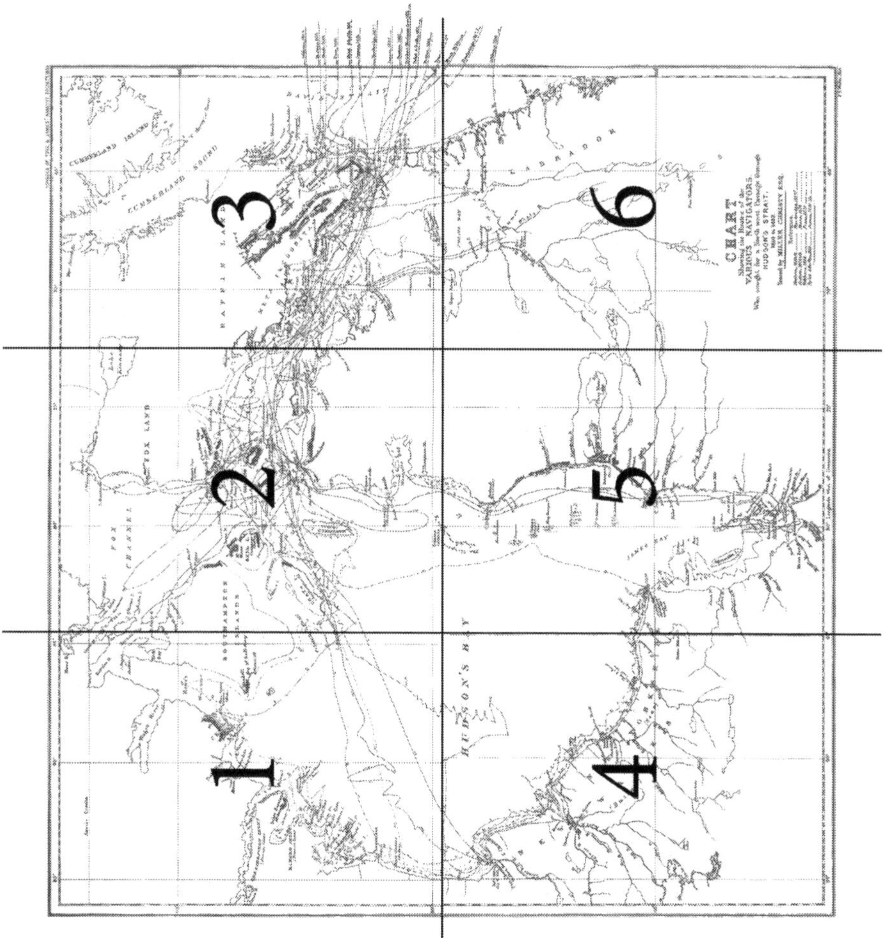

Key to following pages.

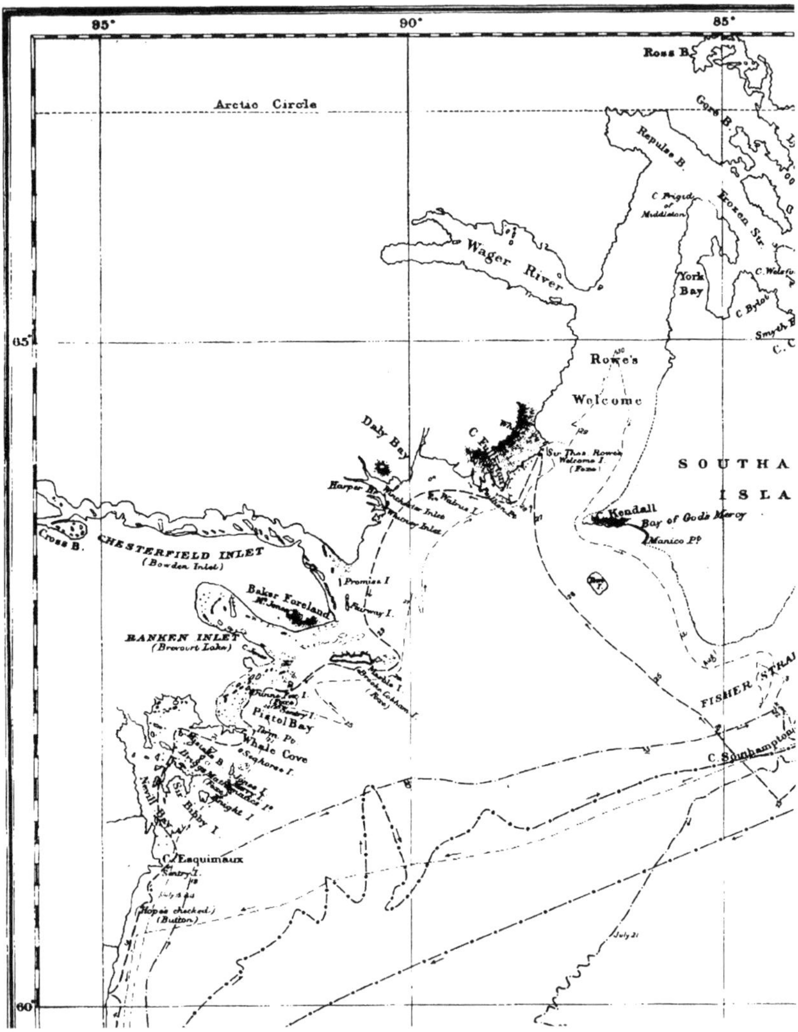

95°
90°
85°
Ross B.
Arctic Circle
Repulse B.
Frozen Str.
Wager River
York Bay
Rowe's
Welcome
Daly Bay
C. Kendall
Bay of God's Mercy
Manico Pt
SOUTHA
ISLA
CHESTERFIELD INLET
(Bowden Inlet)
Cross B.
Baker Foreland
RANKEN INLET
(Brevourt Lake)
Pistol Bay
Whale Cove
FISHER STRAI
C. Southampton
C. Esquimaux
July 21
65°
60°

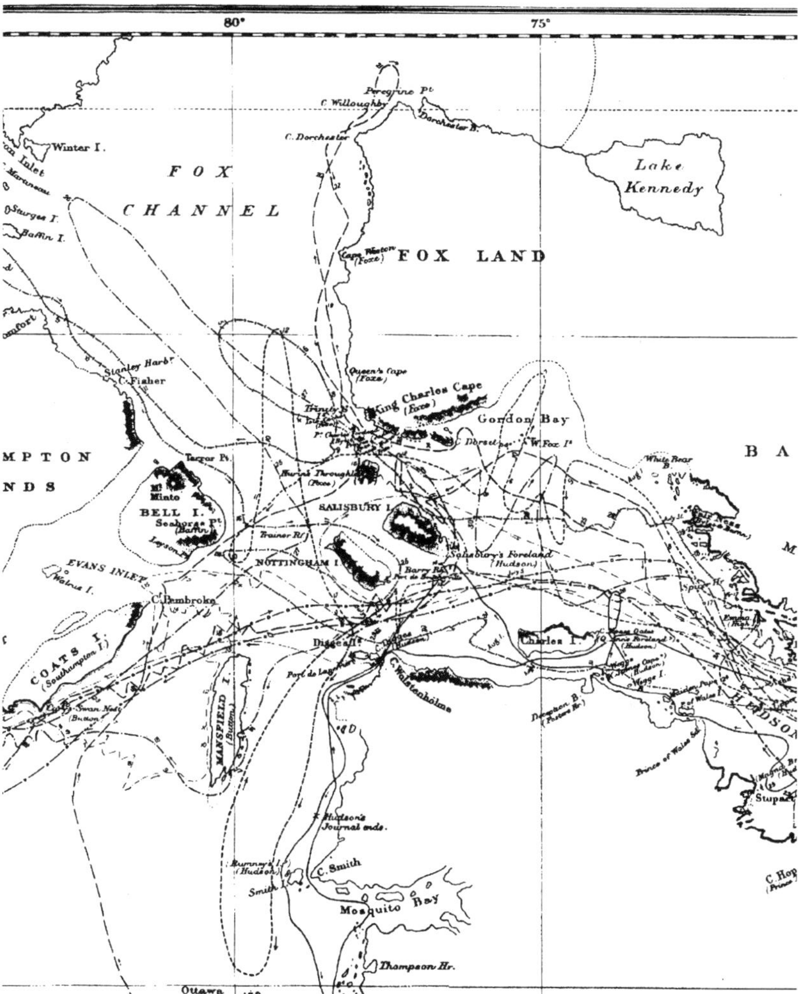
80°
75°
Peregrine Pt
C. Willoughby
Dorchester B.
C. Dorchester
Winter I.
Sturges I.
Baffin I.
FOX
CHANNEL
Lake Kennedy
FOX LAND
Cape Weston (Foxe)
Stanley Harbr
C. Fisher
Queen's Cape (Foxe)
King Charles Cape (Foxe)
Gordon Bay
MPTON
NDS
Mt Minto
BELL I.
Seahorse Pt (Baffin)
EVANS INLET
Walrus I.
C. Pembroke
SALISBURY I.
Trainer Rf
NOTTINGHAM I.
Salisbury's Foreland (Hudson)
White Bear B.
Charles I.
COATS I. (Southampton I.)
Swan Nest (Button)
MANSFIELD I. (Button)
C. Wolstenholme
Prince of Wales Sd.
Stupart
Hudson's Journal ends.
C. Smith
Smith I.
Mosquito Bay
Thompson Hr.
Ottawa Islands

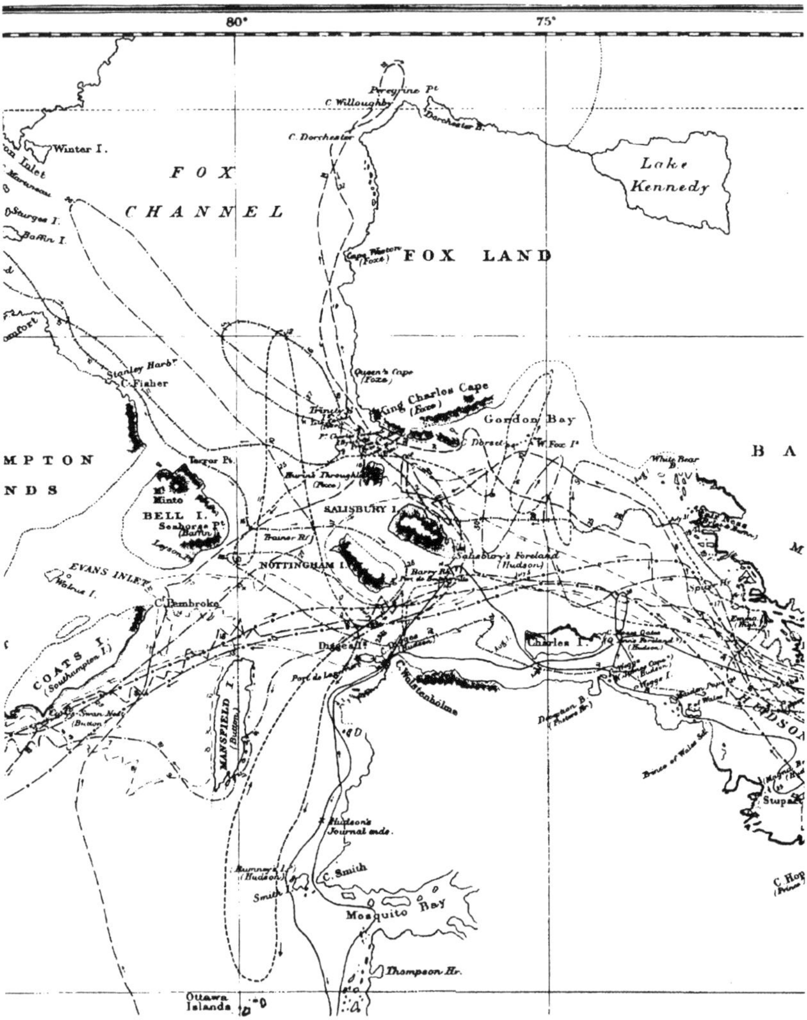

80°
75°
Peregrine Pt
C. Willoughby
Dorchester B.
C. Dorchester
Winter I.
Lake Kennedy
FOX CHANNEL
Sturges I.
Baffin I.
FOX LAND
Stanley Harb
C. Fisher
King Charles Cape
Gordon Bay
MPTON
NDS
Bell I.
SALISBURY I.
NOTTINGHAM I.
Salisbury's Foreland (Hudson)
EVANS INLET
C. Pembroke
Charles I.
COATS I. (Southampton I.)
MANSFIELD I. (Button)
C. Walstenholme
Hudson's Journal ends.
C. Smith
Mosquito Bay
Thompson Hr.
Ottawa Islands
Stupart
B A

2

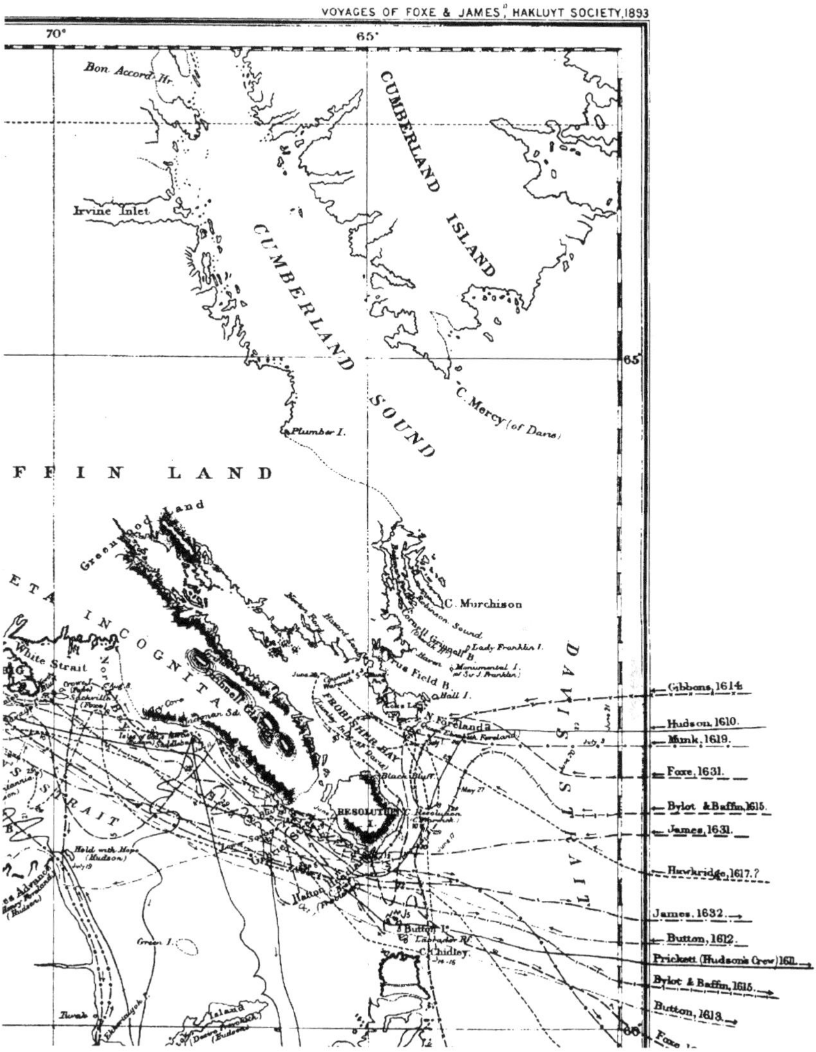
VOYAGES OF FOXE & JAMES, HAKLUYT SOCIETY, 1893
70°
65°
Bon Accord Hr.
CUMBERLAND ISLAND
Irvine Inlet
CUMBERLAND SOUND
65°
C. Mercy (of Davis)
Plumber I.
FFIN LAND
Greenwood Land
ETA INCOGNITA
White Strait
C. Murchison
Lady Franklin I.
Monumental I.
Hall I.
FROBISHER BAY
N. Forelands
DAVIS STRAIT
Black Bluff
RESOLUTION I.
C. Resolution
Hold with Hope (Hudson)
Hatton Hd.
Button Is.
Labrador Rf.
C. Chidley
Green I.
Island
Gibbons, 1614
Hudson, 1610.
Munk, 1619.
Foxe, 1631.
Bylot & Baffin, 1615.
James, 1631.
Hawkridge, 1617.?
James, 1632.
Button, 1612.
Prickett (Hudson's Crew) 1611.
Bylot & Baffin, 1615.
Button, 1613.
60°
Foxe

3

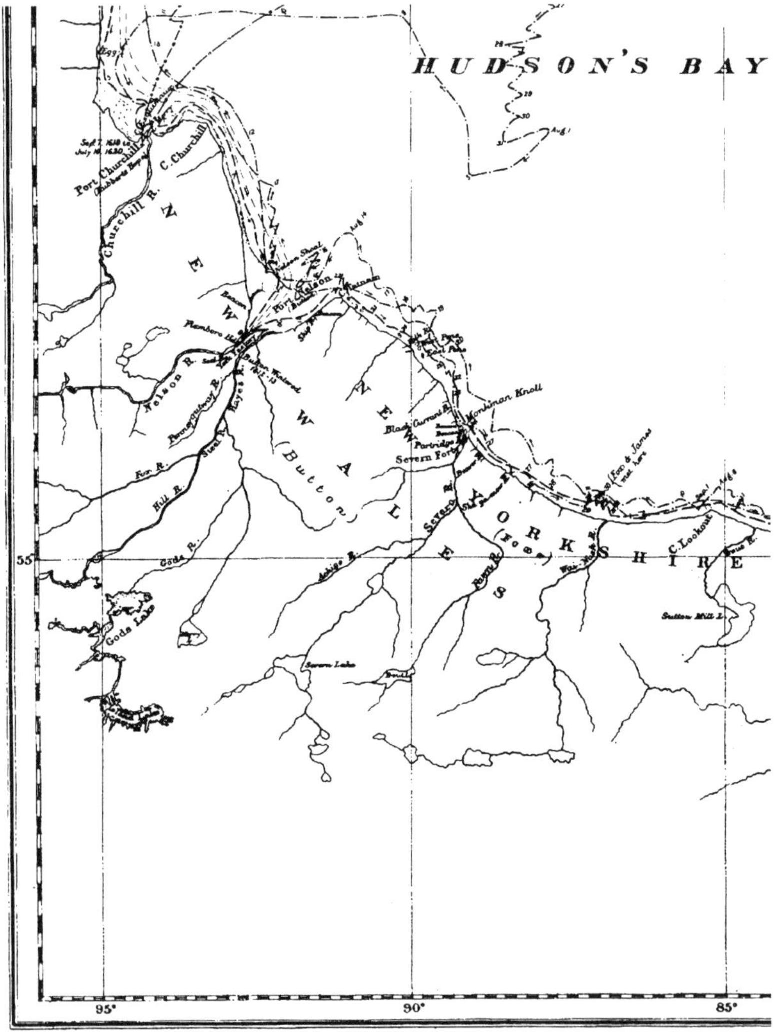
HUDSON'S BAY
Port Churchill
C. Churchill
Churchill R.
NEW WALES
Port Nelson
Nelson R.
Hayes R.
Severn Fort
Monkinan Knoll
Black Currant R.
Partridge
Fox & James met here
NEW WALES (Button)
YORKSHIRE (Foxe)
Severn R.
C. Lookout
Fox R.
Hill R.
Gods R.
Gods Lake
Severn Lake
Sutton Mill L.
55°
95°
90°
85°

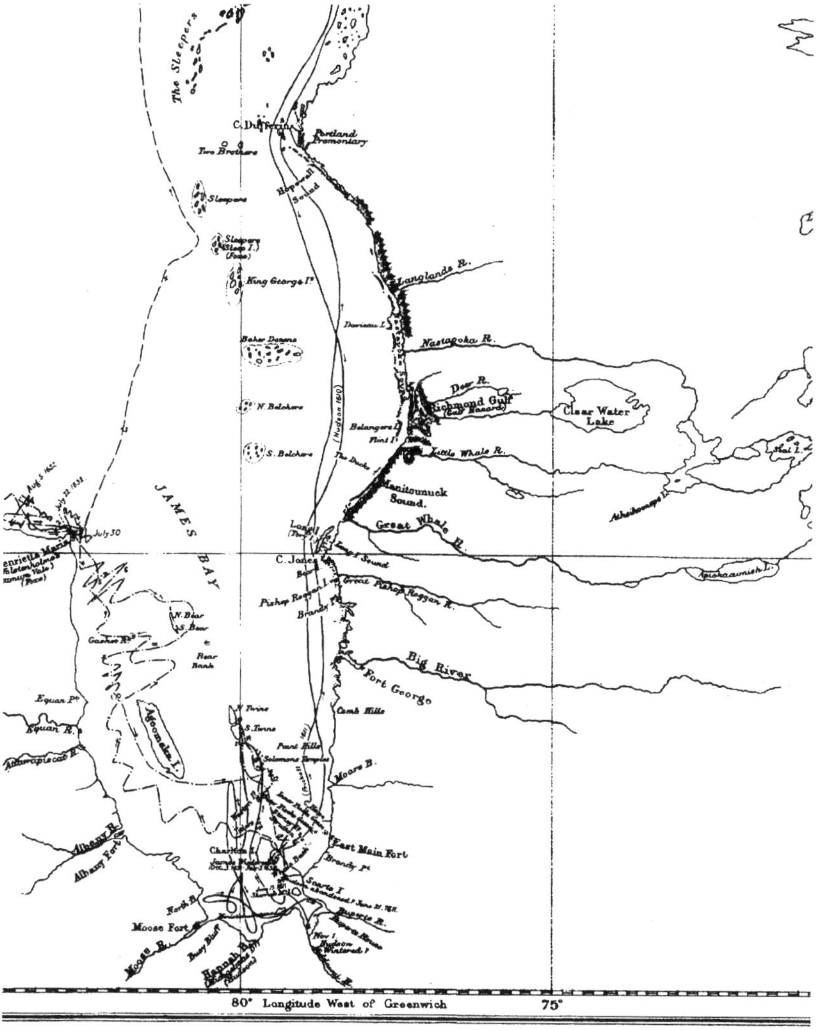
The Sleepers
C. Dufferin
Portland Promontary
Two Brothers
Hopewell Sound
Sleepers
King George Is
Langlands R.
Baker Dozens
Nastapoka R.
Deer R.
Richmond Gulf
Clear Water Lake
N. Belchers
S. Belchers
Belangers I.
Flint I.
The Ducks
Little Whale R.
Manitounuck Sound.
JAMES BAY
Long I.
Great Whale R.
July 30
C. Jones
Long I. Sound
Great Pishop Roggan R.
Pishop Roggan I.
N. Bear
S. Bear
Bear Bank
Big River
Fort George
Equan Pt.
Equan R.
Akimiski I.
Comb Hills
Attawapiscat R.
Moore B.
Albany R.
Albany Fort
Charlton I.
East Main Fort
Moose Fort
Moose R.
Ruperts R.
80° Longitude West of Greenwich
75°

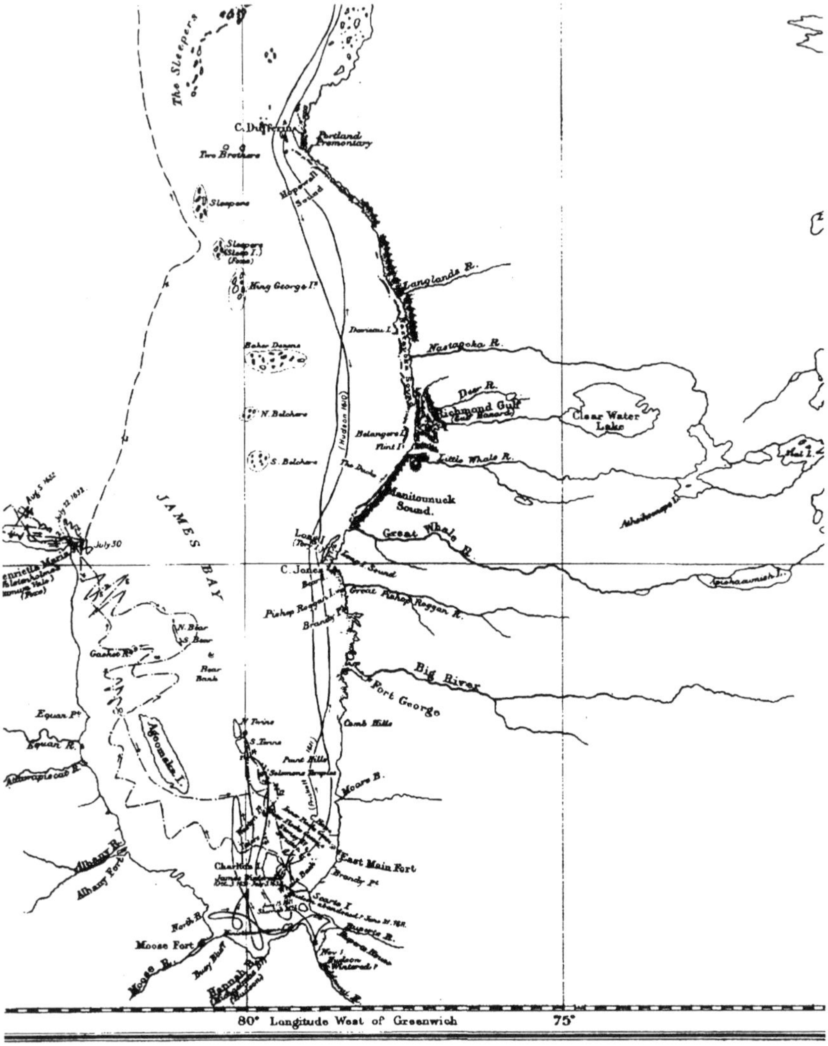
The Sleepers
C. Dufferin
Portland Promontory
Two Brothers
Hopewell Sound
Sleepers
King George Is
Baker Dozens
N. Belchers
S. Belchers
Langlands R.
Nastapoka R.
Deer R.
Richmond Gulf
Clear Water Lake
Belangers I.
Flint I.
Little Whale R.
Manitounuck Sound.
Great Whale R.
JAMES BAY
Long I.
C. Jones
Great Pishop Roggan R.
Big River
Fort George
Camb Hills
Moore B.
Agoomska I.
Equan Pt
Equan R.
Albany R.
Albany Fort
Charlton I.
East Main Fort
Moose Fort
Moose R.
North R.
Rupert R.
80° Longitude West of Greenwich
75°

5

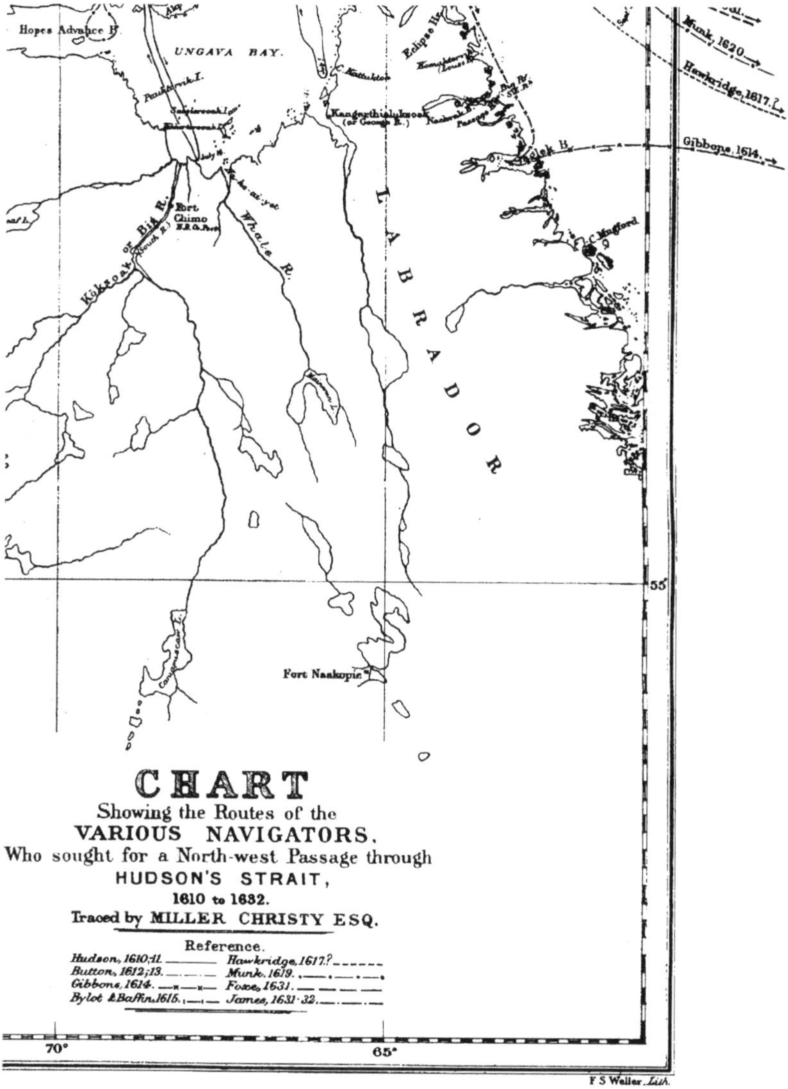
Hopes Advance B.
UNGAVA BAY.
LABRADOR
Fort Chimo
Kôksoak or Big R.
Whale R.
Fort Naskopie
Eclipse Hr.
C. Mugford
Saglek B.
Munk, 1620
Hawkridge, 1617.?
Gibbons, 1614.
55°
CHART
Showing the Routes of the
VARIOUS NAVIGATORS,
Who sought for a North-west Passage through
HUDSON'S STRAIT,
1610 to 1632.
Traced by MILLER CHRISTY ESQ.
Reference.
Hudson, 1610;11.
Button, 1612;13.
Gibbons, 1614.
Bylot & Baffin, 1615.
Hawkridge, 1617.?
Munk, 1619.
Foxe, 1631.
James, 1631·32.
70°
65°
F S Weller, Lith.

Printed in Poland
by Amazon Fulfillment
Poland Sp. z o.o., Wrocław

37466581R00298